Mathematics Study Resources

Volume 7

Series Editors

Kolja Knauer, Departament de Matemàtiques Informàtic,
Universitat de Barcelona, Barcelona, Barcelona, Spain
Elijah Liflyand, Department of Mathematics, Bar-Ilan University,
Ramat-Gan, Israel

This series comprises direct translations of successful foreign language titles, especially from the German language.

Powered by advances in automated translation, these books draw on global teaching excellence to provide students and lecturers with diverse materials for teaching and study.

Marco Hien

Abstract Algebra

Suitable for Self-Study or Online Lectures

 Springer

Marco Hien
Institut für Mathematik
University of Augsburg
Augsburg, Bayern, Germany

ISSN 2731-3824   ISSN 2731-3832   (electronic)
Mathematics Study Resources
ISBN 978-3-662-67973-9   ISBN 978-3-662-67974-6   (eBook)
https://doi.org/10.1007/978-3-662-67974-6

This Springer imprint is published by the registered company Springer-Verlag GmbH, DE, part of Springer Nature.
The registered company address is: Heidelberger Platz 3, 14197 Berlin, Germany

Paper in this product is recyclable.

Preface

This book was created as part of my lecture *Introduction to Algebra* in the winter semester 2020/2021, University of Augsburg. Due to the Corona pandemic, it was conducted in the form of *supported self-study*. The concept of the lecture was that the participants prepare and work on the respective sections on their own. Questions could and should be asked online on the lecture platform (on a wiki page for the next lecture). In the online lectures (2x weekly 90 min), the contents were discussed based on the questions asked.

This manuscript is the basis for this concept. It differs from usual textbooks in so far that more emphasis is placed on derivation, thus facilitating self-study. The highlighted boxes *And now?* and *Check it out!* (see below) are the main tools for this purpose.

The primary goal of the presentation is to represent the definitions and results of each chapter as logical developments of very natural questions. Most propositions arise as answers to obvious questions and their proofs or at least the idea for the proof emerge as if by themselves. My idea of learning mathematics is to always ask oneself:

> What question arises most naturally now and which ideas for an answer would I have developed if I was the first person to do so?

Of course, this does not always lead to the desired goal—many theorems are the result of years or even decades of developments. In the explanations of the *And now?* I often try to give hints on the corresponding ideas, so that one can understand such developments and perceives them as completely natural.

And now?

With *And now?* I start considerations that are supposed to lead to the ideas that will follow, in the sense of: "How do we get to this now? What do we actually want to achieve now?". These paragraphs are highlighted. They introduce the subsequent topic and replace remarks that usually would be added orally in a lecture. They are intended to facilitate the self-study of the contents.

Check it out!
These are suggestions, questions, tasks for the readers, which everyone should think about when reading the book. They hopefully help to better understand the content.

They also served as inspiration for possible questions in advance of the lectures: If it was not clear how to deal with such a task, this was an opportunity to ask a question on the corresponding Wiki page or just leave a remark saying: "I have no idea about the task on page XXX."

Common thread to the previous chapter

After each chapter, I briefly write down in bullet points what I consider the *common thread* of the chapter. What is its purpose? One can recapture whether one has recognized the important steps of the chapter. A good question one could ask oneself after having read a chapter is the following: Could I explain the content of the chapter along the common thread to an imaginary conversation partner who knows just as much as I knew **before** reading this chapter—and answer any questions he/she might ask?

Acknowledgement

I would like to thank the students of the lecture *Introduction to Algebra* in the winter semester 2020/2021 at the University of Augsburg for the numerous comments, questions and especially corrections of errors in the manuscript.

Special thanks also go to Ms. M.Sc. Caren Schinko, who supported and accompanied the lecture. She was in close contact with the students and provided many suggestions for the present book, for which I am very grateful to her.

The basic idea for the lecture concept is known under the name "Just in Time Teaching". I learned about it in the context of a cross-university project (BayernMINT) of the Bavarian Ministry of Science to promote MINT degrees from my colleagues in the Physics department of the Technical University of Rosenheim. I thank them, namely Mr. Prof. Dr. Elmar Junker, Mrs. Prof. Dr. Brigit Naumer and Mrs. Prof. Dr. Claudia Schäfle for their suggestions and the exchange of their experiences.

Many thanks to my wife Beate for her tireless support.

The book is dedicated to my late father, who joyfully followed the beginning of this project, but unfortunately could not witness its completion.

Augsburg   Marco Hien
April 2021

Contents

Motivation and Prerequisites

1.1 Goals

The goals we want to achieve are essentially:

- Getting to know algebraic structures—groups, rings, fields—they will appear in many areas of mathematics: algebraic geometry and number theory, algebraic topology, differential geometry, …
- Galois theory and the question of solving polynomial equations in one variable.

1.1.1 Algebraic Structures

A main focus of high-school mathematics lies in the development of the usual range of numbers, which is developed step by step (perhaps one has not yet seen the complex numbers $\mathbb{C}$ in high-school):

$$\mathbb{N}_0 \hookrightarrow \mathbb{Z} \hookrightarrow \mathbb{Q} \hookrightarrow \mathbb{R} \longrightarrow \mathbb{C}$$

These carry different algebraic structures: $\mathbb{N}_0$ is a so-called monoid, it allows an operation $+$ with a neutral element 0. But the elements $\neq 0$ do not have an inverse with respect to $+$. If one extends the numbers to $\mathbb{Z}$, one introduces these inverses $-x$ to $x \in \mathbb{Z}$. Additionally, one can (already in $\mathbb{N}$) consider the multiplication in $\mathbb{Z}$ and obtain a commutative ring with unit. Most elements, however, do not have a multiplicative inverse. To this end, one again expands the range of numbers , introducing fractions, and obtains the rational numbers $\mathbb{Q}$. These form a field. The next step is less algebraic, but comes from analysis: $\mathbb{Q}$ carries the usual absolute value, but is not complete with respect to this—nestings of intervals do not necessarily have to "converge" to a point. Accordingly, one expands the rational numbers and obtains the real numbers $\mathbb{R}$, a so-called completely valued field. In $\mathbb{R}$, there are

M. Hien, *Abstract Algebra*, Mathematics Study Resources 7, https://doi.org/10.1007/978-3-662-67974-6_1

polynomial equations (like $x^2 + 1 = 0$)without roots. So, once again, there is a reason to extend the range of numbers introducing the complex numbers $\mathbb{C}$. These now give a completely valued (concept of analysis), algebraically closed (concept of algebra) field.

In summary, one observes that in each step, one starts from an range of numbers already constructed before, in which many natural calculations can be executed, but which still lacks some natural requirement. The extensions of the range of numbers are always aimed to achieve this requirement and indeed each time are realised by the simplest, most obvious way to supplement the existing range of numbers so that it fulfills the additional wish. In this book, we want to proceed similarly where applicable: We recognize a natural question, a natural desire, and try to find algebraic structures in a simple way so that this is fulfilled.

Another algebraic concept one often encounters is that of a *group*. They already appear in high-school mathematics—for example in combinatorics/probability : How many ways are there to arrange the numbers $1, \ldots, n$? Each arrangement is a bijective mapping

$$\sigma : \{1, \ldots, n\} \to \{1, \ldots, n\},$$

that assigns to the number j its place $\sigma(j)$. These maps can be composed i.e. executed one after the other and one obtains the *symmetric group*

$$S(n) := \{\sigma : \{1, \ldots, n\} \to \{1, \ldots, n\} \mid \sigma \text{ is bijective}\}.$$

In this book, we want to define and examine several algebraic terms as the ones previously mentioned in an abstract way. What properties does a group possess, how can we construct certain groups, what properties will a commutative ring with unity have, what classes of such rings are there? For example, there is the concept of a *Euclidean ring,* and if R is one of those, many statements that are known for the ring of integers $\mathbb{Z}$ also apply automatically to R.

In algebraic geometry, commutative rings with unity play an essential role. If one wants to study the geometry of the set of zeros

$$V := \{(x, y) \in \mathbf{k}^2 \mid y^2 = x^3 - x\} \subset \mathbf{k}^2$$

for a field $\mathbf{k}$ (e.g., $\mathbf{k} = \mathbb{R}$), it is advisable to consider the ring

$$R := \mathbf{k}[x, y]/(y^2 - x^3 - x)$$

(whose definition we will see in the chapter on rings). The geometric properties of the so-called affine variety V (like *smoothness, dimension,...*) can be read off from R.

In algebraic number theory, one considers (finite) extensions of $\mathbb{Q}$—for example, one might want to calculate with certain roots, without having to make the huge leap to $\mathbb{R}$ or $\mathbb{C}$. You may remember from analysis: $\mathbb{R}$ is uncountable, whereas $\mathbb{Q}$ is countable. If $d \in \mathbb{Z}$ is given, one can define

$$\mathbb{Q}(\sqrt{d}) \subset \mathbb{C}$$

as the smallest field that contains $\mathbb{Q}$ and $\sqrt{d}$ (for $d < 0$ we denote by this any complex solution of $x^2 = d$ - there are two choices possible). It contains a *ring of integers* $\mathcal{O}$ and there is the diagram:

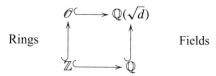

$$\begin{array}{ccc} \mathcal{O} & \hookrightarrow & \mathbb{Q}(\sqrt{d}) \\ \uparrow & & \uparrow \\ \downarrow & & \downarrow \\ \mathbb{Z} & \hookrightarrow & \mathbb{Q} \end{array}$$

Rings   Fields

whose left column consists of commutative rings with unity and whose right column consits of fields. It turns out to be very important which properties the ring $\mathcal{O}$ possesses. For example, in general there might be no unique prime factor decomposition in $\mathcal{O}$. If this were the case, the statement of the *Fermat's conjecture (ca. 1640)* (proven by A. Wiles in 1994)

There are no non-trivial solutions to the equation $x^n + y^n = z^n$ with $x, y, z \in \mathbb{Z}$ for $n \geq 3$.

would have been relatively easy to prove and would not have been an open problem for over 350 years.

1.1.2   Polynomial Equations in One Variable

A classic problem in algebra is to find/examine/understand solutions for polynomial equations in one variable. These are equations of the form

$$a_n X^n + a_{n-1} X^{n-1} + \ldots + a_0 = 0 \tag{1.1}$$

with coefficients a_i in a field **k**. If $a_n \neq 0$, one calls (1.1) an equation of degree n.

For $n = 2$ one has the solution formula (sometimes called *midnight formula* by teachers since their students are supposed to know it by heart at any time of the day or night): The equation $aX^2 + bX + c = 0$ has the solutions

$$x_{1/2} := \frac{-b \pm \sqrt{b^2 - 4ac}}{2a}, \tag{1.2}$$

as soon as there is such a root $\sqrt{b^2 - 4ac}$ in the field **k**. If we start with $\mathbf{k} = \mathbb{Q}$, it may happen that such roots do not exist, but possibly they exist in $\mathbb{R}$ (if not they will exist in $\mathbb{C}$). So, if one passes to $\mathbb{C}$, the solution formula (1.2) can certainly be applied. One could also work in the field

$$\mathbb{Q}(\sqrt{b^2 - 4ac}) \subset \mathbb{C}$$

and could solve the problem remaining in the realm of a field with countably many elements.

One can easily derive the formula (1.2) by completing the square (*quadratic completion*) and never uses the fact that one is working in $\mathbb{R}$ or $\mathbb{C}$, but only that one works in a field. The only point where one has to be careful is when dividing by 2. There are fields where $2 = 0$, and then it is not possible to divide by 2. One can therefore proceed as in the derivation of the formula (including division by 2) in any field where $2 \neq 0$ applies (such fields are then called *fields of characteristic* $\neq 2$). Everything used in this process applies in any field (e.g., the statement $(\pm 1)^2 = 1$). So, in high-school already, the following theorem was proved:

Theorem 1.1 Let **k** be a field of characteristic $\neq 2$ and $a, b, c \in$ **k** elements, such that $a \neq 0$ and such that there is an element $w \in$ **k** with $w^2 = b^2 - 4ac$. We write $\sqrt{b^2 - 4ac} := w$ for such an element. Then the quadratic equation $aX^2 + bX + c = 0$ has the following solutions in **k**:

$$x_{1/2} := \frac{-b \pm \sqrt{b^2 - 4ac}}{2a} \in \mathbf{k}.$$

It is obvious that we are now trying to consider equations of higher degree $n \geq 3$. In doing so, two questions arise, where we again consider a field **k** and try to examine the polynomial equ. (1.1):

1. Can we find a field L with **k** $\subset L$ (being as small as possible), so that (1.1) has a root in L?
2. Do we have a formula analogous to (1.2) for these solutions?

In the case of $n = 2$ the "semi" answer to the first question was: As soon as one finds a field L with **k** $\subset L$ in which $b^2 - 4ac$ has a root, L is a positive answer to the question. For **k** $= \mathbb{Q}$ we could find such a field inside the complex numbers $\mathbb{C}$ if necessary. The answer for a general field **k** is not clear (yet) . We will investigate this in Chap. 10. The answer is positiv and extraordinarily elegant.

The surprising answer to the second question is:

For $n \geq 5$ one can **prove that such a formula cannot exist!** This will be one of the highlights of this book—and will follow from *Galois theory*.

For $n = 3, 4$ there are such formulas, by the way. As an example, I briefly give the formula for $n = 3$. For this, we consider the more specific form of the equation

$$X^3 + pX + q = 0,$$

to which one arrive from the general equation $aX^3 + bX^2 + cX + d = 0$ by division with a and coordinate transformation $X \rightsquigarrow X - \frac{b}{3}$—provided **k** is not of characteristic 3, nor of characteristic 2. Then the **solution formula of Cardano/ Tartaglia, 16th century gives:**

The equation $X^3 + pX + q = 0$ has a solution of the form

$$x = \sqrt[3]{-\frac{q}{2} + \sqrt{\left(\frac{q}{2}\right)^2 + \left(\frac{p}{3}\right)^3}} + \sqrt[3]{-\frac{q}{2} - \sqrt{\left(\frac{q}{2}\right)^2 + \left(\frac{p}{3}\right)^3}},$$

if the specified roots exist in **k**. Here, the first $\sqrt[3]{(1)}$ is any arbitrary 3rd root of the radicand, the second root $\sqrt[3]{(2)}$ must then be chosen such that $3\sqrt[3]{(1)} \cdot \sqrt[3]{(2)} = -p$ holds.

And now?

How should we understand the condition $3\sqrt[3]{(1)} \cdot \sqrt[3]{(2)} = -p$? For sim-
plicity's sake, let's stay within $\mathbb{C}$—then we know that all roots exist. Let's
write $w_1 = \sqrt[3]{(1)}$ for an arbitrary complex number $w_1 \in \mathbb{C}$, such that
$w_1^3 = -\frac{q}{2} + \sqrt{\left(\frac{q}{2}\right)^2 + \left(\frac{p}{3}\right)^3}$. Then we look for another complex number
$w_2 \in \mathbb{C}$ with

$$w_2^3 = -\frac{q}{2} - \sqrt{\left(\frac{q}{2}\right)^2 + \left(\frac{p}{3}\right)^3}. \tag{1.3}$$

There are three complex numbers satisfying this equation (if the right side is
not zero—let's assume that).

If w_2 is one of them, the product $w_1 \cdot w_2$ satisfies

$$(w_1 \cdot w_2)^3 = \left(-\frac{q}{2} + \sqrt{\left(\frac{q}{2}\right)^2 + \left(\frac{p}{3}\right)^3}\right)\left(-\frac{q}{2} - \sqrt{\left(\frac{q}{2}\right)^2 + \left(\frac{p}{3}\right)^3}\right) = -\left(\frac{p}{3}\right)^3.$$

Therefore, (1.3) is equivalent (when $w_1 \neq 0$) to

$$w_2^3 = -\left(\frac{p}{3}\right)^3 \cdot \frac{1}{w_1^3}.$$

Again, there are three solutions, one of which is $w_2 = -\frac{p}{3} \cdot \frac{1}{w_1}$. This is the
choice for Cardano/Tartaglia's formula to hold.

And now?

A small comment on the above formula: To show that it is indeed a solution
formula is almost trivial. One inserts $x = w_1 + w_2$ into the left side of the
equation to be solved $X^3 + pX + q$ and calculates this term. In doing so, one
only uses

- the properties of a field (associativity, distributivity, …),
- the conditions $w_1^3 = -\frac{q}{2} + \sqrt{\left(\frac{q}{2}\right)^2 + \left(\frac{p}{3}\right)^3}$—this is to be read such that

$$\left(w_1^3 + \frac{q}{2}\right)^2 = \left(\frac{q}{2}\right)^2 + \left(\frac{p}{3}\right)^3$$

  applies. Analogously for w_2,
- the condition that $w_1 w_2 = -\frac{p}{3}$, and the result is 0.

So, if one has a candidate for a formula giving a solution, it is trivial to confirm that it holds. The great art is to find such a candidate—either by clever
reshaping (like "completing the square", …) or by "clever guessing". In this
respect, one can understand the despair of all those who have sought such
a formula, for example, for $n = 5$—if only one could find a candidate by

clever reshaping/guessing, one could easily check whether it is correct or not.

The statement that there can be no solution formula of course is of a completely different shape: One must first formalize/abstract the question (when does an equation have a solution formula), and then find a rigorous reason why this is not possible. I give a small idea of what this will be based on.

A group can be assigned to a polynomial equation (the Galois group). If the equation has a solution formula (it will be precisely defined in Sect. 14.2 what is meant by this), one can see that the Galois group then has a certain property (this property is appropriately called *solvable*). However, there are polynomial equations of degree $n = 5$, which do not have a solvable Galois group.

We will need almost the entire content of the book to understand all this. If that's not a motivation!

1.2 Prerequisites

Even though I will briefly give reminders at the appropriate places, I would like to assume that every reader knows/has heard of/is willing to inform him-/herself about the following terms:

- The basic ranges of numbers $\mathbb{Z}, \mathbb{Q}, \mathbb{R}, \mathbb{C}$ and how to calculate within them.
- Basic concepts of linear algebra: Definition of a field, vector space over a field[1], linear mappings $\varphi : V \to W$ between two K-vector spaces, kernel and image thereof.

  Every K-vector space V has a basis and the cardinality of a basis is independent of the choice of the basis—it is called the dimension $\dim_K V$ of V. We will often assume that it is finite.
- Polynomial rings over a field: For a field K, $K[X]$ denotes the polynomial ring in one variable over K, whose elements we refer to as

$$f(X) = a_n X^n + a_{n-1} X^{n-1} + \ldots + a_1 X + a_0 \in K[X]$$

  with $n \in \mathbb{N}_0$ and with coefficients $a_j \in K$. Note: The polynomial in this form is **by definition** exactly described by the tuple of coefficients $(a_n, \ldots, a_0)$—in the sense that: two polynomials are equal if and only if all their coefficients

---

[1] Since in algebra it is often important when speaking about a vector space, which field is underlying, we usually emphasize this and say K-vector space for a vector space over the field K, likewise K-linear mapping, ...

are equal ("comparison of coefficients"). The **degree** of a polynomial $f(X) = a_n X^n + a_{n-1} X^{n-1} + \ldots + a_1 X + a_0 \in K[X]$ is

$$\deg(f) := n, \text{ when } a_n \neq 0.$$

The zero polynomial $f(X) = 0$ has the degree $-\infty$ (this is to be understood in a way that $-\infty < n$ for all $n \in \mathbb{N}_0$ and that $-\infty + n = -\infty$ for all $n \in \mathbb{N}_0 \cup \{-\infty\}$—this avoids case distinctions).

- Division with a remainder of polynomials: If $f, g \in K[X]$ (for a field K) are polynomials and $g \neq 0$, there exist uniquely determined polynomials $q, r \in K[X]$ with

$$f = q \cdot g + r \text{ and } \deg(r) < \deg(g).$$

The proof of its existence is exactly the calculation you hopefully know: If $\deg(f) < \deg(g)$ we are done with $q := 0$ and $r := f$. If $m := \deg(f) \geq \deg(g) =: n$, we write $f(X) = a_m X^m + \ldots + a_0$ and $g(X) = b_n X^n + \ldots + b_0$. We then consider

$$\widetilde{f}(X) = f(X) - \frac{a_m}{b_n} X^{m-n} \cdot g(X) \in K[X].$$

We obtain $\deg(\widetilde{f}) < \deg(f)$ (because we have just eliminated the highest order summand). Now, one can elegantly conduct the proof as an induction on $\deg(f)$ and apply the induction hypothesis to $\widetilde{f}$—or one simply continues successively (this is what is done in the well-known algorithm) with $\widetilde{f}$ and g exactly the same way, until one finally descends below $\deg(g)$.

The uniqueness follows from degree reasons, because

$$qg + r = \widetilde{q}g + \widetilde{r} \Rightarrow (q - \widetilde{q})g = \widetilde{r} - r$$

and the right side has degree $< \deg(g)$ according to our assumption This is only possible if $\widetilde{r} - r = 0$ and then also $\widetilde{q} = q$.

- The direct consequence of this: If K is a field, $f(X) \in K[X]$ is a polynomial and $\alpha \in K$, such that $f(\alpha) = 0$ holds, then one can express $f(X)$ in the form

$$f(X) = (X - \alpha) \cdot g(X)$$

with $g(X) \in K[X]$. It then follows that $\deg(g) = \deg(f) - 1$. In particular, a polynomial of degree n in a field has at most n roots.

- Commutative rings with unity: They will occupy several chapters in this book. The basic definition is usually known from linear algebra. A **commutative ring with unity** is a non-empty set R together with two operations $+$ and $\cdot$ with neutral elements 0 (regarding $+$) and 1 (regarding $\cdot$), which fulfill all axioms of a field, **except for the requirement,** that every $x \in R \setminus \{0\}$ has a multiplicative inverse. A formal definition or reminder can be found in Definition 5.2 in Chap. 5.

Examples: $\mathbb{Z}$, $K[X]$ for a field K. Every field K is also a commutative ring with unity (in which this additional axiom of the multiplicative inverse holds!).

Field Extensions and Algebraic Elements

2

2.1 Field Extensions

The concept of a field is usually encountered in Linear Algebra. Let me remind you that it is a set K with two operations $+$ and $\cdot$, both of which are commutative, associative and together are distributive, addition has a neutral element $0 \in K$, multiplication has a neutral element $1 \in K$, every element $x \in K$ has an additive inverse $-x \in K$ and every $x \in K \setminus \{0\}$ has a multiplicative inverse $x^{-1} \in K$. For a formal definition (or a reminder), see also Remark 5.3.2.

We now consider several fields at the same time:

Definition 2.1 Let L be a field. A subset $K \subset L$ is called a **subfield,** if for all $a, b \in K$ we have $a + b, a \cdot b \in K$ and if $(K, +, \cdot)$ with the restrictions of the operations

$$K \times K \to K, (a, b) \mapsto a + b \text{ or } K \times K \to K, (a, b) \mapsto a \cdot b$$

is again a field—in particular, $0 \in K$ and $1 \in K$ and for every $a \in K$ we have $-a \in K$ and, if $a \neq 0$ also $a^{-1} \in K$.

The pair $K \subset L$ is called a **field extension.** We will write $L|K$ (pronounced "L over K").

And now?

If L is a field and $K \subset L$ a subset such that K is closed under addition and multiplication as above, and hence one can restrict these operations to mappings $K \times K \to K$, note the following:

- The properties of being *commutative,associative* and *distributive* is naturally inherited to the restriction.
- If also $0, 1 \in K$ holds, then these are again the neutral elements with respect to addition and multiplication—they are even neutral for all $z \in L$

To check if K is a subfield, one only needs to show that $-1 \in K$ (which also implies $-x \in K$ for $x \in K$) and that for every $x \in K \setminus \{0\}$ the *multiplicative inverse* $x^{-1} \in L$ (which exists in L) is already in K.

Examples

1. $\mathbb{C}|\mathbb{R}$ or $\mathbb{R}|\mathbb{Q}$ are field extensions.
2. Let $d \in \mathbb{Z} \setminus \{0, 1\}$ be a square-free number (i.e., each prime factor of d appears only in simple power, -1 also counts as square-free). Consider the set

$$\mathbb{Q}(\sqrt{d}) = \{a + b\sqrt{d} \mid a, b \in \mathbb{Q}\} \subset \mathbb{C},$$

where $\sqrt{d} \in \mathbb{C}$ is a complex number with $(\sqrt{d})^2 = d$—we choose one of the two complex numbers with this property.
Claim $K := \mathbb{Q}(\sqrt{d}) \subset \mathbb{C}$ is a subfield.
Proof For $a + b\sqrt{d}, \alpha + \beta\sqrt{d} \in K$ we have:

$$(a + b\sqrt{d}) + (\alpha + \beta\sqrt{d}) = (a + \alpha) + (b + \beta)\sqrt{d} \in K,$$
$$(a + b\sqrt{d}) \cdot (\alpha + \beta\sqrt{d}) = (a\alpha + b\beta d) + (a\beta + b\alpha)\sqrt{d} \in K.$$

Hence we have the operations $+, \cdot$ as mappings $K \times K \to K$. Apparently, $0, 1, -1 \in K$ holds. As discussed in the ***And now?*** above, one only needs to show that every $a + b\sqrt{d} \neq 0$ has a multiplicative inverse in K. Such an inverse exists in $\mathbb{C}$ and this can be transformed as follows—"making the denominator rational":

$$\frac{1}{a + b\sqrt{d}} = \frac{a - b\sqrt{d}}{(a + b\sqrt{d})(a - b\sqrt{d})} = \frac{a - b\sqrt{d}}{a^2 - b^2 d} = \frac{a}{a^2 - b^2 d} + \frac{-b}{a^2 - b^2 d}\sqrt{d} \in K.$$

(Note that $a^2 - b^2 d \neq 0$ must hold, because d is square-free). $\square$
So, $\mathbb{C}|\mathbb{Q}(\sqrt{d})$ is a field extension.
3. The construction from the previous example also gives that $\mathbb{Q}(\sqrt{d})|\mathbb{Q}$ is a field extension.

And now?
Constructions like in the second/third example will now be examined more generally! One could also proceed step by step. For example, try to construct a field that deserves to be called $\mathbb{Q}(\sqrt[3]{7})$.

Check it out!
In analogy to the second example, try to describe a good construction for a subfield $K := \mathbb{Q}(\sqrt[3]{7}) \subset \mathbb{C}$ such that $\mathbb{Q} \subset K$, furthermore $\sqrt[3]{7} \in K$ and K is constructed to be as "simple" as possible! We will discuss this now so you might want to give it a try first. However note that it may not be entirely easy to prove that your attempt yields the correct result—therefore, the task is only about finding the appropriate idea. I will discuss this immediately.

And now?
To complete the task, it is immediately apparent that, analogous to above, all numbers of the form $a + b\sqrt[3]{7} \in \mathbb{C}$ (even in $\mathbb{R}$, but that is irrelevant here) with $a, b \in \mathbb{Q}$ must lie in K (because K is supposed to be closed under $+, \cdot$ and $a, b, \sqrt[3]{7} \in K$ should hold). But this is not enough, because

$$\left(\sqrt[3]{7}\right)^2 = \left(\sqrt[3]{7}\right) \cdot \left(\sqrt[3]{7}\right) \in K$$

must hold also and this number is not of the above form. So the correct approach could be

$$K := \mathbb{Q}(\sqrt[3]{7}) = \{a + b\sqrt[3]{7} + c\left(\sqrt[3]{7}\right)^2 \in \mathbb{C} \mid a, b, c \in \mathbb{Q}\}.$$

It is easy to see that K is closed with respect to $+, \cdot$, and that $0, 1, -1 \in K$ holds. It remains to show that for every $x \in K \smallsetminus \{0\}$ the inverse $x^{-1} \in \mathbb{C}$ (as said, even in $\mathbb{R}$, but that is irrelevant here) also lies in K.

So we need to show that $a + b\sqrt[3]{7} + c\left(\sqrt[3]{7}\right)^2 \in K \smallsetminus \{0\}$ has a multiplicative inverse in K, or that the inverse

$$\frac{1}{a + b\sqrt[3]{7} + c\left(\sqrt[3]{7}\right)^2} \in \mathbb{C}$$

already lies in K! These are calculations that Cardano/Tartaglia had to do in the 16th century to arrive at their solution formula.

1. One can perform this **analogously to above: "make the denominator rational"**—a tedious calculation: Let $\zeta = \exp(2\pi i/3)$. Then, additionally to $\sqrt[3]{7}$ the complex number $\zeta \cdot \sqrt[3]{7}$ also is a third root of 7. In addition to $a + b\sqrt[3]{7} + c\left(\sqrt[3]{7}\right)^2$ we hence consider

$$a + b\zeta\sqrt[3]{7} + c\left(\zeta\sqrt[3]{7}\right)^2 \text{ and } a + b\zeta^2\sqrt[3]{7} + c\left(\zeta^2\sqrt[3]{7}\right)^2$$

(analogous to example (2), where we considered not only $\sqrt{d}$ but also $-\sqrt{d}$ and then expanded the fraction $1/(a + b\sqrt{d})$ by the factor $a + b(-\sqrt{d})$). If we expand with these two factors, we get

$$\frac{1}{a + b\sqrt[3]{7} + c\left(\sqrt[3]{7}\right)^2} \cdot \frac{(a + b\zeta\sqrt[3]{7} + c\left(\zeta\sqrt[3]{7}\right)^2)(a + b\zeta^2\sqrt[3]{7} + c\left(\zeta^2\sqrt[3]{7}\right)^2)}{(a + b\zeta\sqrt[3]{7} + c\left(\zeta\sqrt[3]{7}\right)^2)(a + b\zeta^2\sqrt[3]{7} + c\left(\zeta^2\sqrt[3]{7}\right)^2)}$$

The numerator is in K and we only need to see that the denominator is in $\mathbb{Q}$ and we are done. The calculation of the denominator is a tedious calculation, which we will now omit to carry out. If one computes the denominator one will have to use $\zeta^3 = 1$ as well as $1 + \zeta + \zeta^2 = 0$ —note that $\zeta = \frac{-1+i\sqrt{3}}{2} \in \mathbb{C}$ and $\zeta^2 = \frac{-1-i\sqrt{3}}{2}$. Galois theory will help us avoid this type of calculations!!!

2. One can also try directly to find an inverse in K: we need a candidate $x + y\sqrt[3]{7} + z\sqrt[3]{7}^2 \in K$, such that

$$1 = (a + b\sqrt[3]{7} + c\sqrt[3]{7}^2)(x + y\sqrt[3]{7} + z\sqrt[3]{7}^2)$$

holds. Multiplying the two brackets with each other results in a system of linear equations for x,y,z and one has to prove that it has a solution—this is Linear Algebra. It is very easy to do so with the Gaussian algorithm for explicitely given a,b,c. Example: For if we want to compute the inverse of $1 - 2\sqrt[3]{7} + 5\sqrt[3]{7}^2$ we come up with the equation

$$1 = (1 - 2\sqrt[3]{7} + 5\sqrt[3]{7}^2)(x + y\sqrt[3]{7} + z\sqrt[3]{7}^2)$$
$$= (x + 35y - 14z) + (-2x + y + 35z)\sqrt[3]{7} + (5x - 2y + z)\sqrt[3]{7}^2,$$

and find a solution, if $x, y, z \in \mathbb{Q}$ solves the linear equation

$$\begin{pmatrix} 1 & 35 & -14 \\ -2 & 1 & 35 \\ 5 & -2 & 1 \end{pmatrix} \cdot \begin{pmatrix} x \\ y \\ z \end{pmatrix} = \begin{pmatrix} 1 \\ 0 \\ 0 \end{pmatrix} \tag{2.1}$$

This has as a solution being the first column of the inverse matrix, namely $(x, y, z) = (\frac{71}{6280}, \frac{177}{8280}, \frac{-1}{6280})$, so we finally obtain that

$$y := \frac{71}{6280} + \frac{177}{8280}\sqrt[3]{7} + \frac{-1}{6280}\sqrt[3]{7}^2 \in K \tag{2.2}$$

is the multiplicative inverse to $x = 1 - 2\sqrt[3]{7} + 5\sqrt[3]{7}^2$.

In SAGE, see https://sagemath.org the lines

```
R.<t>=QQ[]
S=R.quotient(t^3-7,'a')
a=S.gen()
x=1-2*a+5*a^2
y=71/6280+177/6280*a-1/6280*a^2
x*y
```

confirm that y is an inverse to x. As a preview/teaser—in the process of this book we will get to know all the following objects and statements: Line 1 defines the polynomial ring over $\mathbb{Q}$ in one variable t. Line 2 defines S as the factor ring (which we will see in Sect. 5.4) modulo the ideal $(t^3 - 7)$. This is (essentially) the field K. In this field, a is a generator and thus takes over the role $a = \sqrt[3]{7}$. The elements x and y are then defined as given/calculated above. The last line calculates $x \cdot y$ in K and if you run that in SAGE, you get the result 1.

The last idea from the *And now?*—using linear algebra—leads to the considerations that will follow shortly. For this, let me recall the (abstract) concept of a **vector space over a field** K. This is a non-empty set V with an addition (commutative and associative), which has a neutral element $0 \in V$ and for every $v \in V$ there is a $-v \in V$ with $v + (-v) = 0$. In addition, there is a **scalar** multiplication (the elements of the field K are often called the **scalars** in linear algebra)

$$K \times V \to V, (x, v) \mapsto xv$$

with further properties (see Linear Algebra 1).

Basic—trivial—observation: If $L|K$ is a field extension, then we have an addition on L which has a neutral element $0 \in L$. Furthermore, there is even a multiplication on L *in the form* $L \times L \to L$. We can restrict this to $K \times L \subset L \times L$—i.e. we only allow elements in K *as the first factor*—and obtain a mapping

$$K \times L \to L, \ (x, z) \mapsto x \cdot z \tag{2.3}$$

(with the $\cdot$ from L, the first factor is by choice in K). Hence we have all the data necessary to describe a K-vector space : the set $V = L$ has an addition and (2.3) can be seen as a scalar multiplication. The conditions that are imposed on a vector space can all be checked rather easily. Hence we have seen:

Lemma 2.2 *If $L|K$ is a field extension, then L with its addition and the restriction of multiplication to $K \times L \to L$ is naturally a K-vector space.*

Therefore, we now have the entire linear algebra available to study field extensions. For example, the concept of the *dimension of a vector space*—which is related to the fact that every K-vector space has a basis and the cardinality of a basis is independent of the choice of the basis.

Definition 2.3 Let $L|K$ be a field extension. It is called **finite** if L is finite-dimensional as a K-vector space. We call the dimension of L as a K-vector space the **degree of the field** and introduce the notation:

$$[L : K] := \dim_K(L).$$

Examples For $d \in \mathbb{Z} \setminus \{0, 1\}$ the field extension $\mathbb{Q}(\sqrt{d})|\mathbb{Q}$ is finite and we have $[\mathbb{Q}(\sqrt{d}) : \mathbb{Q}] = 2$, if d is square-free.

Proof Let d be square-free. The two elements $1, \sqrt{d} \in \mathbb{Q}(\sqrt{d})$ are a $\mathbb{Q}$-basis, as we can easily prove as follows:

- They generate the vector space K which was defined to be the set of all linear combinations

$$K = \{a \cdot 1 + b \cdot \sqrt{d} \in \mathbb{C} \mid a, b \in \mathbb{Q}\}$$

  of 1 and $\sqrt{d}$ with coefficients in $\mathbb{Q}$. So, these two vectors are generators. This step also works for non-square-free d and shows that the dimension is certainly ≤ 2.
- Linearly independent: Let $a, b \in \mathbb{Q}$ be such that $a \cdot 1 + b \cdot \sqrt{d} = 0$ holds. We have to prove that $a = b = 0$. Note that

$$0 = a + b\sqrt{d} \Rightarrow b\sqrt{d} = -a \Rightarrow b^2 \cdot d = a^2.$$

Now, $a, b \in \mathbb{Q}$, let's say $a = \frac{m}{n}$ and $b = \frac{r}{s}$. We deduce that

$$\frac{m^2}{n^2} = \frac{r^2}{s^2}d \Rightarrow (ms)^2 = (rn)^2 d.$$

The latter is an equation in $\mathbb{Z}$. Now use the unique prime factorization in $\mathbb{Z}$. Since every prime factor on the left side appears in an even power, and every prime factor of d either does not appear at all in $(rn)^2$ on the right side or appears in an even power, it must appear in d in an even power also. This is a contradiction to the assumption that d is square-free. $\square$

Short remark: asking for d to be square-free is most natural and harmless. If d is not square-free, we have prime factors with even power and those with odd power in the prime factorization. If we put all even powers of prime factor into a new factor d' (including to it the sign of d), we obtain a representation of d as a product

$$d = b^2 \cdot d'$$

with $b, d' \in \mathbb{Z} \setminus \{0\}, b > 0$ and d' being square-free—note that it could happen that $d' = \pm 1$. Then $\sqrt{d} = b\sqrt{d'}$ and thus we get

$$\mathbb{Q}(\sqrt{d}) = \mathbb{Q}(\sqrt{d'}).$$

The only special case occurs if $d' = 1$, then $\mathbb{Q}(\sqrt{d}) = \mathbb{Q}$ and thus the degree of the field $[\mathbb{Q}(\sqrt{d}) : \mathbb{Q}] = 1$.

And now?
Recall that for a field extension $L|K$ the larger field L is a vector space over the smaller field K. We recommend to take a closer look at the previous example. It is very helpful for what follows.

The difficulty one sometimes encounters is that several algebraic structures occur simultaneously and one should—depending on the question—temporarily *forget* certain of them and *remember* them at the appropriate place. I will try to explain this:

The elements $1, \sqrt{d} \in \mathbb{Q}(\sqrt{d})$ are the *vectors* in the vector space. Let's briefly call them $v_1 = 1$ and $v_2 = \sqrt{d}$. The fact that $K \subset L$ and that the vector $v_1 = 1$ could also be read as an element in K, plays no role here—we simply *forget* this, now we are only talking about L as a K-vector space forgetting any additional structure. When are the two vectors v_1, v_2 in the K-vector space L linearly independent? By definition, if every linear combination of the zero vector

$$a_1 v_1 + a_2 v_2 = 0 \text{ for } a_1, a_2 \in K$$

requires $a_1 = a_2 = 0$. The term $a_1 v_1 + a_2 v_2$ is defined now *remembering*, that $K \subset L$ and that we have the two operations $+, \cdot$ available in L.

We will encounter such situations again soon, and I will then return to the topic of *forgetting* and *remembering*.

We will need a statement from linear algebra, which I would like to briefly recall—the theorem on the rank of a linear map. Start with a K-linear map $\varphi : V \to W$ between two K-vector spaces (i.e. with $\varphi(v_1 + v_2) = \varphi(v_1) + \varphi(v_2)$ and $\varphi(av) = a\varphi(v)$ for $v_1, v_2, v \in V$ and $a \in K$). Let's briefly recall the notion of the **kernel** and the **image** of φ:

$$\ker(\varphi) = \{v \in V \mid \varphi(v) = 0\}, \operatorname{im}(\varphi) = \{\varphi(v) \in W \mid v \in V\}$$

Both are sub-vector spaces in V and W respectively.

The concept of dimension of a vector space can be useful in order to prove the following:

Lemma 2.4 (linear algebra) *If W is a finite-dimensional K-vector space and $U \subset W$ is a subspace with $\dim_K U = \dim_K W$, then $U = W$.*

The following theorem will be important soon:

Theorem 2.5 (Theorem on the rank of a linear map) Let V, W be two K-vector spaces and $\varphi : V \to W$ be a K-linear map. Then the following formula holds:

$$\dim_K(V) = \dim_K(\ker(\varphi)) + \dim_K(\operatorname{im}(\varphi))$$

(The formula has to be interpreted in a meaningful way if $\dim_K V$ is not finite. Then at least one of the two summands on the right is infinite as well.)

Corollary 2.6 Let V be a finite-dimensional K-vector space and $\varphi : V \to V$ a linear map. Then the following holds:

$$\varphi \text{ is injective} \iff \varphi \text{ is surjective}$$

Proof φ is injective if and only if $\ker(\varphi) = \{0\}$, that is, if and only if $\dim_K \ker(\varphi) = 0$. On the other hand, φ is surjective if and only if $\text{im}(\varphi) = V$ and therefore due to the lemma above if and only if $\dim_K \text{im}(\varphi) = \dim_K V$ holds. Since

$$\dim_K \ker(\varphi) = \dim_K V - \dim_K \text{im}(\varphi)$$

the two conditions are obviously equivalent to each other. $\square$

2.2   Intermediate Fields and Algebraic Elements

Let us look at constructions like $\mathbb{Q}(\sqrt{d})$ in a more general way.

Definition 2.7 Given a field extension $L|K$, a subfield $M \subset L$ with $K \subset M \subset L$ is called an **intermediate field** of $L|K$.

Note 2.8 If M, M' are intermediate fields of $L|K$, their intersection $M \cap M'$ is so as well—this is easy to see, one can even generalize this further:

If $M_i \subset L$ is a family of intermediate fields, indexed by a set I, then

$$K \subset \bigcap_{i \in I} M_i \subset L$$

is an intermediate field.

Proof For $a, b \in \bigcap_{i \in I} M_i$ by definition $a, b \in M_i$ for all $i \in I$. Since the M_i are intermediate fields, also $a + b, ab \in M_i$ for all $i \in I$, thus $a + b, ab \in \bigcap_{i \in I} M_i$. Furthermore, we have $K \subset \bigcap_{i \in I} M_i$ and thus $0, 1$ and -1 are elements of the intersection.

If $x \in \bigcap_{i \in I} M_i$, then the inverse $x^{-1} \in L$ is again contained in all M_i, thus $x^{-1} \in \bigcap_{i \in I} M_i$. $\square$

Defining Lemma 2.9 Let $L|K$ be a field extension and $S \subset L$ be any subset. Then there exists a unique intermediate field, written $K(S)$, such that $K \subset K(S) \subset L$,

1. $S \subset K(S)$ and
2. every intermediate field M of $L|K$ with $S \subset M$ satisfies $K(S) \subset M$.

In other words: $K(S)$ is the minimal (in terms of the inclusion relation "$\subset$") intermediate field with $S \subset K(S)$.

We say, $K(S)$ is constructed inside $L|K$ **by adjoining the elements from** S **to** K. If $S = \{\alpha\}$ is a singleton with $\alpha \in L$, we also write more briefly

$$K(\alpha) := K(\{\alpha\}).$$

Similarly for finite sets $S = \{\alpha_1, \ldots, \alpha_r\}$ we will write:

$$K(\alpha_1, \ldots, \alpha_r) := K(\{\alpha_1, \ldots, \alpha_r\}).$$

Proof Uniqueness follows immediately from 2. in the definition: If we consider two candidates $K(S)_1$ and $K(S)_2$, then we deduce from property 2. applied to $K(S)_1$ and $M := K(S)_2$, that $K(S)_1 \subset K(S)_2$. Swapping the roles, we deduce that $K(S)_2 \subset K(S)_1$ also holds. For the existence, consider the set

$$I := \{M \text{ intermediate field of } L|K \text{ with } S \subset M\}$$

as an index set as above and for each index $M \in I$ let M be the corresponding intermediate field. Then

$$K(S) := \bigcap_{M \in I} M \tag{2.4}$$

has the desired properties. $\square$

Brief important remark Further above we had considered the field $\mathbb{Q}(\sqrt{d})$ as an intermediate field of $\mathbb{C}|\mathbb{Q}$, but **not in the way we defined the intermediate field by adjoining elements now,** but we defined it as the 2-dimensional $\mathbb{Q}$-vector space $\mathbb{Q}(\sqrt{d}) = \mathbb{Q} \cdot 1 \oplus \mathbb{Q}\sqrt{d} \subset \mathbb{C}$. We will now see that this gives the same fields and how this kind of ideas generalizes.

And now?

Recall the ***Check it out!*** and the ***And now?*** above, in which we, considered the 3-dimensional $\mathbb{Q}$-vector space

$$\mathbb{Q}(\sqrt[3]{7}) = \mathbb{Q} \cdot 1 \oplus \mathbb{Q} \cdot \sqrt[3]{7} \oplus \mathbb{Q} \cdot \sqrt[3]{7}^2 \subset \mathbb{C} \tag{2.5}$$

(we did not show that $1, \sqrt[3]{7}, \sqrt[3]{7}^2$ are linearly independent over $\mathbb{Q}$—this is an exercise). According to Definition 2.9, the notion used for this field is actually defined as

$$\mathbb{Q}(\sqrt[3]{7}) = \text{ smallest intermediate field in } \mathbb{C}|\mathbb{Q}, \text{ containing } \sqrt[3]{7} \tag{2.6}$$

Of course, I used the same symbol in two different ways because the result will be the same anyway. But what difficulties do we have in proving this?

In the ***And now?*** above not all arguments were completely carried out: the essential issue that (2.5) is indeed a field, i.e. that every $a + b\sqrt[3]{7} + c\sqrt[3]{7}^2 \neq 0$ has an inverse $x + y\sqrt[3]{7} + z\sqrt[3]{7}^2$ was left open in general. If we know this, we are done, because obviously the set defined in (2.5) lies inside of $\mathbb{Q}(\sqrt[3]{7})$ as in (2.6), and if (2.5) is already a field by itself, then it automatically is the smallest intermediate filed containing $\sqrt[3]{7}$

The corresponding condition

$$(a + b\sqrt[3]{7} + c\sqrt[3]{7}^2) \cdot (x + y\sqrt[3]{7} + z\sqrt[3]{7}^2) = 1 \tag{2.7}$$

leads to a linear system of equations, which by multiplying the two brackets in (2.7) and using that $1, \sqrt[3]{7}, \sqrt[3]{7}^2$ is a $\mathbb{Q}$-basis can be written in the form:

$$\begin{pmatrix} a & 7c & 7b \\ b & a & 7c \\ c & b & a \end{pmatrix} \cdot \begin{pmatrix} x \\ y \\ z \end{pmatrix} = \begin{pmatrix} 1 \\ 0 \\ 0 \end{pmatrix}$$

We expect this system of equations to have a unique solution (since there should be a unique inverse), so the matrix, let's call it A, should be invertible. We can try to prove this by hand, for example by computing $\det(A) = a^3 + 7b^3 + 49c^3 - 21abc$. Writing a,b,c as fractions, multiplying with the least common denominator, considering prime factorizations, one can indeed show that $\det(A) \neq 0$, whenever $(a, b, c) \neq (0, 0, 0)$ —this is a very tedious computation and unfortunately not easily generalizable to arbitrary intermediate fields of the form $K(\alpha)$!!!

The question now is: Can one find an elegant (and ideally immediately generalizable) argument where we do not have to calculate such a determinant in a complicated way -- or at least prove its non-vanishing? The answer is Yes, and we will evoke our ideas of of *Forgetting* and *Remembering* at the appropriate place. The situation is the following: Let's write $V := (2.5)$ (so by definition it is the vector space with basis $1, \sqrt[3]{7}, \sqrt[3]{7}^2$) and consider

$$0 \neq \alpha := a + b\sqrt[3]{7} + c\sqrt[3]{7}^2 \in \mathbb{Q}(\sqrt[3]{7}).$$

We *remember* now that V is not only a $\mathbb{Q}$-vector space, but also that V is included in the complex numbers, $V \subset \mathbb{C}$, that $\mathbb{C}$ is a field and V is closed with respect to addition and multiplication. Therefore, *multiplication with α is a map*

$$\varphi : V \to V, \ v \mapsto \alpha \cdot v.$$

It is $\mathbb{Q}$-linear—easy to check, follows from the commutativity of $\cdot$ and the distributivity in $\mathbb{C}$. How do we compute the matrix B which describes φ with respect to the basis we are considering, namely $1, \sqrt[3]{7}, \sqrt[3]{7}^2$? To this end, let

$$\psi : \mathbb{Q}^3 \to V$$

be the basis isomorphism $e_1 \mapsto 1, e_2 \mapsto \sqrt[3]{7}$ and $e_3 \mapsto \sqrt[3]{7}^2$ (with e_i denoting the standard basis vectors in $\mathbb{Q}^3$). Then the matrix B is exactly the one that makes the following diagram commutative:

$$
\begin{array}{ccc}
V & \xrightarrow{\ \varphi\ } & V \\
\psi \big\uparrow \cong & & \psi \big\uparrow \cong \\
\mathbb{Q}^3 & \xrightarrow{\ B\ } & \mathbb{Q}^3
\end{array}
$$

and it turns out that B is **exactly the matrix A we considered above:** $B = A$ –*check this directly!!!*. We want to prove that A is an isomorphism. To do this, it is sufficient to show that A is injective—Corollary 2.6. But this is clear from our *remembering idea*: We have $V \subset \mathbb{C}$ and multiplication with an element $0 \neq \alpha$ in $\mathbb{C}$ is injective, thus it is also injective when restricted to V. There is nothing more to prove. To add some more explanations consider the commutative diagram:

$$
\begin{array}{ccc}
\mathbb{C} & \xrightarrow{\;\alpha\cdot\;} & \mathbb{C} \\
\uparrow & & \uparrow \\
V & \xrightarrow{\;\varphi\;} & V
\end{array}
$$

where the upper line is injective (even bijective with inverse mapping $\frac{1}{\alpha} \cdot \_$), and therefore φ is injective as well, thus also bijective due to corollary (2.6). You can also put it this way: $1 \in V$ lies in the image of φ, hence there exists a $v \in V$ with $\alpha \cdot v = 1$.

Now, guided by the example from the *And now?*, let us carry out the general situation: Let $L|K$ be a field extension.

- First, let's consider the polynomial ring $K[X]$. We have the injective map

$$K[X] \to L[X], \ f(X) \mapsto f(X).$$

- If $\alpha \in L$ is an element, then we have the **homomorphism given by inserting the element:**
$$L[X] \to L, \ f(X) \mapsto f(\alpha).$$

- The concatenation of these two maps yields the **homomorphism on $K[X]$**

$$K[X] \hookrightarrow L[X] \to L, \ f(X) \mapsto f(\alpha).$$

Definition 2.10 If $L|K$ is a field extension and $\alpha \in L$, we denote by $K[\alpha]$ the image of this homomorphism:

$$K[\alpha] := \{ f(\alpha) \in L \mid f(X) \in K[X] \} \subset L.$$

Note 2.11 Obviously, $K[\alpha] \subset L$ is a K-subvector space. But it has an additional structure, it is a **subring**—see the next definition.

Definition 2.12 Let R be a commutative ring with unity. A **subring** is a subset $S \subset R$, which is closed under addition and multiplication, and such that $0, 1 \in S$ holds and for every $x \in S$ we have $-x \in S$.

Remark 2.13 Every subring S is itself a commutative ring with unity, the addition and multiplication being the restrictions of the original ones on R.

Let's consider the situation from Definition 2.10: We consider some $\alpha \in L$ for a field extension $L|K$. We associate to this element the following two objects:

1. The subring $K[\alpha]$ of L. It is a commutative ring with unity and at the same time a K-vector space.
2. The intermediate field $K(\alpha)$ of $L|K$. It is by itself a field extension $K(\alpha)|K$ and thus also a K-vector space. Furthermore, it is by definition a field.

And now?
A bit more informally, I would like to phrase it like this: The two objects $K[\alpha]$ and $K(\alpha)$ have different "qualities" or "flavors".

- $K[\alpha]$ is *easily accessible*, we know the elements immediately, they are exactly the polynomial expressions

$$a_n\alpha^n + a_{n-1}\alpha^{n-1} + \ldots + a_0 \in L$$

with $a_j \in K$. For example, the structure as a K-vector space is very easy to understand, the (although countably many, but that doesn't matter for now) elements

$$1, \alpha, \alpha^2, \alpha^3, \ldots$$

are a generating system.[1] over K. Therefore, the structure as a vector space is given in *very explicit terms.*. **On the other hand side**, it is not easy to investigate general properties—in particular, considering the question whether $K[\alpha]$ is a field (i.e. if the "missing" axiom on the multiplicative inverse also holds)—as we have seen in the example $\mathbb{Q}(\sqrt[3]{7})$.
- $K(\alpha)$ is, by definition, equipped with the nice property of being a field. **On the other hand side**, we do not a priori know its elements explicitly (see the definition (2.4)). Although one can show that

$$K(\alpha) = \left\{ \frac{f(\alpha)}{g(\alpha)} \in L \mid f(X), g(X) \in K[X] \text{ and } g(\alpha) \neq 0 \right\}$$

---

[1] if this was not mentioned in Linear Algebra: An infinite tuple $(v_l)_{l \in I}$ (with index set I) in a K-vector space V is called a **generating system,** if for every $v \in V$ there are finitely many indices $i_1, \ldots, i_r \in I$ and corresponding elements $a_{i_v} \in K$, such that $v = a_{i_1}v_{i_1} + \ldots + a_{i_r}v_{i_r}$ holds.

holds (see below), this is far from being as nice as in the situation of $K[\alpha]$
—for example, there are several representations of the same element:
$\frac{f(\alpha)}{g(\alpha)} = \frac{h(\alpha)}{\ell(\alpha)}$, we have to understand the condition $g(\alpha) \neq 0$, and we don't
easily see generators of $K(\alpha)$ as a K-vector space.
In summary:

	K-Vector space structure	Field?
$K[\alpha]$	easy to access	difficult to decide
$K(\alpha)$	hard to access	trivial by definition

The beauty of the next Theorem we will prove below lies in the fact that in
the situations we are mostly interested in (Definition 2.14 below) both constructions yield the same result. We have already discussed this in the example $\mathbb{Q}(\sqrt[3]{7})$.

First, let us give a definition, which $\alpha \in L$ will be of special interest to us.

Definition 2.14 Let $L|K$ be a field extension. An element $\alpha \in L$ is called **algebraic
over** K, if there is a polynomial $f(X) \in K[X] \setminus \{0\}$ such that $f(\alpha) = 0$ holds.
   If $\alpha \in L$ is not algebraic over K, it is called **transcendental over** K.

Remark 2.15

1. Every $\alpha \in K \subset L$ is algebraic over K, because one can choose
   $f(X) = X - \alpha \in K[X]$ as such a polynomial.
2. The element $\sqrt[3]{7} \in \mathbb{C}$ is algebraic over $\mathbb{Q}$, because one can choose
   $f(X) = X^3 - 7 \in \mathbb{Q}[X]$ as such a polynomial.
3. The number $\pi \in \mathbb{C}$ is transcendental over $\mathbb{Q}$—but this is not easy to show and is
   usually referred to as Lindemann's theorem.

Defining Lemma 2.16 If $L|K$ is a field extension and $\alpha \in L$ is algebraic over K,
then there exists a unique monic[2] polynomial $p(X) \in K[X]$ of minimal degree with
$p(\alpha) = 0$. It is called **the minimal polynomial of α over** K.
   If $f(X) \in K[X]$ is an arbitrary polynomial with $f(\alpha) = 0$, then $p(X)$ is
a divisor of $f(X)$—written: $p(X)|f(X)$—i.e., there exists a $q(X) \in K[X]$ with
$f(X) = p(X) \cdot q(X)$.

Proof The existence of such a polynomial $\neq 0$ with minimal degree is easy to
see: The set of polynomials $f(X) \neq 0$ with $f(\alpha) = 0$ is not empty, because α is

---

[2]i.e. of the form $f(X) = X^n + a_{n-1}X^{n-1} + \ldots + a_0$ with leading coefficient 1. Also note: If you
have any polynomial $q(X)$ of degree n with $q(\alpha) = 0$, you can divide by the leading coefficient
and obtain a monic polynomial $p(X)$ of the same degree n with $p(\alpha) = 0$.

algebraic over K, so there is one among them with minimal degree. We can divide by the leading coefficient and obtain a monic polynomial.

If $f(X) \in K[X]$ is arbitrary with $f(\alpha) = 0$, without loss of generality $f \neq 0$, then we perform division with remainder:

$$f(X) = q(X)p(X) + r(X) \text{ with } \deg(r) < \deg(p).$$

Inserting α and using $f(\alpha) = 0$ and $p(\alpha) = 0$ we obtain that $r(\alpha) = 0$. However, since $\deg(r) < \deg(p)$ and p has minimal degree among all polynomials$\neq 0$ with $f(\alpha) = 0$, we deduce that $r(X) = 0$.

This immediately also gives the uniqueness, because if there are two candidates $p(X)$ and $\widetilde{p}(X)$ with minimal degree, then $p(X)|\widetilde{p}(X)$ and vice versa. Since both are monic, they are already equal. $\square$

Note 2.17 If $\alpha \in L$ and $f(X) \in K[X]$ with $f(\alpha) = 0$, we call α a **root of f in L**.

Now to the promised theorem:

Theorem 2.18 Let $L|K$ be a field extension and $0 \neq \alpha \in L$ be any element. Then the following statements are equivalent:

1. α is algebraic over K.
2. The ring $K[\alpha]$ is a field.
3. We have: $K[\alpha] = K(\alpha)$.
4. The dimension $[K(\alpha) : K]$ is finite.

And now?
Just a quick note: To show that all four statements are equivalent, it is sufficient to make a circular conclusion, i.e., to show that

$$(1) \Rightarrow (2) \Rightarrow (3) \Rightarrow (4) \Rightarrow (1).$$

Check it out!
Before we write down the proof, let us briefly note that we already know some of the implications, or some of them are clear, or let us prepare some easy steps first:

- $(2) \Leftrightarrow (3)$: this is clear. Why?
- Preparation for $(1) \Rightarrow (2)$: Let α be algebraic over K and let $p(X) = X^n + a_{n-1}X^{n-1} + \ldots + a_0 \in K[X]$ be the minimal polynomial. Convince yourself that

$$1, \alpha, \alpha^2, \ldots, \alpha^{n-1} \in K[\alpha]$$

is a generating system of $K[\alpha]$!!! (Hint: why is the element $\alpha^n \in K[\alpha]$ a linear combination (coefficients in K) of the elements $1, \alpha, \ldots, \alpha^{n-1}$? That's the real trick! Think about $p(\alpha) = 0$. For general $f(\alpha) \in K[\alpha]$, use polynomial division with remainder. The following proof will execute this.)

Proof We show the individual implications as follows:

(1) $\Rightarrow$ (2) (solves the above **Check it out!**). So let $\alpha \in L$ be algebraic. We first show that $K[\alpha]$ then is a **finite-dimensional** K-vector space. Let $p(X) = X^n + a_{n-1}X^{n-1} + \ldots + a_0 \in K[X]$ be the minimal polynomial. Then

$$1, \alpha, \alpha^2, \ldots, \alpha^{n-1}$$

is a generating system of $K[\alpha]$ as a K-vector space, because if $f(\alpha) \in K[\alpha]$ is any element with $f(X) \in K[X]$, then by polynomial division with remainder, we have

$$f(X) = q(X)p(X) + r(X) \quad \text{with} \quad \deg(r) < \deg(p),$$

and $\deg(r) \leq n - 1$. If we insert α, we get

$$f(\alpha) = r(\alpha) \in \operatorname{span}_K\{1, \alpha, \ldots, \alpha^{n-1}\}.$$

Addendum: in this situation, $1, \alpha, \ldots, \alpha^{n-1}$ is even a K-basis of $K[\alpha]$, because if

$$c_0 1 + c_1 \alpha + \ldots + c_{n-1}\alpha^{n-1} = 0$$

with $c_j \in K$, we deduce for $r(X) := c_{n-1}X^{n-1} + \ldots + c_1 X + c_0 \in K[X]$, that $r(\alpha) = 0$. But since $\deg(r) \leq n - 1 < \deg(p)$ and p is the minimal polynomial of α over K, it follows that $r(X) = 0$, so all $c_j = 0$. Therefore, $1, \alpha, \ldots, \alpha^{n-1}$ are linearly independent over K. So we have shown:

$$\left.\begin{array}{l} \alpha \in L \text{ algebraic over } K \\ \text{with minimal polynomial } p(X) \in K[X] \end{array}\right\} \implies \dim_K K[\alpha] = \deg(p). \quad (2.8)$$

Our actual goal, however, is to show that $K[\alpha]$ is a field for algebraic α.

Check it out!
We have almost done that. Do you see it? We have done this in the example $K = \mathbb{Q}$ and $\alpha = \sqrt[3]{7}$ in such a way that it can immediately be generalized to the general situation. Try to convince yourself that we have already seen the rest of (1) $\Rightarrow$ (2) there!

We now know that $V := K[\alpha]$ is a finite-dimensional K-vector space, let's say again $n := \dim_K K[\alpha]$. Let $x \in K[\alpha] \smallsetminus \{0\}$ be an arbitrary element. Then consider the K-linear mapping "multiplication with x":

$$\varphi : K[\alpha] \to K[\alpha], \ v \mapsto x \cdot v$$

(remembering that $K[\alpha]$ in addition to being a K-vector space, is a commutative ring with unity as well). This map is injective because multiplication with x in the field L is injective and φ is the restriction of this injective map.

But as an injective linear mapping $\varphi : V \to V$ of the **finite-dimensional** K-vector space V into itself, φ is also surjective due to Corollary 2.6. In particular, $1 \in \text{Bild}(\varphi)$, i.e., there exists a $v \in K[\alpha]$ with $x \cdot v = 1$, so x has a multiplicative inverse in $K[\alpha]$.

(2) $\Leftrightarrow$ (3) This is indeed clear. We always have $K[\alpha] \subset K(\alpha)$. Now, $K(\alpha)$ is the smallest intermediate field in $L|K$, which contains α. Therefore, if $K[\alpha]$ is a field, it must already be this smallest one. Conversely, if $K[\alpha] = K(\alpha)$, then the left side is a field, because the right one is.

(3) $\Rightarrow$ (4) If $K[\alpha] = K(\alpha)$ applies, so $K[\alpha]$ is a field. Then, for every element in $K[\alpha] \smallsetminus \{0\}$ there is a multiplicative inverse, especially there is one for $\alpha \in K[\alpha]$. Therefore, a polynomial $g(X) \in K[X]$ exists and thus an element $g(\alpha) \in K[\alpha]$, so that

$$\alpha \cdot g(\alpha) = 1.$$

Then, $f(X) := X \cdot g(X) - 1 \in K[X] \smallsetminus \{0\}$ is a polynomial that has α as a root. Therefore, α is algebraic over K. (Note that we now also have shown "(3)$\Rightarrow$(1)").

We have already shown above that for algebraic α, we know that $K[\alpha]$ is a finite-dimensional K-vector space. Therefore, finally

$$[K(\alpha) : K] = \dim_K K(\alpha) = \dim_K K[\alpha] < \infty$$

if $K[\alpha] = K(\alpha)$ (and thus α is algebraic over K, as we have just seen).

(4) $\Rightarrow$ (1) If $[K(\alpha) : K] < \infty$, then, since $K[\alpha] \subset K(\alpha)$ is a subspace, we know that $\dim_K K[\alpha] < \infty$ as well. Let us denote its dimension by n. Then consider the elements

$$1, \alpha, \alpha^2, \ldots, \alpha^n \in K[\alpha].$$

These are $n + 1$ vectors in the n-dimensional K-vector space $K[\alpha]$, hence they are linearly dependent over K. Therefore, there exist $a_j \in K$, not all 0, such that

$$0 = a_0 1 + a_1 \alpha + \ldots + a_n \alpha^n.$$

Then

$$f(X) := a_0 + a_1 X + \ldots + a_n X^n \in K[X] \smallsetminus \{0\}$$

is a polynomial with α as a root and thus α is algebraic over K. $\square$

From the proof, we draw the **extremely important conclusion**—it will be almost the only method to determine degrees in a field extension:

Corollary 2.19 If $\alpha \in L$ is algebraic over K and if $p(X) \in K[X]$ is the minimal polynomial over K, then:

$$[K(\alpha) : K] = \deg(p).$$

Proof Follows from (2.8). $\square$

Definition 2.20 A field extension $L|K$ is called **algebraic** if every $\alpha \in L$ is algebraic over K.

Corollary 2.21 (to Theorem 2.18) If $[L : K] < \infty$, then $L|K$ is algebraic.

> **Check it out!**
> Why does this immediately follow from Theorem 2.18? If you have no idea, take a quick look at Eq. (2.9) below—do you have an idea then?

Proof If $\alpha \in L$, then one has

$$K(\alpha) \subset L \qquad (2.9)$$

as a K-subvector space. This implies that $\dim_K K(\alpha) \leq \dim_K L < \infty$, so $[K(\alpha) : K] < \infty$ and according to theorem 2.18, α is algebraic over K. $\square$

The converse of the corollary does not hold: A field extension $L|K$ can be infinite and algebraic. We will see an important example shortly. First, we prove an important formula:

Theorem 2.22 (Degree Formula) Let $K \subset M \subset L$ be field extensions (thus M is an intermediate field of $L|K$). Then the **Degree Formula**

$$[L : K] = [L : M] \cdot [M : K],$$

holds, even in the case that $L|K$ is infinite, if it is read accordingly: Then at least one of the factors on the right hand side is infinite—and vice versa.

Proof If one of the two factors on the right side is infinite, then $[L : K] = \infty$—a finite-dimensional vector space cannot contain an infinite-dimensional subspace. Note that L is both a K- and an M-vector space.

We therefore assume that $[L : M] < \infty$ and $[M : K] < \infty$ applies. Let's say $r := [L : M]$ and $s := [M : K]$ for $r, s \in \mathbb{N}_0$. Then we can choose

- a basis $m_1, \ldots, m_s$ of M as a K-vector space and
- a basis $\ell_1, \ldots, \ell_r$ of L as an M-vector space.

> **Check it out!**
> Give it a shot: Without thinking too much about a justification etc., suggest a candidate for a basis
>
> $$v_1, \ldots, v_{rs}$$
>
> consisting of $r \cdot s$ elements of L as a K-vector space! For this, of course, you use the s elements $m_j \in M \subset L$ and the r elements $\ell_j \in L$.

A candidate for a (hopefully it is one) K-basis of L is quite simple:

$$\underbrace{m_1\ell_1, m_1\ell_2, \ldots, m_1\ell_r}, \underbrace{m_2\ell_1, \ldots, m_2\ell_r}, \ldots, \underbrace{m_s\ell_1, \ldots, m_s\ell_r} \in L,$$

i.e. the family $(m_i\ell_j)_{i=1,\ldots,s;j=1,\ldots,r}$. This is indeed a K-basis of L, because:

- **Linearly independent:** Let $a_{ij} \in K$, such that

$$\sum_{i=1}^{s}\sum_{j=1}^{r} a_{ij}m_i\ell_j = 0.$$

We need to show that all $a_{ij} = 0$. If we sort the sum accordingly (and use distributivity), we get:

$$0 = \sum_{j=1}^{r} \underbrace{\left(\sum_{i=1}^{s} a_{ij}m_i \right)}_{\in M} \ell_j$$

and since $\ell_1, \ldots, \ell_r$ are linearly independent over M, it follows that for all $j = 1, \ldots, r$:

$$0 = \sum_{i=1}^{s} a_{ij}m_i.$$

Since $m_1, \ldots, m_s$ are linearly independent over K, it follows that $a_{ij} = 0$ for all i,j.
- **Generating system:** The process is analogous: Let $\ell \in L$ be arbitrary. Then one can write ℓ as an M-linear combination in the M-basis $\ell_1, \ldots, \ell_r$ of L:

$$\ell = \sum_{j=1}^{r} \mu_j\ell_j$$

with $\mu_j \in M$. For a fixed $j = 1, \ldots, r$, one can write μ_j again as a K-linear combination in the K-basis. $m_1, \ldots, m_s$ of M, so that

$$\mu_j = \sum_{i=1}^{s} a_{ij}m_i$$

with coefficients $a_{ij} \in K$. In total, then

$$\ell = \sum_{j=1}^{r}\sum_{i=1}^{s} a_{ij}m_i\ell_j.$$

In particular, we also see that $[L:K]$ is finite, when the two factors on the right are. Then, everything is shown. $\square$

Defining Lemma 2.23 Consider the subset.

$$\widetilde{\mathbb{Q}} := \{z \in \mathbb{C} \mid z \text{ is algebraic over } \mathbb{Q}\} \subset \mathbb{C}$$

This is an intermediate field, which is called the **algebraic closure of $\mathbb{Q}$ in $\mathbb{C}$**. The field extension $\overline{\mathbb{Q}}|\mathbb{Q}$ is algebraic, but not finite.

Proof Apparently, $0, 1, -1 \in \overline{\mathbb{Q}}$. What needs to be shown is that for $z, w \in \overline{\mathbb{Q}} \setminus \{0\}$ the numbers $z + w, zw$ and z^{-1} are also contained in $\overline{\mathbb{Q}}$. This is not so obvious and I would like to briefly comment on this first $\square$

And now?

So the question is: Considering a field extension $L|K$ in which the elements $z, w \in L$, $z \neq 0$, are algebraic over K, why are $z + w, zw, z^{-1}$ also algebraic over K? Let's consider $z + w$: Since both are algebraic over K, we have

- a polynomial $0 \neq f(X) \in K[X]$ with $f(z) = 0$ and
- a polynomial $0 \neq g(X) \in K[X]$ with $g(w) = 0$.

The task now is to find a polynomial $0 \neq h(X) \in K[X]$ such that $h(z + w) = 0$ holds. This is anything but easy!!! One cannot "construct" h from f and g in an easy way. How do we proceed? The answer lies in theorem 2.18—it provides criteria for when an element (like $z + w$) is algebraic. Additionally, using corollary 2.21, even if $L|K$ itself is not finite, together with the degree formula of Theorem 2.22, the argument is a nice trick—I formulate it as ***Check it out!***, even though it is not so easy.

Check it out!

Do you see a trick on how to use Theorem 2.18, Corollary 2.21 and the degree formula of Theorem 2.22 to see that $z + w$ is algebraic for given algebraic z,w ?

The trick is to consider the intermediate field

$$\mathbb{Q}(z, w) \subset \overline{\mathbb{Q}}.$$

We then have the field extensions $\mathbb{Q}(z, w)|\mathbb{Q}(z)|\mathbb{Q}$. Since z is algebraic over $\mathbb{Q}$, the degree $[\mathbb{Q}(z) : \mathbb{Q}] < \infty$ is a finite number (Theorem 2.18). Since w is algebraic over $\mathbb{Q}$, the element w is certainly algebraic over $\mathbb{Q}(z)$ –*think about it for a moment!*. So, $[\mathbb{Q}(w, z) : \mathbb{Q}(z)] < \infty$ (same theorem). The degree formula (Theorem 2.22) shows that

$$[\mathbb{Q}(z, w) : \mathbb{Q}] = [\mathbb{Q}(z, w) : \mathbb{Q}(z)] \cdot [\mathbb{Q}(z) : \mathbb{Q}] < \infty,$$

so according to Corollary 2.21 the field extension $\mathbb{Q}(z, w)|\mathbb{Q}$ is algebraic. But $z + w, zw, z^{-1} \in \mathbb{Q}(z, w)$, thus they are all algebraic. $\square$

And now?

Arguments like the last ones will be used very often. Let us add a comment: Given field extensions

$$K \subset M \subset L,$$

we see that L is both a K-vector space and an M-vector space. Depending on the question to attack, one of these structures should be exploited (and the other *forgotten*). Sometimes you even look for other intermediate fields M' to incorporate these into the arguments.

For example—as used in the proof of the Defining Lemma 2.23: If $L|K$ is a field extension and $z, w \in L$. Then we can introduce the intermediate field

$$K \subset K(z, w) \subset L,$$

and $M := K(z, w)$ is a K-vector space. For the above argument, it was helpful to consider the intermediate field $M' := K(z)$. We then have

$$K \subset K(z) \subset K(z, w)$$

and we can consider $K(z,w)$ also as a $K(z)$ -vector space. Why is this helpful? Because we know very little about the degree $[F : E]$ of a field extension $F|E$, only **in the special case,** that $F = E(\alpha)$ for a $\alpha \in F$. Then we have:

$$[E(\alpha) : E] < \infty \Leftrightarrow \alpha \text{ algebraic über } E$$

and then we even have the explicit formula:

$$[E(\alpha) : E] = \deg(\text{Minimal polynomial of } \alpha \text{ over } E)$$

when α is algebraic over E.

Now our $K(z,w)|K$ is usually **not** of the form as in the **special case,** but if we introduce $K(z)$ as an intermediate step, we have the situation of the special case twice! To this end, note that

$$K(z, w) = K(z)(w)$$

trivially holds (*do you understand this statement and is it really "trivially" clear to you?*). With this, we have

$$[K(z, w) : K] = \underbrace{[K(z, w) : K(z)]}_{\text{special case}} \cdot \underbrace{[K(z) : K]}_{\text{special case}}$$

Often you have to construct or find suitable intermediate fields M' (like the field $K(z)$ in the example above)—this is some kind of creative act which you are asked to perform in order to solve the problem and this kind of creative procedures are often perceived as difficult—at least they become easier if one is aware that such a creative act could be the solution.

We conclude the chapter with a few examples (and a final proposition, which mainly concerns infinite, algebraic extensions— note, however, that in the future we will mostly consider finite extensions).

Examples
1. Let's consider the polynomial $f(X) = X^4 - 2 \in \mathbb{Q}[X]$. It has the roots $\pm\sqrt[4]{2}, \pm i\sqrt[4]{2} \in \mathbb{C}$. We want to consider the field $\mathbb{Q}(\sqrt[4]{2}, i\sqrt[4]{2})$. This is apparently the smallest field in $\mathbb{C}$ that contains all roots of $f(X)$ (we will later call this field the **splitting field** of $f(X)$). Which degree does it have over $\mathbb{Q}$? As indicated in the last *And now?*, we are looking for a suitable intermediate field. This can be done in various ways—a creative act. Note that

$$\mathbb{Q}(\sqrt[4]{2}, i\sqrt[4]{2}) = \mathbb{Q}(\sqrt[4]{2}, i) \tag{2.10}$$

applies—*think about it for a moment!* So, we have the *tower of fields*[3]

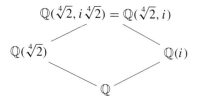

Let's consider the field extension $\mathbb{Q}(i)|\mathbb{Q}$ (the bottom right inclusion in the diagram). Since $i \notin \mathbb{Q}$, we have $[\mathbb{Q}(i) : \mathbb{Q}] \geq 2$. Therefore, the minimal polynomial of i over $\mathbb{Q}$ has at least degree 2. Since $X^2 + 1$ is a polynomial that has i as a root, it is already the minimal polynomial and therefore $[\mathbb{Q}(i) : \mathbb{Q}] = 2$.
We come to a usually difficult question: We know that $f(X) = X^4 - 2$ is a polynomial that has $\sqrt[4]{2}$ as a root, but is it the minimal polynomial? We will develop tricks for this in a later chapter. Then we will see at a glance that $X^4 - 2$ is the minimal polynomial—keyword: Eisenstein. Now we have to proceed in a different way. Let's denote by $p(X) \in \mathbb{Q}[X]$ the minimal polynomial of $\sqrt[4]{2}$ over $\mathbb{Q}$. It is certainly of degree

$$2 \leq \deg(p) \leq 4,$$

the first inequality holds, because $\sqrt[4]{2} \notin \mathbb{Q}$, the second one holds, because $X^4 - 2$ is indeed a polynomial with root. $\sqrt[4]{2}$. According to the Defining Lemma 2.16 we know that $p(X)$ is a divisor of $X^4 - 2$, so

$$X^4 - 2 = p(X) \cdot q(X) \in \mathbb{Q}[X]$$

---

[3] I will use these diagrams more often. They are always to be read in such a way that the connecting lines between two bodies, read from bottom to top, denote an inclusion ("below" ⊂ "above").

for a $q(X) \in \mathbb{Q}[X]$. But we know the roots of $f(X)$ in $\mathbb{C}$ and therefore we have the splitting of the polynomials **now in the polyomial ring** $\mathbb{C}[X]$ over the complex numbers:

$$p(X) \cdot q(X) = X^4 - 2 = (X - \sqrt[4]{2})(X + \sqrt[4]{2})(X - i\sqrt[4]{2})(X + i\sqrt[4]{2}).$$

In $\mathbb{C}[X]$ the polynomial $p(X)$ also factorizes into linear factors associated to its roots, so $p(X)$ has the form

$$p(X) = (X - \alpha_1)(X - \alpha_2) \text{ or } (X - \alpha_1)(X - \alpha_2)(X - \alpha_3)$$
$$\text{or } (X - \alpha_1)(X - \alpha_2)(X - \alpha_3)(X - \alpha_4)$$

for pairwise different $\alpha_1, \alpha_2, \alpha_3, \alpha_4 \in \{\pm\sqrt[4]{2}, \pm i\sqrt[4]{2}\}$. If one checks the first two possibilities for all combinations of the α_j, one finds that $(X - \alpha_1)(X - \alpha_2)$ and $(X - \alpha_1)(X - \alpha_2)(X - \alpha_3)$ do **not** lie in $\mathbb{Q}[X]$. Thus, only the last possibility remains and it is equivalent to $p(X) = X^4 - 2$. Therefore, Corollary 2.19 provides that $[\mathbb{Q}(\sqrt[4]{2}) : \mathbb{Q}] = 4$ holds.

Now we know the field degrees of the two lower inclusions in the diagram. What about the upper inclusions? The easiest one is for the field extension $\mathbb{Q}(\sqrt[4]{2}, i) | \mathbb{Q}(\sqrt[4]{2})$ (top left). This is again a special case as in Corollary 2.19, i.e., we know:

$$[\mathbb{Q}(\sqrt[4]{2}, i) : \mathbb{Q}(\sqrt[4]{2})] = \deg(\text{Minimal polynomial of } i \text{ over } \mathbb{Q}(\sqrt[4]{2})).$$

Which polynomial is this minimal polynomial? There is a candidate having i as a root, namely

$$X^2 + 1 \in \mathbb{Q}[X] \subset \mathbb{Q}(\sqrt[4]{2})[X].$$

The question of whether it is the minimal polynomial is (which is usually difficult to answer) has an easy answer here, because $\mathbb{Q}(\sqrt[4]{2}) \subset \mathbb{R}$ and $i \notin \mathbb{R}$, therefore $[\mathbb{Q}(\sqrt[4]{2}, i) : \mathbb{Q}(\sqrt[4]{2})] \geq 2$ must hold, thus $X^2 + 1$ is the minimal polynomial of i also over $\mathbb{Q}(\sqrt[4]{2})$.

Hence, we finally know all the field degrees that occur in the diagram and due to the degree formula we obtain $[\mathbb{Q}(\sqrt[4]{2}, i) : \mathbb{Q}] = 8$. The "intermediate degrees" are drawn in the next diagram next to the inclusion line—we will often use this convention in the following:

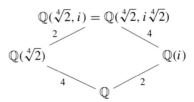

2. We stick with the above example: We can make a very important observation. For the field extension in the top left of the diagram, we determined the degree of the extension in (1) by using $K := \mathbb{Q}(\sqrt[4]{2})$, $L := \mathbb{Q}(\sqrt[4]{2}, i)$ and considering L in the form $L = K(i)$. But assume we had not observed (2.10)—that

observation was a creative act from our part—we could also have tried to compute the degree as follows: Obviously, $L = K(i\sqrt[4]{2})$ holds and thus (Corollary 2.19) gives:

$$[\mathbb{Q}(\sqrt[4]{2}, i\sqrt[4]{2}) : \mathbb{Q}(\sqrt[4]{2})] = \deg(\text{Minimal polynomial of } i\sqrt[4]{2} \text{ over } \mathbb{Q}(\sqrt[4]{2})).$$

Let's call this minimal polynomial $p(X) \in \mathbb{Q}(\sqrt[4]{2})[X]$. We have $\alpha := i\sqrt[4]{2} \in \mathbb{C}$ as a root and we consider its two various minimal polynomials:

- the minimal polynomial $X^4 - 2 \in \mathbb{Q}[X]$ over $\mathbb{Q}$ and
- the minimal polynomial $p(X) \in \mathbb{Q}(\sqrt[4]{2})[X]$ over $\mathbb{Q}(\sqrt[4]{2})$ (that's why I always emphasized **minimal polynomial of ... over ...**). There is a relation between these: The former can be read in the bigger polynomial ring

$$X^4 - 2 \in \mathbb{Q}[X] \subset \mathbb{Q}(\sqrt[4]{2})[X]$$

and due to the Defining Lemma 2.16, we deduce that $p(X)$ is a divisor of $X^4 - 2$ in $\mathbb{Q}(\sqrt[4]{2})[X]$:

$$X^4 - 2 = p(X)q(X) \in \mathbb{Q}(\sqrt[4]{2})[X].$$

From the calculations of the field degrees above, we know that $\deg(p) = 2$ must hold! So it is a proper divisor $\neq X^4 - 2$.

As above, we write $X^4 - 2$ in $\mathbb{C}[X]$ as a product of the linear factors

$$p(X)q(X) = X^4 - 2 = (X - \sqrt[4]{2})(X + \sqrt[4]{2})(X - i\sqrt[4]{2})(X + i\sqrt[4]{2}).$$

Now it can be seen that the product of the last two factors

$$(X - i\sqrt[4]{2})(X + i\sqrt[4]{2}) = X^2 + \left(\sqrt[4]{2}\right)^2 \in \mathbb{Q}(\sqrt[4]{2})[X]$$

lies in the corresponding polynomial ring and $i\sqrt[4]{2}$ is one of its roots. Therefore, $p(X) = X^2 + (\sqrt[4]{2})^2 = X^2 + \sqrt{2} \in \mathbb{Q}(\sqrt[4]{2})[X]$ is the minimal polynomial of $i\sqrt[4]{2}$ over $\mathbb{Q}(\sqrt[4]{2})$.

And now?

Let me comment on an important issue: When making a statement like *"If α is algebraic over K, then $K(\alpha)|K$ is finite"*, the larger field L does not explicitly appear in the formulation of the statement itself. The reason is that the statement forgot to mention where the element α comes from. However, such a statement only makes sense in a given field extension $L|K$. I emphasize this because we want to deal with the reverse problem later: Given some $f(X) \in K[X]$ without a larger field (which could then be called L) available in which one could look for roots how can we construct a suitable field L which contains a root $\alpha \in L$ of f? This will be an important step in the whole theory.

If you think of $K = \mathbb{Q}$, it's easy because you can always start with $L = \mathbb{C}$ and find the roots there. But think of $K = \mathbb{F}_5$, the field with 5 elements— which you hopefully have already seen in linear algebra, or in the exercises. There you might not see (I dare to assume) a direct candidate for a field L with $\mathbb{F}_5 \subset L$, in which a given polynomial has a root.

Let us add one last theorem for completeness:

Theorem 2.24 Let $K \subset M \subset L$ be field extensions. If $L|M$ and $M|K$ are algebraic, then so is $L|K$.

Proof If $L|M$ and $M|K$ are both finite, this follows immediately from the degree formula and Corollary 2.21. For infinite extensions, this is reduced to the finite case with a trick.

Onsider an arbitrary α in L. Since $L|M$ is algebraic, α is algebraic over M, so there is a polynomial $\neq 0$:

$$f(X) = \mu_m X^m + \mu_{m-1} X^{m-1} + \ldots + \mu_0 \in M[X]$$

with $f(\alpha) = 0$. Now consider the intermediate field $K(\mu_0, \ldots, \mu_m)$ in $L|K$. We have

$$[K(\mu_0, \ldots, \mu_m) : K] < \infty$$

(by successive application of the degree formula and Corollary 2.19, since all $\mu_j \in M$ are algebraic over K). Furthermore, α is also algebraic over $K(\mu_0, \ldots, \mu_m)$, so we have

$$[K(\mu_0, \ldots, \mu_m, \alpha) : K(\mu_0, \ldots, \mu_m)] < \infty.$$

Due to the degree formula, we also have

$$[K(\mu_0, \ldots, \mu_m, \alpha) : K] < \infty,$$

and thus $K(\mu_0, \ldots, \mu_m, \alpha)$ and in particular α is algebraic over K. $\square$

Red thread to the previous chapter
We have seen in this chapter,

- what is called a field extension $L|K$,
- that in such a situation L is naturally a K-vector space,
- that $\alpha \in L$ is algebraic over K if it is a root of a polynomial $f(X) \in K[X] \setminus \{0\}$,
- that an algebraic α has a unique minimal polynomial $p(X) \in K[X]$,
- that for an $\alpha \in L$ between $K \subset L$ the two objects
  - the subring $K[\alpha]$ and
  - the intermediate field $K(\alpha)$ exist,

and that $K \subset K[\alpha] \subset K(\alpha) \subset L$ holds,

- that exactly then $K[\alpha] = K(\alpha)$ holds if and only if α is algebraic over K, and if and only if $[K(\alpha) : K] := \dim_K K(\alpha) < \infty$,
- that for algebraic α we have a formula:

$$[K(\alpha) : K] = \deg(\text{minimal polynomial of } \alpha \text{ over } K)$$

- that for a situation of two field extensions $K \subset M \subset L$ (let's say all finite), the degree formula applies:

$$[L : K] = [L : M] \cdot [M : K].$$

- that for every subset $S \subset L$ there is a smallest intermediate field $K \subset K(S) \subset L$.

Excercises:

2.1 Consider the real number $\alpha := \sqrt{2} + \sqrt{3}$.

(a) Determine the minimal polynomial $f(X) \in \mathbb{Q}[X]$ of α over $\mathbb{Q}$,
(b) as well as the minimal polynomial $g(X) \in \mathbb{Q}(\sqrt{6})[X]$ of α over $\mathbb{Q}(\sqrt{6})$.

2.2 Let $\alpha := e^{i\frac{\pi}{4}} = \frac{1}{\sqrt{2}}(1 + i)$. Determine

(a) the minimal polynomial of α over $\mathbb{Q}$,
(b) the minimal polynomial of α over $\mathbb{Q}(\sqrt{2})$.

2.3 Consider the field $L := \mathbb{Q}(\sqrt[5]{3}, \sqrt{7})$. Show that $L = \mathbb{Q}(\sqrt[5]{3} \cdot \sqrt{7})$ applies.

2.4 Roots of Unity:

(a) Show that the polynomial $X^n - 1 \in \mathbb{Q}[X]$ is a multiple of $X - 1$ and calculate $\frac{X^n - 1}{X - 1}$.

(b) Let $\zeta := \exp(2\pi i/n) \in \mathbb{C}$ (apparently a root of $X^n - 1$). Consider

$$\mathbb{Q}[\zeta] := \{g(\zeta) \in \mathbb{C} \mid g(X) \in \mathbb{Q}[X]\} \subset \mathbb{C},$$

the set of all polynomial expressions in ζ with coefficients in $\mathbb{Q}$. Show that

$$1, \zeta, \zeta^2, \dots, \zeta^{n-1} \tag{2.11}$$

is a generating system over $\mathbb{Q}$.

(c) Show that in the case of $n = 4$ the generating system (2.11) is not a basis over $\mathbb{Q}$.

2.5 Consider the polynomial $f(X) := X^3 - 3X + 1 \in \mathbb{Q}[X]$ and a real root $\alpha \in \mathbb{R}$ (exists according to the mean value theorem of analysis).

(a) Show that $f(X)$ has no root in $\mathbb{Q}$ and deduce from this that $f(X)$ is the minimal polynomial of α.
(b) Represent the element $(\alpha^2 - 2\alpha + 1)^{-1} \in \mathbb{Q}[\alpha]$ as a $\mathbb{Q}$-linear combination of the basis $1, \alpha, \alpha^2$ of $\mathbb{Q}[\alpha]$.

2.6 Let $L|K$ be a finite field extension, $K \subset M_1 \subset L$ and $K \subset M_2 \subset L$ are two intermediate fields. Prove the following: If $[M_1 : K]$ and $[M_2 : K]$ are coprime, then

$$[M_1 : K] \cdot [M_2 : K] \text{ divides } [L : K].$$

2.7 The roots of the polynomial $f(X) = X^3 - 7 \in \mathbb{Q}[X]$ are in the field $Z := \mathbb{Q}(\sqrt[3]{7}, \zeta)$ with $\zeta := \exp(\frac{2\pi i}{3}) \in \mathbb{C}$. You may use that $f(X)$ is the minimal polynomial of $\sqrt[3]{7}$ over $\mathbb{Q}$.[4] Show the following:

(a) The polynomial $\Phi(X) := X^2 + X + 1$ is the minimal polynomial of ζ over $\mathbb{Q}$.
(b) The same polynomial, now read as $\Phi(X) \in \mathbb{Q}(\sqrt[3]{7})[X]$, is also the minimal polynomial of ζ over $\mathbb{Q}(\sqrt[3]{7})$ and $[Z : \mathbb{Q}] = 6$. (Hint: Exercise 2.6).

---

[4] We will be able to easily deduce this later with theorem 9.12.

Groups

3

And now?

We are now making a "cut" and start a new topic—we are examining algebraic structures and starting with the one with very few axioms, the groups. We can practice dealing formally with algebraic structures with this notion. In addition, we will encounter two important aspects for later:

- the concept of a *group,* properties, examples, theorems—not least, the Galois group of a polynomial, which is one of the main goals of the book, is a *group* and this fact alone, together with principal theorems about groups, allows many statements about the relation among the roots of the polynomial, which can be deduced without being able to calculate the roots explicitly.
- Quotients by an equivalence relation and factor groups—we will see this construction again when we will consider factor rings. You may already know this concept from linear algebra—quotient vector spaces. The construction is extremely important, but (honestly) often perceived as difficult by students. I will try to explain the difficulties—this will be the topic of the subsequent chapter.

3.1 General Definition and Consequences

Definition 3.1 A **group** is a triple $(G, \cdot, e)$ consisting of a non-empty set G, a map (the **group operation**)

$$\cdot : G \times G \to G, \ (g, h) \mapsto g \cdot h$$

and an element $e \in G$, such that the following properties hold:

M. Hien, *Abstract Algebra,* Mathematics Study Resources 7, https://doi.org/10.1007/978-3-662-67974-6_3

1. For all $g, h, k \in G$ we have $(g \cdot h) \cdot k = g \cdot (h \cdot k)$ (Associativity),
2. For all $g \in G$ we have $e \cdot g = g$ (left-neutral element),
3. For each $g \in G$ there exists a $h \in G$, such that $h \cdot g = e$ holds (left-inverse element).

Note 3.2 We usually write gh instead of $g \cdot h$.

And now?
Sometimes it is helpful to imagine such a definition as a game: Player A versus Player B.

Player A wants to present a group to Player B, he has won if he succeeds. The game then proceeds as follows:

1. Player A presents Player B the data required by the definition of a group: The set G, the operation $\cdot$ and the element $e \in G$.
2. Then it's Player B's turn to check whether this data fulfills the properties of the definition.

After Player B's check, it is clear whether the data $(G, \cdot, e)$, coming from Player A, form indeed a group or if Player A made a mistake.

If you look at it this way, you sort out which kind of data has to be given (by Player A) and what are the tests one has to run on this data (by Player B).

A statement like "Here is the set G, this is a group." is meaningless because Player A also has to provide additional data, namely the operation $\cdot$ and the special element $e \in G$, so that Player B then can start his tests on these data. Of course, sometimes the operation and the neutral element are clear from the context and might not be recalled explicitly by Player A, but even then Player A gives the initial data to Player B by agreeing that they are the 'natural ones' everybody (especially both players) agrees on from the context.

Examples
1. $(G, \cdot, e) = (\mathbb{Z}, +, 0)$. Here, the operation is written as $+$ and the (left-)neutral element as 0.
2. $(G, \cdot, e) = (\mathbb{R}_{>0}, \cdot, 1)$.
3. $(G, \cdot, e) = (GL_n(\mathbb{R}), \cdot, 1)$ with the unit matrix 1 as neutral element.

And now?
The question arises as to why we have considered **left**-neutral or **left**-inverse elements. In all examples above, e is also **right**-neutral in the obvious sense, ... We will soon show that a left-neutral e as in the definition

is automatically right-neutral. So why don't we demand it right away in the definition and instead of (2) introduce the following axiom into the definition.

(2*variant*) For all $g \in G$ we have $e \cdot g = g \cdot e = g$

The reason is that one wants to put as **few** axioms as possible into a definition. Every time you have a candidate for a group $(G, \cdot, e)$ at hand, you have to check all of them. In doing so, one wants to avoid unnecessary axioms to be checked.

In a similar way, with (2) and (3) as in the definition, the question arises whether the neutral e is the only element with this property, similarly the h for a given g in (3) -- see the following lemma.

Lemma 3.3 *Let $(G, \cdot, e)$ be a group. Then the following holds:*

1. *the left-neutral element e is also right-neutral, i.e. $g \cdot e = g$ for all $g \in G$.*
2. *The element e is the only left-neutral element.*
3. *If $g \in G$, then the left-inverse $h \in G$ is uniquely determined and it is also a right-inverse to $g : g \cdot h = e$.*

Proof We prove the statements almost simultaneously. The "left-right" question can be shown as follows:

Let $g \in G$ and let $h \in G$ be a left-inverse of g. Then $h \in G$ also has a left-inverse, let's call it $k \in G$. Thus, (always exploiting the associativity):

$$gh = egh = (kh)gh = k(hg)h = k(eh) = kh = e.$$

Furthermore, still with $g, h \in G$ as chosen before:

$$ge = g(hg) = (gh)g = eg = g.$$

For uniqueness: If $e, e' \in G$ are left-neutral elements, according to the above, both are also right-neutral. Therefore,

$$e' = ee' = e$$

where the first equality holds, because e is left-neutral, and the second one holds because e' is right-neutral.

If $h, h' \in G$ are inverse elements to $g \in G$ (according to the above both left and right), the following holds:

$$h' = eh = (hg)h = h(gh) = he = h.$$

Remark 3.4 Since we now know that the inverse to an element $g \in G$ is uniquely determined by g, we will write $g^{-1} \in G$ for this element.

Remark 3.5

- Usually, one often shortly writes: *Let G be a group, ...* This means that one is considering a group $(G, \cdot, e)$, but the data $\cdot$ and e are not specifically mentioned—because they are clear from the context.
- Whenever the operation is denoted with $\cdot$ (one then also talks of a *multiplicative group*), the (unique) neutral element is often also denoted by $1 \in G$.
  If the group is *written additively*, i.e., the operation is denoted as $+$, the neutral element is often also written as $0 \in G$. The inverse to g is then written as $-g \in G$.
- In a *multiplicatively written* group, one uses the following abbreviation for $g \in G$ and $k \in \mathbb{N}$

$$g^k := \underbrace{g \cdot g \cdots g}_{k \text{ times}} \in G$$

and for $k < 0$ likewise

$$g^k := \underbrace{g^{-1} \cdot g^{-1} \cdots g^{-1}}_{|k| \text{ times}} \in G$$

In an *additively written* group we will in an analogous way write $kg = g + \ldots + g$ (with k summands) for $k > 0$ and $kg = (-g) + \ldots + (-g)$ (with $|k|$ summands) for $k < 0$.

Lemma 3.6 *Let G be a group. For every $g \in G$, we have*

$$(g^{-1})^{-1} = g. \tag{3.1}$$

Check it out!
Do you understand what the statement (3.1) means? What do you need to show in order to prove it? Think about this question before you continue reading. Recall what the notion $\square^{-1}$ means for a given element $\square \in G$.

You should conclude that in order to prove (3.1) (once we already believe Lemma 3.3) there is actually really nothing deep to show!

Proof As we have agreed above, $\square^{-1} \in G$ is the (uniquely determined) inverse (left and right) to $\square \in G$. This means the left side of (3.1) is by definition (set $\square := g^{-1}$):

$$(g^{-1})^{-1} := \text{ the uniquely determined inverse to } g^{-1} \in G.$$

The statement (3.1) can therefore be equivalently formulated as:

$$(g^{-1})^{-1} = g \iff \text{ the uniquely determined inverse to } g^{-1} \text{ is } g.$$

We therefore only need to convince ourselves that g is the inverse to g^{-1} is, so that $gg^{-1} = e$ applies. But this is the case as we already know from Lemma 3.3, so there is nothing more to do. $\square$

Check it out!
Very similar to this, one considers that in a group G we always have

$$(gh)^{-1} = h^{-1}g^{-1}.$$

What does this mean? Show this! Once you have understood what the statement means, there is again nothing really deep to do!

Definition 3.7 A group G is called **abelian,** if its operation is commutative, i.e., if for all $g, h \in G$ we have

$$gh = hg.$$

Remark 3.8 Whenever a group G is written additively, it is (almost always) abelian. I have never seen an additively written, non-abelian group. But one can also write abelian groups multiplicatively, for example $(\mathbb{R}_{>0}, \cdot, 1)$.

Examples of abelian groups
1. Obviously, $(\mathbb{Z}, +, 0)$ is an abelian group.
2. If $0 \neq m \in \mathbb{Z}$, we consider the set $\mathbb{Z}/m\mathbb{Z} := \{[0], [1], \dots, [m-1]\}$—right now this is just a notation, later we will see that it is an example of a general construction. We define the operation $[a] + [b] := [r]$, when division with remainder $a + b = qm + r$ yields the remainder $0 \leq r \leq m$.
   One can easily check that $(\mathbb{Z}/m\mathbb{Z}, +, [0])$ is an abelian group: The operation is associative, the element $[0]$ is a neutral element and for any $[k] \in \mathbb{Z}/m\mathbb{Z}$ the element $[m - k] \in \mathbb{Z}/m\mathbb{Z}$ is an inverse of $[k]$.

An important example of a group For $n \in \mathbb{N}$ let

$$S(n) := \{\sigma : \{1, \dots, n\} \to \{1, \dots, n\} \mid \sigma \text{ is bijective}\}$$

be the set of bijective self-maps of the set $\{1, \dots, n\}$ with n elements. The operation is the concatenation:

$$\sigma\tau := \sigma \circ \tau$$

of mappings. The neutral element is the identity map id .

Defining Lemma 3.9 These data form a group, it is called the **symmetric group to the number** n.

Proof The axioms are easily verified. Since $\sigma \in S(n)$ is bijective, the inverse mapping $\sigma^{-1} \in S(n)$ exists and it is the inverse in the sense of the definition of a group. $\square$

Remark 3.10 The cardinality of S(n) is #$S(n) = n!$.

One often uses the following notation for the elements in S(n):

$$\sigma = \begin{pmatrix} 1 & 2 & 3 & \cdots & n \\ \sigma(1) & \sigma(2) & \sigma(3) & \cdots & \sigma(n) \end{pmatrix} \in S(n).$$

With this notation, the group operation can also be easily determined explicitly, as illustrated in the example:

$$\begin{pmatrix} 1 & 2 & 3 & 4 & 5 \\ 2 & 4 & 5 & 1 & 3 \end{pmatrix} \cdot \begin{pmatrix} 1 & 2 & 3 & 4 & 5 \\ 2 & 5 & 1 & 3 & 4 \end{pmatrix} = \begin{pmatrix} 1 & 2 & 3 & 4 & 5 \\ 4 & 3 & 2 & 5 & 1 \end{pmatrix} \in S(5).$$

Another notation exists for special elements, the **cycles.** These are elements $\sigma \in S(n)$ of the following form:

Definition 3.11 Let $k \leq n$. A **k-cycle in** S (n) is an element $\sigma \in S(n)$, such that there is a tuple $(a_1, \ldots, a_k)$ of pairwise different $a_j \in \{1, \ldots, n\}$ such that the following applies:

1. For each $j < k$ we have $\sigma(a_j) = a_{j+1}$ and $\sigma(a_k) = a_1$.
2. For all $a \in \{1, \ldots, n\} \setminus \{a_1, \ldots, a_k\}$ we have $\sigma(a) = a$.

We will write $\sigma = (a_1\, a_2\, \ldots\, a_k)$.

Note 3.12 Two observations are in order:

1. Note that the entries of a cycle can be permuted cyclically without changing the cycle:

$$(a_1\, a_2\, a_3\, \ldots\, a_k) = (a_k\, a_1\, a_2\, \ldots\, a_{k-1}) = (a_{k-1}\, a_k\, a_1\, \ldots\, a_{k-2}) = \ldots$$

   A good representation of a cycle is the diagram:

2. For a k-cycle τ we have: $\tau^k = 1$.

Example The element

$$\begin{pmatrix} 1 & 2 & 3 & 4 & 5 & 6 \\ 1 & 4 & 5 & 3 & 2 & 6 \end{pmatrix} = (2\,4\,3\,5)$$

is a 4-cycle. Of course, cycles can also be composed with each other—the result usually is not a cycle any more—for example

$$(2\,4\,3\,5) \cdot (1\,6) = \begin{pmatrix} 1 & 2 & 3 & 4 & 5 & 6 \\ 6 & 4 & 5 & 3 & 2 & 1 \end{pmatrix}$$

Defining Lemma 3.13 Let $n \in \mathbb{N}$. Two cycles $\sigma = (a_1, \ldots, a_k)$, $\tau = (b_1, \ldots, b_\ell)$ in $S(n)$ are called **disjoint,** if

$$\{a_1, \ldots, a_k\} \cap \{b_1, \ldots, b_\ell\} = \emptyset$$

holds. Disjoint cycles commute $\sigma\tau = \tau\sigma$.

Every element $\eta \in S(n)$ allows a decomposition as a product of pairwise disjoint cycles which is unique up to ordering the cycles.

Proof There is not really anything to say about the commutativity of disjoint cycles. Regarding the decomposition into pairwise disjoint cycles, first note that every $1 \le j \le n$ starts a cycle *within* η: For this, let

$$j \overset{\eta}{\mapsto} \eta(j) \overset{\eta}{\mapsto} \eta^2(j) \overset{\eta}{\mapsto} \ldots \eta^\ell(j)$$

be the maximum chain such that all occurring elements are pairwise different (every such chain is finite, because the occurring elements come from the finite set $\{1, \ldots, n\}$). Then we have

$$\eta^{\ell+1}(j) \in \{j, \eta(j), \eta^2(j), \ldots, \eta^\ell(j)\}.$$

Hence $\eta^{\ell+1}(j) = j$ holds, because η is injective and each $\eta^\nu(j)$ has the only preimage $\eta^{\nu-1}(j)$, if $\nu \ge 2$. Thus, we have the ℓ-cycle

$$\tau_1 := (j \; \eta(j) \; \eta^2(j) \; \ldots \; \eta^\ell(j)) \in S(n). \tag{3.2}$$

We now introduce the following notation for $\sigma \in S(n)$:

$$\mathrm{Fix}(\sigma) := \{a \in \{1, \ldots, n\} \mid \sigma(a) = a\}$$

being the set of elements fixed under σ. Then the following holds:

$$\mathrm{Fix}(\tau_1^{-1}\eta) = \mathrm{Fix}(\eta) \cup \{j, \eta(j), \eta^2(j), \ldots, \eta^\ell(j)\}.$$

Check it out!
Convince yourself that the last equation is correct!

Note:

$$\#\mathrm{Fix}(\eta) = n \Leftrightarrow \eta = 1.$$

Keeping this in mind, one can easily prove the claim successively: If $\eta \ne 1$, then there is a $j \notin \mathrm{Fix}(\eta)$. With the above construction, one then has a cycle τ_1, so that

$$\#\mathrm{Fix}(\tau_1^{-1}\eta) > \#\mathrm{Fix}(\eta).$$

If $\tau_1^{-1}\eta = 1$ is true, we are done, since then $\eta = \tau_1$ holds. Otherwise, the same argument now can be applied to $\tau_1^{-1}\eta$ and a cycle τ_2 can be found with

$$\#\mathrm{Fix}(\tau_2^{-1}\tau_1^{-1}\eta) > \#\mathrm{Fix}(\tau_1^{-1}\eta) > \#\mathrm{Fix}(\eta).$$

Note that the cycle τ_2 only moves elements that are outside $\mathrm{Fix}(\tau_1^{-1}\eta)$, so τ_1 and τ_2 are disjoint.

After finally many steps, one has found pairwise disjoint cycles $\tau_1, \ldots, \tau_m$ with

$$\#\mathrm{Fix}(\tau_m^{-1} \cdots \tau_2^{-1}\tau_1^{-1}\eta) = n$$

from which $\eta = \tau_1 \cdot \tau_2 \cdots \tau_m$ follows.

The uniqueness is clear, because the cycle in the decomposition that contains an element j must be the cycle of the form (3.2). $\square$

Remark The proof has presented a successive argument. One could have also done this more elegantly as an induction according to the number of non-fixed elements. On the other hand side, the proof directly shows how to determine the decomposition!

> **Check it out!**
> Convince yourself that the proof gives a recipe how to compute the decomposition by applying it to some example!

> **And now?**
> What should next steps into a theory of groups consist of? As it is the case with many algebraic structures, we will continue to
>
> - ask ourselves what the appropriate notion of a sub-object should be, i.e. looking for(smaller) groups that lie within a given (larger) group.
> - investigate maps between these structures: What are the properties for a map $\varphi : G \to H$ between two groups to be *well-behaved* with respect to the group structures?
> - define quotients *modulo* a subgroup—this will become an extremely important term.

3.2   Subgroups and Group Homomorphisms

Definition 3.14 Let G be a group. A subset $H \subset G$ is called a **subgroup,** if the following holds:

1. $e \in H$,
2. for $h_1, h_2 \in H$ we have $h_1 h_2 \in H$,
3. for $h \in H$ we have $h^{-1} \in H$.

We obeserve the following lemma:

Lemma 3.15 *If $H \subset G$ is a subgroup, restriction of the operation to $H \times H$ yields a map*

$$H \times H \to H, \ (h_1, h_2) \mapsto h_1 h_2$$

and $(H, \cdot, e)$ is again a group.

Proof Easy to see. Due to the axioms of a subgroup, the restriction of the operation ends up in H as claimed. The neutral element is in H and for every $h \in H$ there exists an inverse in H, namely the inverse in $h^{-1} \in G$, which according to the axioms automatically lies in H. $\square$

Example The subset $H := \mathrm{SL}_n(\mathbb{Q}) := \{A \in M(n \times n, \mathbb{Q}) \mid \det(A) = 1\} \subset G := \mathrm{GL}_n(\mathbb{Q})$ is a subgroup. *Verify this!*

Definition 3.16 Let G be a group and $g \in G$ be an element. Then

$$\langle g \rangle := \{g^k \mid k \in \mathbb{Z}\} \subset G$$

is a subgroup, the **cyclic subgroup** *generated by g.*

More generally, for a subset $S \subset G$ the **subgroup generated by** S is the subgroup

$$\langle S \rangle := \bigcap_{\substack{H \subset G \text{ subgroup} \\ S \subset H}} H$$

And now?

In general, the elements of $\langle S \rangle$ do not have a simple description as in the case of $S = \{g\}$. There is a description as follows: A **word with letters in** S is called an element in G of the form

$$s_1^{m_1} \cdot s_2^{m_2} \cdots s_k^{m_k}$$

with $s_j \in S$, $k \in \mathbb{N}_0$ and $m_j \in \mathbb{Z}$ (for $k = 0$ we read this as the element e). Then $\langle S \rangle$ is the set of all words in G with letters in S. **But:** Given two words, it is not at all easy to decide whether they are the same element in G or not. Therefore, this description of elements in $\langle S \rangle$ is usually not very useful at all.

Definition 3.17 Let $g \in G$ be an element in a group G. Then

$$\mathrm{ord}\,(g) := \#\langle g \rangle \in \mathbb{N} \cup \{\infty\}$$

is called the **order of** g **in** G.

Lemma 3.18 *Let $g \in G$ be an element of finite order. Then*

$$\mathrm{ord}\,(g) = \min\{k \in \mathbb{N} \mid g^k = e\}.$$

If $\mathrm{ord}\,(g) = \infty$, then $g^k \neq e$ for all $k \in \mathbb{N}$.

Proof It is clear that $\{g^k \mid k \in \mathbb{N}\} \subset \langle g \rangle$ and if $\#\langle g \rangle = \mathrm{ord}\,(g) < \infty$, this subset also has only finitely many elements. Hence there are $k \neq \ell \in \mathbb{N}$ with $g^k = g^\ell$, WLOG $\ell > k$. Then $g^{\ell-k} = e$ and thus the set $\{k \in \mathbb{N} \mid g^k = e\}$ is not empty. Therefore, the minimum of this set exists.

Let κ be this minimum. Then it is clear that $e, g, g^2, \ldots, g^{\kappa-1} \in G$ are pairwise different in G, so

$$\{e, g, g^2, \ldots, g^{\kappa-1}\} \subset \langle g \rangle \tag{3.3}$$

is a subset with cardinality κ, thus $\kappa \leq \mathrm{ord}\,(g)$.

Now, let $\ell \in \mathbb{Z}$ be arbitrary. By division with remainder, we have $\ell = q\kappa + r$ with $0 \leq r < \kappa$ and we deduce that

$$g^\ell = g^{q\kappa+r} = (g^\kappa)^q \cdot g^r = g^r \in \{e, g, g^2, \ldots, g^{\kappa-1}\},$$

so (3.3) is an equality. $\square$

And now?

Note that in general it is not so easy to determine $\mathrm{ord}\,(g)$ —if one does not want compute all elements $e, g, g^2, g^3, g^4, \ldots$ or one does not have enough time to do so. In particular, one has to be careful that given $g, h \in G$ for which we knoword (g) and $\mathrm{ord}\,(h)$, we usually cannot simply compute $\mathrm{ord}\,(gh)$ in general. Note that in order to understand $\mathrm{ord}\,(gh)$ one has to compute $e, gh, (gh)^2, (gh)^3, \ldots$ and

$$(gh)^k = ghghghgh \cdots gh \text{ (all in all } k \text{ pairs } gh)$$

If the group is not abelian, this is generally **not the same** as $g^k h^k$—of course, this is clear to you, but it is still a common mistake.

In particular for the symmetric group, where we decomposed $\sigma \in S(n)$ into **pairwise disjoint** cycles

$$\sigma = \tau_1 \cdot \tau_2 \cdots \tau_N$$

the requirement that they are pairwise disjoint is very helpful. It allows them to commute with each other and then we have

$$\sigma^k = \tau_1^k \cdot \tau_2^k \cdots \tau_N^k, \text{ because all factors commute with each other!}$$

Definition 3.19 A group is called **cyclic,** if there is an element $g \in G$ such that

$$G = \langle g \rangle$$

holds. Such an element (which in general is not uniquely determined) is then called a **generator.**

Now we come to the notion of *good* mappings between groups:

Definition 3.20 Let G, G' be groups. We will write e for each of the neutral elements, since in the formulas below it is usually clear which group will be considered. A map $\varphi : G \to G'$ is called a **group homomorphism**, if the following holds:

1. $\varphi(e) = e$,
2. $\varphi(g \cdot h) = \varphi(g) \cdot \varphi(h)$ for all $g, h \in G$.

And now?
In both axioms, one should be aware that the symbols e and $\cdot$ on the respective left hand side represent the data of G (actually $(G, \cdot, e)$) and on the right side those of G' (which we again wrote as $(G', \cdot, e)$). Usually, there should be no risk of confusing these, because for example $\varphi(g), \varphi(h) \in G'$, so the $\cdot$ on the right hand side of (2) must be the one of G'.

And now?
Again, one can see from the definition that redundant information is not specifically written as additional axioms if it can be deduced from the given axioms. For a group, in addition to the operation and the neutral element, there is an inverse for each $g \in G$. Why don't we additionally require for a group homomorphism that $\varphi(g^{-1}) = (\varphi(g))^{-1}$ holds? Because it already follows from the given axioms.

Check it out!
Read the *And now?* and consider the last statement—prove the given formula! This will be the next lemma, so you might want to continue reading after you have completed this task.

Lemma 3.21 *For a group homomorphism $\varphi : G \to G'$ we have $\varphi(g^{-1}) = (\varphi(g))^{-1}$ for all $g \in G$.*

Proof Let $g \in G$. The right side of the assertion is by definition the (unique!) element in G', that is inverse to $\varphi(g) \in G'$. The formula to be shown can therefore be equivalently reformulated as follows:

$$\varphi(g^{-1}) = (\varphi(g))^{-1} \iff \text{the element } \varphi(g^{-1}) \in G' \text{ is the inverse of } \varphi(g) \in G'.$$

This is immediately clear because

$$\varphi(g^{-1}) \cdot \varphi(g) \overset{(2)}{=} \varphi(g^{-1} \cdot g) = \varphi(e) \overset{(1)}{=} e$$

holds. $\square$

And now?

Of course one could write the proof in a different way: We have

$$\varphi(g^{-1}) \cdot \varphi(g) \overset{(2)}{=} \varphi(g^{-1} \cdot g) = \varphi(e) \overset{(1)}{=} e,$$

and if you multiply both sides from the right with $(\varphi(g))^{-1}$, it remains a correct equation, namely

$$\varphi(g^{-1}) \cdot \underbrace{\varphi(g) \cdot (\varphi(g))^{-1}}_{=e} = e \cdot (\varphi(g))^{-1}.$$

We deduce that $\varphi(g^{-1}) = (\varphi(g))^{-1}$. I find the first proof more elegant because it refers back to the definition and uniqueness of the inverse.

Examples

1. (**an abstract example—very important!**) Let G be a group and $g \in G$ a fixed element. Then

$$\varphi = \mathrm{conj}_g : G \to G, x \mapsto \mathrm{conj}_g(x) := gxg^{-1}$$

   is a group homomorphism, the so-called **conjugation with** g.
   Note that conj_g is bijective, because $\mathrm{conj}_{g^{-1}}$ is an inverse bijection to it.

2. (reading the abstract example even more abstractly—*an expert example: Do not despair if you find this difficult to read the first time—later you will find it easy when we have more practice in group theory.*)
   If G is a group, then

   $$\mathrm{Aut}(G) := \{\varphi : G \to G \mid \varphi \text{ is a bijective group homomorphism}\}$$

   together with the concatenation as the multiplication and the identity as a neutral element is itself a group *(think about it)!*. With this notation,

   $$\mathrm{conj} : G \to \mathrm{Aut}(G), \ g \mapsto \mathrm{conj}_g$$

   is a group homomorphism. **Check this!**

3. For $m \in \mathbb{N}$ the mapping

   $$\varphi : \mathbb{Z} \to \mathbb{Z}/m\mathbb{Z}, \ x \mapsto [x]$$

   is a group homomorphism of abelian groups.

And now?

For additively written (abelian) groups, of course, you have to replace $\cdot$ with $+$. There can also be situations where one group is written additively, the other multiplicatively, as in the next example:

4. Let $c \neq 1$ be a positive real number. Then

$$\varphi : \mathbb{Z} \to \mathbb{R}_{>0}, \; n \mapsto c^n$$

is a group homomorphism, because

- We have $\varphi(0) = c^0 = 1$ (and on the left hand side, 0 is the neutral element; on the right hand side, 1 is the neutral element).
- For $n, m \in \mathbb{Z}$ we have

$$\varphi(n + m) = c^{n+m} = c^n \cdot c^m = \varphi(n) \cdot \varphi(m)$$

(and on the left hand side, $+$ is the group operation; on the right hand side, it is the multiplication $\cdot$).

5. The map

$$\det : \mathrm{GL}_n(\mathbb{Q}) \to \mathbb{Q}^\times, \; A \mapsto \det(A)$$

is a group homomorphism *(verify this!)*, where $\mathbb{Q}^\times := \mathbb{Q} \setminus \{0\}$ denotes the multiplicative group $(\mathbb{Q}^\times, \cdot, 1)$.

Defining Lemma 3.22 The **kernel** of a group homomorphism $\varphi : G \to G'$ is the subgroup

$$\ker(\varphi) := \{g \in G \mid \varphi(g) = e\} \subset G.$$

Proof It needs to be shown that $\ker(\varphi) \subset G$ is a subgroup. This follows because

1. by definition $\varphi(e) = e$, thus $e \in \ker(\varphi)$,
2. for $g, h \in \ker(\varphi)$ also $\varphi(gh) = \varphi(g)\varphi(h) = e$ holds, thus $gh \in \ker(\varphi)$ and
3. for $g \in \ker(\varphi)$ according to Lemma 3.21 also $\varphi(g^{-1}) = (\varphi(g))^{-1} = e^{-1} = e$, thus $g^{-1} \in \ker(\varphi)$.$\square$

Example
1. The kernel of $\varphi : \mathbb{Z} \to \mathbb{Z}/m\mathbb{Z}, \; x \mapsto [x]$ is the subgroup $m\mathbb{Z} = \{my \in \mathbb{Z} \mid y \in \mathbb{Z}\}$.
2. The kernel of $\det : \mathrm{GL}_n(\mathbb{Q}) \to \mathbb{Q}^\times$ is the subgroup

$$\ker(\det) = \mathrm{SL}_n(\mathbb{Q}) := \{A \in \mathrm{GL}_n(\mathbb{Q}) \mid \det(A) = 1\}.$$

And now?
We now want to show an application of the term *kernel*. I would like to comment on this briefly: We now know what a group homomorphisms

$$\varphi : G \to G'$$

is and of course we want to investigate these and for example ask ourselves whether a given φ is **injective**. This is a definition for set-theoretical maps and one generally has to test that for $g, h \in G$ the conclusion

$$\varphi(g) = \varphi(h) \Longrightarrow g = h$$

is fulfilled. If you want to check this individually, you basically have to calculate $\varphi(g)$ for all $g \in G$ and check whether they are all pairwise different.
But then you have not used at all that G and G' are not just sets (but have
group structure) and that φ is not just a mapping between sets (but fulfills the
homomorphism properties).

The additional structures should indeed be useful for something. If one
considers the kernel $\ker(\varphi)$, one immediately sees:

$$\varphi \text{ is injective} \implies \ker(\varphi) = \{e\}$$

(do you see that?). The next lemma shows that it is even an equivalence of
statements. So, for group homomorphisms, one only needs to calculate the
kernel and can read off a potential injectivity from the knowledge of the
kernel.

Lemma 3.23 *For a group homomorphism $\varphi : G \to G'$ the following holds:*

$$\varphi \text{ is injective} \iff \ker(\varphi) = \{e\}.$$

Check it out!
Prove it!—if you want/need a hint, just read Eq. (3.4) below.

Proof Let $g, h \in G$ given. Note that

$$\varphi(g) = \varphi(h) \iff \varphi(g) \cdot \varphi(h)^{-1} = e \tag{3.4}$$

holds (this is highlighted as a formula so that it can serve as a hint for the above
task). We continue with the calculation:

$$\varphi(g) = \varphi(h) \iff e = \varphi(g) \cdot \varphi(h)^{-1} = \varphi(gh^{-1}) \iff gh^{-1} \in \ker(\varphi).$$

From this, the assertion follows immediately: If φ is injective, then $\ker(\varphi) = \{e\}$
(obviously clear). Conversely, if $\ker(\varphi) = \{e\}$ and $g, h \in G$ with $\varphi(g) = \varphi(h)$, then
$gh^{-1} \in \ker(\varphi) = \{e\}$, thus $gh^{-1} = e$, from which $g = h$ follows. $\square$

Lemma 3.24 *The image of a group homomorphism $\varphi : G \to G'$ is a subgroup in G'.*

Proof Easy to verify—***Check it out!*** $\square$

Definition 3.25 A group homomorphism $\varphi : G \to G'$ is called an **isomorphism,**
if there is a group homomorphism

$$\psi : G' \to G$$

such that $\varphi \circ \psi = \text{id}_{G'}$ and $\psi \circ \varphi = \text{id}_G$ holds.

We then call G and G' **isomorphic groups** and write $G \cong G'$.

Remark 3.26 1. One often also calls it more precisely a **group isomorphism.**
2. If $\varphi : G \to G'$ is a group isomorphism, then φ is bijective.
3. Conversely, every bijective group homomorphism is an isomorphism.
4. If $\varphi : G \to G'$ is an **injective** group homomorphism, then

$$\varphi : G \to \text{Image}(\varphi)$$

is an isomorphism from G to the subgroup $\text{Image}(\varphi) \subset G'$.

Proof (of (3)) If φ bijective, then an inverse map $\psi : G' \to G$ exists. What remains to be shown is that ψ is then automatically a group homomorphism. To this end: Since $\varphi(e) = e$, $\psi(e) = e$ must hold as well. If $g_1', g_2' \in G'$ are given, then we have

$$\varphi(\psi(g_1') \cdot \psi(g_2')) = \varphi\psi(g_1') \cdot \varphi\psi(g_2') = g_1' \cdot g_2'.$$

If both sides are inserted into ψ, one obtains

$$\psi(g_1') \cdot \psi(g_2') = \psi(\varphi(\psi(g_1') \cdot \psi(g_2'))) = \psi(g_1' \cdot g_2').$$

And now?

Writing $G \cong G'$ is the statement that there is a group isomorphism $\varphi : G \to G'$ (which does not have to be uniquely determined).

What does it mean for groups to be isomorphic? It means that they are essentially the same groups, but in different notations: There is a **translator,** namely an isomorphism

$$\varphi : G \overset{\cong}{\to} G'$$

with **inverse translator** φ^{-1}. Every statement that holds in the group G and only refers to the group properties, also holds in G' (one only needs to transition with φ to G'). Example: If G has an element of order 16—let's say g, then G' also does, namely $\varphi(g)$. If there is an injective homomorphism $\alpha : G \hookrightarrow S(17)$, then also $G' \hookrightarrow S(17)$, namely $\alpha \circ \varphi^{-1}$.

Be careful: If you just say $G \cong G'$, you know that such translators exist, but they are usually not unique. You then know that G' behaves exactly like G, but you still have no translator between these two. You can/must first **choose** such a translator to really use the above reasoning.

Example The two groups $\mathbb{Z}/4\mathbb{Z}$ and $\mathbb{Z}/2\mathbb{Z} \times \mathbb{Z}/2\mathbb{Z}$ are **not isomorphic.** The latter group is defined by component-wise multiplication.

Proof The group $\mathbb{Z}/4\mathbb{Z}$ (written additively) has an element of order 4, namely $[1]$. The group $\mathbb{Z}/2\mathbb{Z} \times \mathbb{Z}/2\mathbb{Z}$ does not have such an element. $\square$

Thread to the Previous Chapter
In this chapter, we have seen

- how a group is defined,
- what a subgroup $H \subset G$ is and
- what group homomorphisms $\varphi : G \to G'$ are,
- how to calculate explicitly with pairwise disjoint cycles in the symmetric group $S(n)$,
- that the injectivity of a group homomorphism $\varphi : G \to G'$ can be read off the kernel: φ is injective if and only if $\ker(\varphi) = \{e\}$,
- what a group isomorphism is.

Excercises:
3.1 In the group $GL_2(\mathbb{Q})$ choose $a = \begin{pmatrix} 0 & -1 \\ 1 & 0 \end{pmatrix}$ and $b = \begin{pmatrix} 0 & 1 \\ -1 & 1 \end{pmatrix}$.

(a) Show that a and b have finite order, and determine these orders.
(b) Show that $c = ab$ does not have a finite order.

3.2 Let G be a group and $H_1, H_2 \subset G$ be subgroups. We define (only for this exercise) $H_1^{\text{naiv}} \cdot H_2 := \{h_1 h_2 \in G \mid h_1 \in H_1, h_2 \in H_2\}$. Show by means of a counterexample that $H_1^{\text{naiv}} \cdot H_2$ is not necessarily a subgroup.

3.3 Symmetric Group:

(a) Calculate the product $\rho := \pi\sigma^{-1} \in S(7)$ of

$$\pi := \begin{pmatrix} 1\ 2\ 3\ 4\ 5\ 6\ 7 \\ 6\ 7\ 2\ 3\ 5\ 1\ 4 \end{pmatrix} \text{ and } \sigma := \begin{pmatrix} 1\ 2\ 3\ 4\ 5\ 6\ 7 \\ 2\ 4\ 5\ 6\ 3\ 7\ 1 \end{pmatrix}.$$

(b) Calculate the order $\operatorname{ord}(\pi)$ of the element $\pi \in S(7)$ from (a).

3.4 Let $\varphi : G \to H$ be a group homomorphism between the groups G and H. Let $g \in G$ and $m := \operatorname{ord}(\varphi(g))$. Justify that $g^m \in \ker(\varphi)$ holds.
3.5 Consider the symmetric group $S(n)$. Show

(a) that every element $\sigma \in S(n)$ can be written as a product of finitely many 2-cycles:

$$\sigma = \tau_1 \ldots \tau_m.$$

Hint: Induction after n. In the induction step: Multiply σ suitably with 2-cycles to obtain a new σ' that fulfills $\sigma'(n) = n$.
(b) that the number m of 2-cycles in the above representation is not unique, but their parity $m \mod 2$ is.

Group Quotients and Normal Subgroups

4

4.1 Equivalence Relations

We are now preparing the important concept of the quotient of a group *modulo* a subgroup.

> **And now?**
>
> In the following i will try to accompany the text with a non-mathematical example "running" alongside the mathematical examples. The reason is that the concept of constructing the quotient modulo an equivalence relation is often perceived as difficult and the non-mathematical example provides a situation which is fairly easy to understand. The example will follow us in the sections entitled **And now?** and will deal with football players and football teams.
>
> An important trick to understanding the quotients will again be the usual game of "forgetting some structure" and "remembering the structure at the right place", as it has already occurred in other **And now?** before. This will become clearer when reading the following **And now?**.

We start with the set-theoretical concept that you (perhaps) already know:

Definition 4.1 Let M be a non-empty set. An **equivalence relation on** M is a subset $R \subset M \times M$ of pairs $(x, y) \in M \times M$, where we will use the following notation

$$x \sim y :\Longleftrightarrow (x, y) \in R \text{ (see also the \textbf{\textit{Andnow?}} below),}$$

© The Author(s), under exclusive license to Springer-Verlag GmbH, DE, part of Springer Nature 2024
M. Hien, *Abstract Algebra,* Mathematics Study Resources 7,
https://doi.org/10.1007/978-3-662-67974-6_4

such that the following axioms holds:

1. For every $x \in M$ we have $x \sim x$ (in words: $\sim$ is reflexive),
2. We have $x \sim y \Longrightarrow y \sim x$ ($\sim$ is symmetric) and
3. For $x, y, z \in M$ with $x \sim y$ and $y \sim z$, we also have $x \sim z$ ($\sim$ is transitive).

And now?

The correct formulation of the definition with help of the subset $R \subset M \times M$ is the simplest and fastest. However, one should perhaps understand an equivalence relation in such a way that it is defined by a relation $\sim$ which decides whether two elements $x, y \in M$ are **equivalent (with respect to $\sim$)** or not. For each pair $x, y \in M$ we get:

Either

$x \sim y$ (thus x and y are considered equivalent to each other)

or

$x \not\sim y$ (thus x and y are not equivalent to each other).

The requirements in the definition are quite natural! x should be equivalent to itself (clearly: "equality implies equivalence!"), the order does not matter $x \sim y \Leftrightarrow y \sim x$ and transitivity is also quite natural: If x is equivalent to y and y is equivalent to z, then all three are equivalent to each other.

  The promised (non-mathematical) example: Let M be the set of all Bundesliga (or any other soccer league of your preference) football players. Then we define

$x \sim y :\Leftrightarrow x$ and y play in the same team.

This fulfills the requirements *(check this briefly!)*.

Definition 4.2 Let $\sim$ be an equivalence relation on the non-empty set M. Then we define:

- For each $x \in M$ let

$$[x] := \{y \in M \mid y \sim x\} \subset M$$

  be the subset of all elements equivalent to x. We call $[x]$ the **equivalence class of** x (often just the **class of** x).
- The **quotient of M modulo** $\sim$ is the set

$$M / \sim := \{[x] \mid x \in M\}$$

  of all classes of elements in M.

And now?

One of the difficulties is that $M/\sim$ is a set, the elements of which are themselves sets (more precisely subsets of M). However, at some points, it is better to simply *forget* this—one does not need to *remember* at all times that $[x]$ is by definition a set, but only whenever it is important by itself.

In our example (football player), for a Bundesliga player x, the class

and thus
$$[x] = \text{team in which } x \text{ plays,}$$

$$M/\sim \overset{1-1}{\leftrightarrow} \{\text{FCA, Bayern München, 1. FC Köln, Borussia Dortmund, ...}\}$$

is in bijection to the set of all Bundesliga teams—I pedantically do not write $=$ but $\overset{1-1}{\leftrightarrow}$, because by definition $M/\sim$ is the set of equivalence classes, which is naturally bijective to the set of Bundesliga clubs.

Perhaps you can see what I mean by *forgetting* and *remembering*: When someone says, "Bayern Munich won the match versus FC Augsburg yesterday," he/she has *forgotten* that both teams consist of their players (i.e., are sets of players). When the conversation partner then asks: "Did Harry Kane[1] score a goal?" he/she has *remembered* that the team consists of players.

The axioms of an equivalence relation have an important implication for the classes:

Lemma 4.3 *Let $\sim$ be an equivalence relation on M. Then for every $x \in M$, we have $x \in [x]$. Furthermore, for $x, y \in M$:*

$$[x] \cap [y] \neq \emptyset \Longleftrightarrow x \sim y \Longleftrightarrow [x] = [y].$$

In other words, if two classes (as subsets of M) have a non-empty intersection, they are already equal.

Proof The first statement follows from the reflexivity of $\sim$.

Regarding the second statement: $[x] \cap [y] \neq \emptyset$ holds if and only if there exists a $z \in [x] \cap [y]$. But then, $z \sim x$ and $z \sim y$, so due to transitivity $x \sim y$. Conversely, if $x \sim y$, then $x \in [y]$ and $x \in [x]$, hence $x \in [x] \cap [y]$. This proves the first $\Longleftrightarrow$.

Secondly: If $x \sim y$ and $w \in [x]$, then $w \sim x$. Since $x \sim y$ and due to transitivity, we deduce $w \sim y$ and thus $w \in [y]$. Due to symmetry, we apply the same arguments with x, y interchanged, and deduce that $[x] = [y]$. Conversely, if $[x] = [y]$, then $x \in [x] = [y]$, thus $x \sim y$. $\square$

Note 4.4 An element $y \in [x]$ is also called **a representative of the class**.

---

[1] For those who don't follow German soccer: Harry Kane is a football player who, at the time of the book's translation, is a player for Bayern München.

Fig. 4.1 The equivalence relation from the canoe example in the *And now?* in Sect. 4.1. The points/canoeists in one of the ellipses form a class, for example the far left class is Canoe 1 = [E] = [A]

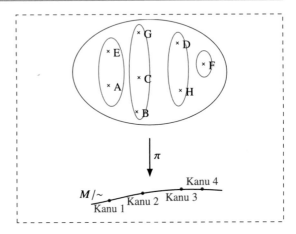

What now?

Apparently, x is a representative of the class $[x]$. One should better understand this as follows: If you write $[x]$ for a class, you have already chosen a representative (namely x), i.e. you already **selected a representative** with the help of which you determine/describe the class. But you could have chosen any other $y \in [x]$ because then $[x] = [y]$ applies—as we have just seen in the lemma.

On the other hand, one could simply write as follows: *Let $\xi \in M/\sim$ be a class.* Then you have not yet chosen a representative—and you will choose one whenever you'll need one.

We will consider another non-mathematical example to illustrate the construction in the following **What now?**. A sketch of the situation can be found in Fig. 4.1 (in Sect. 4.2).

What now?

A group of friends $M = \{A, B, C, D, E, F, G, H\}$ meet for a canoe trip. They divide themselves into the canoes as follows:

Canoe 1	A,E
Canoe 2	B,C,G
Canoe 3	D,H
Canoe 4	F

The equivalence relation is

$$x \sim y :\Leftrightarrow x \text{ and } y \text{ are in the same canoe,}$$

so the quotient set is

$$M/\sim \overset{1\text{-}1}{\leftrightarrow} \{\text{ Canoe 1, Canoe 2, Canoe 3, Canoe 4 }\}.$$

One interesting aspect in this example is that we see that the classes $[x]$ are not as homogeneous as in the other examples. Here, for example

$$[F] = \{F\} \text{ and } [B] = \{B, C, G\}$$

have different numbers of elements.

With group quotients, the classes will be very homogeneous again, but you should appreciate this fact as something particularly nice for group quotients!

Remark 4.5 If $\sim$ is an equivalence relation on M, then we have the **natural projection**

$$\pi : M \to M/\sim, \; x \mapsto [x].$$

It is obviously surjective.

Well-definedness of maps: The well-definedness of maps from a quotient $M/\sim$ to another set often is an issue that has to be handled with great care, more precisely, whenever a map is defined as follows (N being another set):

$$f : M/\sim \to N, \; [x] \mapsto \varphi(x), \tag{4.1}$$

i.e., one specifies what $f([x])$ should be by giving an element assigned to x (see examples below). The problem, however, is that the representative x is generally **not determined by the class** $[x]$.

And now?

In the example with the canoeists, for instance, B,C and G are each representatives of the class

$$\text{Canoe } 2 = [B] = [C] = [G].$$

If one tried to define a map that indicates the age of the canoeist (let's assume that, for example, B and C are not of the same age):

$$f : M/\sim \to \mathbb{N}, \; [x] \mapsto \text{Age}(x),$$

then this is **not well-defined**, because we consider the class *Canoe 2* as the argument of the map f and we have *Canoe 2*=$[B] = [C]$, but $\text{Age}(B) \neq \text{Age}(C)$.

The question of well-definedness therefore always arises when the class
$[x] \in M/\sim$ is represented by a representative $x \in M$ and the actual map

$$[x] \mapsto ???$$

is defined in terms of the representative x (and not directly in terms of the class
$[x]$). This means that a **choice** is involved (namely the choice of the representative)
and one must prove that the right side does not depend on the choice:

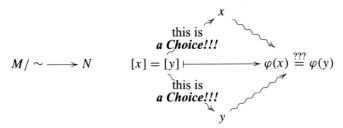

Remark 4.6 two remarks:

- It is possible to formulate the question of well-definedness in pure mathematical terms: A map $\varphi : M \to N$ induces a **well-defined mapping** $M/\sim \to N$,
  $[x] \mapsto \varphi(x)$, if there is a mapping $f : M/\sim \to N$ (i.e., a map defined on the
  classes themselves) such that the diagram

$$M \xrightarrow{\ \varphi\ } N$$
$$\pi \downarrow \quad \nearrow f$$
$$M/\sim$$

  commutes,[2] i.e., such that $f \circ \pi = \varphi$ holds.
- The question of well-definedness naturally also arises with candidates for maps

$$M/\sim \times M/\sim \to N, \quad ([x],[y]) \to \varphi(x,y).$$

  or more general constructions involving quotients on the source side of the
  map.

And now?

Examples of issues of well-definedness in the **football example**:

- The attempt to define the age of a player as a map on the classes is of
  course nonsense:

---

[2] A diagram **commutes**, if for all objects A and B in the diagram all possible combinations leading from A to B yield the same mapping.

$$f : M/\sim \to \mathbb{N}, \; [x] \mapsto age(x)$$

is **not well-defined**—unless all players of each team by chance are of the same age.

- The map which indicates the color of a player's home jersey induces a **well-defined map**

$$f : M/\sim \to \{\text{all colors}\}, \; [x] \mapsto \text{Jersey color}(x),$$

because if $[y] = [x]$, then x and y play in the same team, so they wear the same jersey color (we agree that the goalkeepers also wears a jersey of the same color).

- The map which indicates the rank in the standings list after the 5th match is certainly also **well-defined**: $g : M/\sim \to \{1, 2, \ldots, 18\}, \; [x] \mapsto \text{rank in the standings after 5 gamedays}(x)$.

The last two (well-defined) examples are of different nature with respect to well-definedness because the last example (rank) is directly defined as a map on the set of teams (when you look at the standings list, you *forget* that the teams consist of players, it is only the team you are interested in. The list anyway contains the names of the teams, not of the players).

The example before that (color of the jersey) is more interesting because the image of a team/class $[x]$, i.e. $f([x])$ is given by the following procedure:

$$[x] \sim_{Choice!!} \rightsquigarrow \text{Representative } x \rightsquigarrow \text{Shirt colour}(x)$$

—if you want to determine the jersey color, you choose a player from the team (e.g., if you attend the match in the stadium or watch it on TV) and take a look at his jersey. No matter which player of the team you have chosen, you will always ome up with the same color of the jersey.

It is often advantageous to choose a representative for each class a priori. The following definition gives a precise meaning to this:

Definition 4.7 If $\sim$ is an equivalence relation on the set M, a subset $S \subset M$ is called a **system of representatives** if there is *exactly one* element $x \in S$ for each class $\xi \in M/\sim$ with $\xi = [x]$.

In other words: $S \subset M$ is a system of representativesif the restriction of the projection

$$\pi|_S : S \to M/\sim, \; x \mapsto [x]$$

is a bijection.

And now?
In the football example, S could be the set of captains of each team.
But also any other subset $S \subset M$ of all Bundesliga players, such that exactly one player for each team belongs to S, is a system of representatives.

Lemma 4.8 *If $S \subset M$ is a system of representatives, M decomposes into the pairwise disjoint union*

$$M = \bigsqcup_{s \in S} [s].$$

(4.2)

And now?
Looking at equation (4.2), one must remember that $[s] \in M/\sim$ is by definition a subset $[s] \subset M$—only with this interpretation (4.2) is a meaningful expression.

For the cardinalities involved we deduce the following:

Corollary 4.9 If the set M is finite, we have

$$\#M = \sum_{s \in S} \#[s].$$

4.2   Group Quotients

We do not want to consider arbitrary equivalence relations further on, but those that are induced by subgroups H in a group G:

Defining Lemma 4.10 Let $H \subset G$ be a subgroup of the group G. Then

$$g \sim g' :\Longleftrightarrow g'^{-1} \cdot g \in H$$

defines an equivalence relation on G. For each $g \in G$ the associated class is given by

$$[g] = gH := \{gh \mid h \in H\}.$$

gH is also called the **left coset of** g.
The **group quotient of** G **modulo** H is defined to be the quotient set with respect to this equivalence relation:

$$G/H := G/\sim.$$

Fig. 4.2 Illustration of the
group quotient G/H

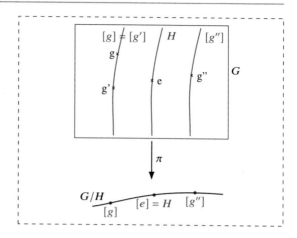

Proof The axioms of an equivalence relation are easy to verify (only the third axiom requires a common trick):

- $g \sim g$, because $g^{-1}g = e \in H$.
- If $g'^{-1} \cdot g \in H$ holds, then its inverse element satisfies $(g'^{-1}g)^{-1} = g^{-1}g' \in H$, from which $g' \sim g$ follows.
- If

$g \sim g_1 \sim g_2$, thus by definition $g_1^{-1}g \in H$ and $g_2^{-1}g_1 \in H$. Then

$$g_2^{-1}g = g_2^{-1}\underbrace{g_1 \cdot g_1^{-1}}_{=e}g = (g_2^{-1}g_1) \cdot (g_1^{-1}g) \in H.$$

Note 4.11 The definition can also be formulated as follows. For $g, g' \in G$ we have:

$$g \sim g' \iff \text{there exists an } h \in H \text{ with } g' = g \cdot h.$$

This is often convenient in applications because one can then directly work with $g' = gh$.

And now?
The trick in proving the third property of Lemma 4.10 is to insert a factor $e \in G$ into the equation and then write it in the form $e = g_1 \cdot g_1^{-1}$. We will see this kind of trick again soon.

Examples

1. Let $H = \mathrm{SL}_2(\mathbb{Q}) \subset \mathrm{GL}_2(\mathbb{Q}) = G$. Then for two invertible matrices $A, B \in G$:

$$[A] = [B] \Longleftrightarrow B^{-1}A \in \mathrm{SL}_2(\mathbb{Q}) \Longleftrightarrow \det(B^{-1}A) = 1 \Longleftrightarrow \det(A) = \det(B).$$

In particular,

$$\mathbb{Q}^\times \to G/H, \ a \mapsto \begin{pmatrix} a & 0 \\ 0 & 1 \end{pmatrix}$$

is a bijective mapping.

> **Check it out!**
> Prove the last statement!!!

2. For $m \geq 2$ let $G = \mathbb{Z}$ and $H = m\mathbb{Z}$. Then we have the following statement in $\mathbb{Z}/m\mathbb{Z}$:

$$[a] = [b] \Longleftrightarrow -b + a \in m\mathbb{Z} \Longleftrightarrow \text{there exists a } k \in \mathbb{Z} \text{ with } b = a + km$$

(note the additive notation!!!). In particular,

$$\{0, 1, 2, \ldots, m - 1\} \to \mathbb{Z}/m\mathbb{Z}, \ a \mapsto [a]$$

is a bijective mapping.

Remark 4.12 For a group quotient G/H, the individual classes $[g] = gH$ are homogeneous (as opposed to the general set theoretic situation, cp. the examples above, the canoe example for instance). See Fig. 4.1 and 4.2—more precisely: If $[g], [g'] \in G/H$ are two classes, then one can choose representatives $g \in [g]$ and $g' \in [g']$ and the following map is a bijection

$$[g] \xrightarrow{\text{1-1}} [g'], \ x \mapsto g'g^{-1}x$$

In particular, $\#[g] = \#[g']$ holds.

> **Check it out!**
> Do you understand the bijection in the remark? What is the inverse map? Do you understand that the bijection depends on the choices of representatives?

Important addition to the remark: The above remark shows that two classes $[g]$ and $[g']$ are bijectively related to each other as sets—but the bijection depends on choices and is therefore not *natural* or *canonical*.

4.3 The Lagrange Theorem

Now we can prove an extremely important theorem (by Lagrange) in group theory. It concerns the cardinality of groups and subgroups, so we assume that $\#G < \infty$. Then we call G a **finite group**. Let $H \subset G$ be a subgroup.

If one chooses a representative system $S \subset G$ for G/H (and one can always find such), then the restriction of the projection

$$\pi|_S : S \to G/H,\ g \mapsto [g]$$

is a bijection. Note that since G is finite, the quotient set G/H again only has finitely many elements. Thus, we consider the following cardinalities:

1. The group quotient G/H has $\#(G/H) = \#S$ elements and
2. each class $[g]$ has exactly as many elements as the class $[e] = H$, so for each class $[g]$, we have $\#[g] = \#H$.

Applying corollary 4.9 we immediately deduce the theorem

Theorem 4.13 Let $H \subset G$ be a subgroup in the finite group G, then

$$\#G = \#(G/H) \cdot \#H.$$

Lagrange's theorem is a direct consequence of this:

Theorem 4.14 (Lagrange's Theorem) Let G be a finite group and $H \subset G$ be a subgroup. Then $\#H$ is a divisor of $\#G$:

$$\#H \mid \#G.$$

In particular, $\mathrm{ord}\,(g) | \#G$ holds for every $g \in G$.

Corollary 4.15 If G is a finite group and $g \in G$, then $g^{\#G} = 1$.

Proof The assertion follows because $\#G$ according to Lagrange's Theorem is a multiple of $\mathrm{ord}\,(g)$.□

And now?
Lagrange's theorem often serves as an argument as to why a group, for example, cannot have a given fixed group as a subgroup.

For instance, there can be **no injective** group homomorphism

$$\varphi : \mathbb{Z}/7\mathbb{Z} \hookrightarrow S(5)$$

because if there were, $\text{im}(\varphi) \subset S(5)$ would be a subgroup isomorphic to $\mathbb{Z}/7\mathbb{Z}$ (see remark 3.26) and thus would have 7 elements, but

$$7 \nmid \#S(5) = 5! = 120.$$

Such arguments often appear in exercises.

4.4   Normal Subgroups and Factor Groups

Let's go back to the initial situation: G is a group and $H \subset G$ is a subgroup.

Check it out!
We then have the group quotient G/H (as a set) and the surjective map of sets:

$$\pi : G \to G/H, \; g \mapsto [g] = gH.$$

What question arises immediately in this situation? What are you asking yourself about G/H?

And now?
In response to the ***Check it out!*** : You are (hopefully) asking yourself whether G/H is a group again!

   Initially, however, the question is not very meaningful, because a group includes $\cdot$ and e. In addition, the group structure on G/H (if there is a natural one) should have something to do with the group structure of the group G. What does that mean? Well, one certainly has a candidate for the group operation/multiplication:

A natural candidate for a multiplication on G/H is the following:

$$\cdot_{\text{candidate}} : G/H \times G/H \to G/H, \; ([g_1], [g_2]) \mapsto [g_1] \cdot_{\text{candidate}} [g_2] := [g_1 \cdot g_2]$$
$$(4.3)$$

where the latter multiplication, i.e. the one in $g_1 \cdot g_2 \cdot$ is that from G.

> **Check it out!**
> Do you see the problem that arises?—think about this question before going
> on with the text!

The problem is the question of well-definedness, because (4.3) again includes
choices (each time a choice of g_i as a representative of $[g_i]$ is made):

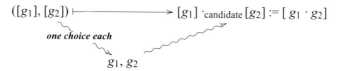

(Don't be fooled: On the right hand side we find the class $[g_1 g_2] \in G/H$, but this
arises from the classes $[g_1], [g_2]$ **only after the choice of representatives**:

$$([g_1], [g_2]) \sim_{choices} \rightsquigarrow g_1, g_2 \longmapsto g_1 \cdot g_2 \longmapsto [g_1 \cdot g_2].$$

One chooses g_1, g_2, then multiplies those to obtain $g_1 \cdot g_2$ and then takes the class
of this element $[g_1 \cdot g_2] \in G/H$ as the candidate for the product.)
  Now you might expect that one has to/can show the well-definedness, but:
  In general, (4.3) is **not a well-defined** mapping.

Example: Let's consider the example $H = \mathrm{Diag}_2(\mathbb{Q}) \subset GL_2(\mathbb{Q}) = G$ of the
invertible diagonal matrices

$$H = \{\begin{pmatrix} a & 0 \\ 0 & b \end{pmatrix} \mid ab \neq 0\}$$

Briefly check that this is a subgroup!
The attempt to try to define a multiplication is:

$$G/H \times G/H \to G/H, \ ([A], [B]) \mapsto [AB],$$

where AB on the right is the matrix product. Take the following pair of classes

$$([A], [B]) := (\left[\begin{pmatrix} 1 & 1 \\ 0 & 1 \end{pmatrix}\right], \left[\begin{pmatrix} 1 & 1 \\ 0 & 1 \end{pmatrix}\right])$$

We have

$$\left[\begin{pmatrix} 1 & 1 \\ 0 & 1 \end{pmatrix}\right] = \left[\begin{pmatrix} 1 & 2 \\ 0 & 2 \end{pmatrix}\right] = \left[\begin{pmatrix} 1 & 3 \\ 0 & 3 \end{pmatrix}\right],$$

because

$$\begin{pmatrix} 1 & 1 \\ 0 & 1 \end{pmatrix}^{-1} \cdot \begin{pmatrix} 1 & a \\ 0 & a \end{pmatrix} = \begin{pmatrix} 1 & -1 \\ 0 & 1 \end{pmatrix} \cdot \begin{pmatrix} 1 & a \\ 0 & a \end{pmatrix} = \begin{pmatrix} 1 & 0 \\ 0 & a \end{pmatrix} \in H$$

holds. Hence, we observe the following:

- If we choose the representatives $A = B = \begin{pmatrix} 1 & 1 \\ 0 & 1 \end{pmatrix}$, we obtain $A \cdot B = \begin{pmatrix} 1 & 2 \\ 0 & 1 \end{pmatrix}$.

- If we choose the representatives

$$A' = \begin{pmatrix} 1 & 2 \\ 0 & 2 \end{pmatrix}, \quad B' = \begin{pmatrix} 1 & 3 \\ 0 & 3 \end{pmatrix},$$

we obtain $A' \cdot B' = \begin{pmatrix} 1 & 9 \\ 0 & 6 \end{pmatrix}$

and since

$$(AB)^{-1}(A'B') = \begin{pmatrix} 1 & 2 \\ 0 & 1 \end{pmatrix}^{-1} \cdot \begin{pmatrix} 1 & 9 \\ 0 & 6 \end{pmatrix} = \begin{pmatrix} 1 & -2 \\ 0 & 1 \end{pmatrix} \cdot \begin{pmatrix} 1 & 9 \\ 0 & 6 \end{pmatrix} = \begin{pmatrix} 1 & 3 \\ 0 & 6 \end{pmatrix} \notin H$$

we end up with two different classes: $[A \cdot B] \neq [A' \cdot B']$.

And now?
One might think it is a pity that the construction of a group structure on G/H does not work in general. On the other hand, that's just the way it is and such a (negative) realization leads to further considerations, which then again bring forth interesting new notions and theorems.

   The solution to this issue is to define an additional property for subgroups $H \subset G$ so that for the subgroups with this new property, the above construction works. The question is therefore: Given a subgroup $H \subset G$, can we understand under which conditions the candidate (4.3) for a group multiplication on G/H is well-defined?

We try to prove the (not always valid) well-definedness of (4.3) to see where the problems occur: Let $g_1, g_2 \in G$ be given and choose other representatives $g_1', g_2' \in G$ of the same classes:

$$[g_1] = [g_1'] \text{ and } [g_2] = [g_2'].$$

Then there are $h_1, h_2 \in H$, such that $g_1' = g_1 \cdot h_1$ and $g_2' = g_2 \cdot h_2$. We now have to compare the respective right hand side of (4.3):

$$[g_1 \cdot g_2] \in G/H \text{ and } [g_1' \cdot g_2'] \in G/H.$$

The second expression yields:

$$[g_1' \cdot g_2'] = [g_1 \cdot h_1 \cdot g_2 \cdot h_2].$$

Now we have located the problem: If we could swap h_1 and g_2, we would be done:

$$\text{if } h_1 g_2 = g_2 h_1 \text{ holds, we have } [g_1 \cdot h_1 \cdot g_2 \cdot h_2] = [g_1 g_2 h_1 h_2] = [g_1 g_2], \quad (4.4)$$

the last equation holds since $h_1 h_2 \in H$. In general, however, h_1 and g_2 do not commute.

And now?

One could already find a way out of the dilemma by assuming that all $h \in H$ commute with all $g \in G$. This indeed even has a name: A subgroup $H \subset G$ is called **central**, if $gh = hg$ holds for all $g \in G$ and all $h \in H$. Then we have proven the lemma above (with 4.4):

Lemma 4.16 *If $H \subset G$ is a central subgroup, then* (4.3) *is well-defined and thus G/H is a group.*

This is nice, but too restrictive. If we take another look at the term causing the problem:

$$[g_1 h_1 g_2 h_2]$$

it is not really necessary that $h_1 g_2 = g_2 h_1$ holds, because it is sufficient to demand that one finds an $h \in H$ such that one can proceed as follows:

$$[g_1 h_1 g_2 h_2] = [g_1 g_2 h h_2] = [g_1 g_2] \in G/H$$

Therefore, if we are able to achieve the following:

Given any $g_2 \in G$ and $h_1 \in H$ we can find an $h \in H$ such that $h_1 g_2 = g_2 h$ holds.
$$(4.5)$$

we are done as well (see the subtle difference to the condition $h_1 g_2 = g_2 h_1$ we had before: On the right hand side of such an equation, it is not necessary to have exactly the element $h = h_1$, it is sufficient to find any $h \in H$ instead). This leads exactly to the notion of a normal subgroup that will be defined now. How can we formulate the condition (4.5) in a more elegant way? To this end, let us observe the following:

$$h_1 g_2 = g_2 h \text{ for an } h \in H \Leftrightarrow$$
$$h = g_2^{-1} h_1 g_2 \text{ for an } h \in H \Leftrightarrow$$
$$g_2^{-1} h_1 g_2 \in H,$$

(the purpose of this observastion is that we have avoided to introduce a letter (i.e., h) for the element $h \in H$, because the only important fact is that it is any element in H). This condition must hold for all $g_2 \in G$ and all $h_1 \in H$ (and we therefore omit the indices). It has also become customary to write the inverse element $()^{-1}$ on the other hand side—this causes no problem: If $g^{-1}hg \in H$ for all $g \in G$ holds, then so does $ghg^{-1} = (g^{-1})^{-1}h(g^{-1}) \in H$.

The considerations in the last *And now?* result in the following definition:

Definition 4.17 A subgroup $H \subset G$ is called a **normal subgroup**—in notation $H \lhd G$—if the following holds:

$$\text{For every } g \in G \text{ we have the inclusion } gHg^{-1} \subset H$$

Note that $gHg^{-1} := \{ghg^{-1} \mid h \in H\} \subset G$.
    Let's summarize our considerations again:

Theorem 4.18 If $H \lhd G$ is a normal subgroup, then

$$G/H \times G/H \;\rightarrow\; G/H$$
$$([g_1] \;,\; [g_2]) \;\mapsto\; [g_1 \cdot g_2]$$

defines a well-defined multiplication, so that G/H together with this multiplication and $[e] \in G/H$ as a neutral element becomes a group. G/H is then also called the **factor group** to the normal subgroup H.
    The projection $\pi : G \twoheadrightarrow G/H$ is a surjective group homomorphism with kernel H.

Proof Since the well-definedness was worked out in several steps above, here is a short summary: Let $g_1, g_1', g_2, g_2' \in G$ be elements with $[g_1] = [g_1']$ and $[g_2] = [g_2']$. Then there are $h_1, h_2 \in H$, such that $g_1' = g_1 \cdot h_1$ and $g_2' = g_2 \cdot h_2$. With this,

$$[g_1'g_2'] = [g_1 h_1 g_2 h_2] = [g_1 \cdot g_2 \cdot \underbrace{\underbrace{g_2^{-1} h_1 g_2}_{\in H} h_2}_{\in H}] = [g_1 g_2] \in G/H.$$

The group properties and the statements about the projection π are very easy to see. $\square$

Hence, we have seen that a normal subgroup $H \lhd G$ leads to a group G/H and a group homomorphism $\pi : G \rightarrow G/H$, whose kernel is exactly H. The reverse also holds:

Lemma 4.19 *If $\varphi : G \rightarrow G'$ is a group homomorphism, then $\ker(\varphi)$ is a normal subgroup in G.*

Proof Let $h \in \ker(\varphi)$ and $g \in G$ be arbitrary. Then the following holds

$$\varphi(ghg^{-1}) = \varphi(g)\varphi(h)\varphi(g)^{-1} = \varphi(g)e\varphi(g)^{-1} = e,$$

thus $ghg^{-1} \in \ker(\varphi).\square$

One last definition dealing with normal subgroups:

Definition 4.20 A group G is called **simple** if it contains no normal subgroup other than $\{1\}$ and G.

And now?

The idea behind this definition is that one might hope to better understand a—let's say finite—group G if it has a normal subgroup $N \lhd G$ because then we have the surjective homomorphism

$$\pi : G \twoheadrightarrow G/N =: G'$$

with kernel N. One might want to use this to argue that G looks as if it arises from the (smaller) group G/N and the (also smaller) group N. Then the *simple* groups would be something like the indivisible Lego blocks which cannot be deconstructed into smaller groups.

This is not entirely correct, because for given groups N and G' there are in the general situation several groups G that fit into the scheme:

there exists a surjective $\pi : G \to G'$ with kernel N

—perhaps this also fits well with the "Lego blocks image", as you can assemble two Lego blocks in different ways.

We will get to know a series of simple groups:

Defining Lemma 4.21 Let $n \in \mathbb{N}$. Then there exists a unique group homomorphism

$$\text{sgn} : S(n) \to \mu_2 := \{\pm 1\},$$

such that $\text{sgn}(\tau) = -1$ for every transposition (=2-cycle) $\tau = (ij)$. The kernel of sgn is the **Alternating Group** $A(n) := \ker(\text{sgn})$.

Proof The map sgn and its properties are already known from Linear Algebra, they necessarily appear when studying determinants (Leibniz formula). $\square$

Without proof, let us state the following theorem:

Theorem 4.22 $A(n)$ is simple for $n \geq 5$.

The proof is not really difficult, but somewhat *cumbersome* and does not provide any really interesting insight for us in the moment.

And now?
We clearly have $\#A(n) = \frac{n!}{2}$, so there are arbitrarily large simple groups—or building blocks, if you will.

But there are also quite extraordinary simple groups that are very large and not part of a series (as the $A(n)$ are), for example there is a simple group called the **Monster group** with

$$2^{46} \cdot 3^{20} \cdot 5^9 \cdot 7^6 \cdot 11^2 \cdot 13^3 \cdot 17 \cdot 19 \cdot 23 \cdot 29 \cdot 31 \cdot 41 \cdot 47 \cdot 59 \cdot 71 \approx 8 \cdot 10^{53}$$

elements.

4.5   The Fundamental Homomorphism Theorem for Groups

We now prove the Fundamental Homomorphism Theorem for groups. There are analogous theorems for rings (later), vector spaces (Linear Algebra, where this theorem is the reason for the formula on the rank of linear maps), …. The theorem is extremely important and actually quite easy to prove:

Theorem 4.23 (Fundamental Homomorphism Theorem for Groups) Let be a group homomorphism. Then φ induces the following **injective** group homomorphism

$$\overline{\varphi} : G/\ker(\varphi) \to G' \; [g] \mapsto \varphi(g).$$

Proof Note that the first result to prove is for the map to be well-defined:

- **Well-definedness:** If $[g_1] = [g_2]$, then there exists an $h \in \ker(\varphi)$ such that $g_2 = g_1 h$. Then

$$\varphi(g_2) = \varphi(g_1 h) = \varphi(g_1)\varphi(h) = \varphi(g_1),$$

  so $\overline{\varphi}$ is well-defined.
- **Group homomorphism:** This is trivial, because φ is one.
- **Injectivity:** The kernel of $\overline{\varphi}$ has to be determined. Let $[g] \in \ker(\overline{\varphi})$. Then the following holds:

$$e = \overline{\varphi}([g]) \stackrel{\text{Def}}{=} \varphi(g),$$

  so $g \in \ker(\varphi)$ and thus $[g] = e \in G/\ker(\varphi)$.

$\square$

Check it out!
Read the proof of the injectivity carefully and make sure that the well-definedness (which was shown before this step) plays some role here—in the proof of the injectivity we (silently) made a a choice. I explain this briefly in the following *And now?*.

And now?
In the proof we wrote: Let $[g] \in \ker(\overline{\varphi})$ -- here a choice of a representative $g \in [g]$ is implicitly given. If you are not yet confident enough with group quotients, sort out exactly where which elements (or subsets) live, either in G or in $G/\ker(\varphi)$: I will elaborate on this in detail (and to emphasize the distinction i will call everything that plays in G to be **upstairs**, everything in $G/\ker(\varphi)$ to be **downstairs**), this comes from my habit to write/sketch the projection as vertical:

$$G$$
$$\downarrow \pi$$
$$G/\ker(\varphi).$$

Now back to the proof of the injectivity. Contained in the statement *"Let $[g] \in \ker(\overline{\varphi})$"* are the following considerations:

- The kernel of $\overline{\varphi}$ is $\ker(\overline{\varphi}) \subset G/\ker(\varphi)$ a normal subgroup in the factor group $G/\ker(\varphi)$, so it lives **downstairs**.
- An element $\xi \in \ker(\overline{\varphi})$ is an element $\xi \in G/\ker(\varphi)$ and also lives **downstairs**.
- The elements **downstairs** can always be described by giving a representative from **upstairs** (this is the surjectivity of π).
- So let's describe the $\xi \in \ker(\overline{\varphi})$ by choosing a representative $g \in G$ from **upstairs**, i.e. $\xi = [g]$. This involves the choice of a representative $g \in [g]$.

Then $e = \overline{\varphi}(\xi) = \overline{\varphi}([g])$ (this happens **downstairs**), and by definition $\overline{\varphi}([g]) = \varphi(g)$—and here the choice of representative doesn't matter, because we already know the well-definedness of the map. It follows that the representative g (which lives **upstairs**) lies in the kernel $\ker(\varphi) \subset G$ **upstairs**. Therefore, its class $\xi = [g]$ is the class $[e]$ **downstairs**, and we are done.

One can also write down the situation of the fundamental homomorphism theorem as a diagram:

$$
\begin{array}{ccc}
G & \xrightarrow{\;\varphi\;} & G' \\
{\scriptstyle\pi}\big\downarrow & \nearrow{\scriptstyle\overline{\varphi}} & \\
G/\ker(\varphi) & &
\end{array}
$$

Corollary 4.24 If $\varphi : G \to G'$ is a surjective group homomorphism, then

$$\overline{\varphi} : G/\ker(\varphi) \overset{\cong}{\longrightarrow} G'$$

is a group isomorphism.

Example Let $n \in \mathbb{N}$ and $\zeta = \exp(2\pi i/n) \in \mathbb{C}$. Then $\zeta^n = 1$ and even $\mathrm{ord}\,(\zeta) = n$ as an element in the multiplicative group $\mathbb{C}^\times = \mathbb{C} \smallsetminus \{0\}$. Let's write

$$\mu_n(\mathbb{C}) := \langle \zeta \rangle \subset \mathbb{C}^\times$$

for the subgroup generated by this complex number —the group of n-th roots of unity in $\mathbb{C}$. We have the surjective group homomorphism

$$\varphi : \mathbb{Z} \to \mu_n(\mathbb{C}), \; k \mapsto \zeta^k$$

with source the additively written group $\mathbb{Z}$. We have $\ker(\varphi) = n\mathbb{Z}$, so

$$\overline{\varphi} : \mathbb{Z}/n\mathbb{Z} \overset{\cong}{\to} \mu_n(\mathbb{C}), \; [k] \mapsto \zeta^k$$

is an isomorphism (note that the group on the left hand side is written additively and the one on the right hand side multiplicatively).

4.6   Finite Cyclic Groups

We conclude this chapter with three considerations on finite cyclic groups.
     The first statement is a generalization of the last example.

Lemma 4.25 *Any two finite cyclic groups of the same cardinality are isomorphic to each other.*
     In more detail: If G and H are cyclic groups with $\#G = \#H < \infty$, then there is an isomorphism $G \overset{\cong}{\to} H$.

Proof Let G be cyclic with $\#G = n$. Then, by choosing a generator $g \in G$ the homomorphism

$$\varphi : \mathbb{Z} \to G, \; k \mapsto g^k$$

is surjective (this is the definition of *being cyclic with generator g*). The kernel is obviously $n\mathbb{Z}$ and thus we have an isomorphism

$$\overline{\varphi} : \mathbb{Z}/n\mathbb{Z} \xrightarrow{\cong} G,$$

(which depends on the choice of the generator). The same applies to H. $\square$

Corollary 4.26 For every number $n \in \mathbb{N}$ there exists, up to isomorphism, exactly one cyclic group of order $\#G = n$.

Additionally, we have:

Lemma 4.27 *If G is a cyclic group with infinitely many elements, then $G \cong \mathbb{Z}$.*

Proof If $g \in G$ is a generator, then $\varphi : \mathbb{Z} \to G, k \mapsto g^k$ is obviously bijective. $\square$

The second statement is the following:

Lemma 4.28 *If G is a finite group and $\#G$ is a prime number p, then G is cyclic and thus $G \cong \mathbb{Z}/p\mathbb{Z}$.*

Proof Let $g \in G$ be an element $g \neq e$. Then consider the subgroup

$$\langle g \rangle \subset G,$$

with $\text{ord}(g)$ elements. According to Lagrange's theorem, $\text{ord}(g) \mid \#G = p$ and since $1 \neq \text{ord}(g)$, necessarily $\text{ord}(g) = p$ holds. But then $\langle g \rangle = G$. $\square$

The third statement is:

Lemma 4.29 *If G is a cyclic group of order n, and $d|n$ any positive divisor, there is exactly one subgroup of order d. Furthermore, all subgroups and all factor groups of G are cyclic again.*

Proof We fix a generator of G, so $G = \langle \gamma \rangle$. Then for $d|n$, let's say $n = dm$, the element γ^m has order d, so $H := \langle \gamma^m \rangle$ is a subgroup with d elements.

Conversely, if H is any subgroup in G, consider the minimal $k \in \mathbb{N}$ with $\gamma^k \in H$. If then $h \in H$ is any element, then $h = \gamma^\ell$ for an $\ell \in \mathbb{Z}$. After division with remainder, we have $\ell = qk + r$ with $0 \leq r < k$. We deduce that

$$h = \gamma^\ell = (\gamma^k)^q \cdot \gamma^r \Rightarrow \gamma^r = (\gamma^k)^{-q} \cdot \gamma^\ell \in H,$$

from which, due to the minimality of k, we obtain $r = 0$. Therefore, $H = \langle \gamma^k \rangle$.

This shows that every subgroup in G is cyclic and that every subgroup H can be written as $H = \langle \gamma^k \rangle$ with the minimal $k \in \mathbb{N}$ with $\gamma^k \in H$. Then $\#H = \text{ord}(\gamma^k)$ is a divisor of n, from which $k|n$ follows and finally $\#H = \frac{n}{k}$. Thus, there is exactly the subgroup $\langle \gamma^{n/d} \rangle$ of order d.

Is $G' = G/N$ for a subgroup $N \triangleleft G$, then $[\gamma]$ is obviously a generator of G/N. $\square$

And now?

These considerations are helpful whenever you have a group of a given order and need to determine which group it is (up to isomorphism).

Lemma 4.28 itself is the direct answer to such a question: *How many groups of order p(a prime number) exist up to isomorphism?* Answer: Exactly one, namely the cyclic group $\mathbb{Z}/p\mathbb{Z}$.

Note that it does not matter whether the group is written additively or multiplicatively: a group is a group.

Red thread to the previous chapter

In this chapter, we have seen

- what an equivalence relation $\sim$ on a set M is and how the quotient $M/\sim$ is defined,
- that every subgroup $H \subset G$ induces a group quotient G/H as a set—namely via the equivalence relation

$$g \sim g' \Leftrightarrow \text{ there exists a } h \in H \text{ with } g' = gh,$$

- that one must require that $H \lhd G$ is a normal subgroup if one wants to obtain a group structure on G/H via $[g_1][g_2] := [g_1g_2]$,
- that a subgroup $H \subset G$ is called a normal subgroup, if $gHg^{-1} \subset H$ holds for all $g \in G$,
- that usually the well-definedness of maps $\psi : G/H \to ??$ is an issue that has to be proven,
- that the fundamental homomorphism theorem holds,
- that, given a subgroup H in a finite group G, the equation

$$\#G = \#(G/H) \cdot \#H$$

holds and thus in particular (Lagrange's theorem) $\#H \mid \#G$ holds as well,

- that applying these considerations to cyclic groups, the latter are completely understood up to isomorphism.

Excercises

4.1 Justify the statement: if G *is abelian*, every subgroup is a normal subgroup (this is actually rather easy -- too easy for an exercise, but nevertheless worth mentioning!)

4.2 Consider the group $G := \mathrm{GL}_n(\mathbb{Q})$ of invertible $n \times n$-matrices over $\mathbb{Q}$ for $n \geq 2$. Let

$$D := \{A \in G \mid A \text{ is an upper triangular matrix.}\}$$
$$S := \{A \in G \mid \det(S) = 1\}$$

Check whether D and S are normal subgroups in G.

4.3 Let H be a subgroup of a (not necessarily finite) group G with finite index $[G : H] = n \in \mathbb{N}$.

(a) Show that for all $g \in G$ there exists a $j \in \mathbb{N}$ such that $1 \leq j \leq n$ and $g^j \in H$ holds. (Hint: Consider the cosets $g^i H$,$0 \leq i \leq n$.) (Note on the note: There are $n + 1$ natural numbers i with $0 \leq i \leq n$.)
(b) Solve the exercise with the additional assumotion that H is a normal divisor, **in one line**.

4.4 Let G be a cyclic group with $n = \#G$ elements and let $g \in G$ be a generator. Show that

$$\mathrm{ord}(g^k) = \frac{n}{\mathrm{ggT}(k, n)}.$$

4.5 Let $n \in \mathbb{N}$. Show that for a $[a] \in \mathbb{Z}/n\mathbb{Z}$ the greatest common divisor $\mathrm{ggT}(a, n)$ is well-defined. Also show that:

(a) the set

$$(\mathbb{Z}/n\mathbb{Z})^\times := \{[a] \in \mathbb{Z}/n\mathbb{Z} \mid \mathrm{ggT}(a, n) = 1\}$$

together with the multiplication $[a] \cdot [b] := [ab]$ (well-defined?) and the neutral element $[1] \in (\mathbb{Z}/n\mathbb{Z})^\times$ forms a group.[3]
(b) the map

$$\psi : \mathrm{Aut}(\mathbb{Z}/n\mathbb{Z}, +) \to (\mathbb{Z}/n\mathbb{Z})^\times, \; \varphi \mapsto \varphi([1])$$

is a group isomorphism.

4.6 For an abelian group G (written additively), let

$$\mathrm{Tor}(G) := \{g \in G \mid \text{there exists a } n \in \mathbb{N} \text{ such that } ng = 0\}$$

be the set of torsion elements (where $ng := g + \ldots + g$ with n summands). Show that $\mathrm{Tor}(G)$ is a subgroup and that

$$\mathrm{Tor}\big(G/\mathrm{Tor}(G)\big) = \{e\}$$

is the trivial group with only one element.
4.7 You find the following formulation in a handwritten script, which you cannot read at one place:

Let G be a group and let $H, K \subset G$ be subgroups. Assume that $H \not\subseteq K$. Then the mapping

$$G/H \to G/K, \; [g] \mapsto [g]$$

---

[3] The reason for the notation $(\mathbb{Z}/n\mathbb{Z})^\times$ will be seen later in the topic *Rings*.

is well-defined.

Justify whether $\rightleftarrows$ stands for $\subset$ or $\supset$.

4.8 Let G be a finite group and $U \subset G$ be a subgroup, such that $\#(G/U) =: n \geq 3$. Let $S(G/U) := \{\varphi : G/U \to G/U \mid \varphi$ is a bijective mapping.$\}$. Show:

(a) If $\circ$ is the composition of maps, then $(S(G/U), \circ, \mathrm{id})$ is a group.

(b) If $g \in G$ is given, then $\mu_g : G/U \to G/U$, $[x] \mapsto [gx]$ is a well-defined, bijective map and thus $\mu_g \in S(G/U)$.

(c) The map $\psi : G \to S(G/U)$, $g \mapsto \mu_g$ is a group homomorphism.

(d) If ψ is injective, it follows that $\#G \mid n!$.

Rings and Ideals

5

And now?

We are now considering another algebraic structure, commutative rings with unity. At certain places we can rely on our considerations about groups, because similar concepts occur and often it is enough to convince oneself that one can carry out proofs and ideas analogously. The cornerstones are also the same:

- Definition of a ring and subrings
- Ring homomorphisms
- Ring quotients
- Ideal (which roughly play the role of normal subgroups played for groups) and factor rings
- Fundamental Homomorphism Theorem

In principle, one could now write down the definition of a commutative ring and then give it as an exercise to carry out the details of these tasks—that would certainly be too much to ask for an exercise but let me stress out that the solution will be very straightforward and -- as with groups -- considering what we want to achieve for each entry in the list above, every construction will be very natural. I will comment on this remark at several places in the following.

5.1 Commutative Rings with Unity

For the sake of completeness, I am writing down the definition again although we have already used this notion (cp. with the prerequisites at the beginning of this book).

© The Author(s), under exclusive license to Springer-Verlag GmbH, DE, part of
Springer Nature 2024
M. Hien, *Abstract Algebra*, Mathematics Study Resources 7,
https://doi.org/10.1007/978-3-662-67974-6_5

Definition 5.1 A **ring** is a tuple $(R, +, \cdot, 0)$ consisting of a set R, two operations (called addition and multiplication—we often write the latter again $xy := x \cdot y$)

$$+, \cdot : R \times R \to R$$

and an element $0 \in R$, so that the following axioms apply:

1. The triple $(R, +, 0)$ is an abelian group.
2. $\cdot$ is associative: $x(yz) = (xy)z$ for all $x, y, z \in R$.
3. The distributive laws apply:

$$x \cdot (y + z) = xy + xz, \quad (x + y) \cdot z = xz + yz$$

for all $x, y, z \in R$.

We will usually demand even more. We also write again the statement *Let R be a ring* and assume that the additional data $+, \cdot, 0$ is also given (i.e. should be clear from the context).

Definition 5.2

- A ring R is called **commutative**, if the equation $xy = yx$ holds for all $x, y \in R$.
- A ring **has a unity (or unit element)**, if there is an element $1 \in R$ with $x \cdot 1 = x = 1 \cdot x$ for all $x \in R$. We assume that $1 \neq 0$ holds.
- A **commutative ring with unity** is exactly what the words say.

Remark 5.3

1. $\mathbb{Z}$ is a commutative ring with unity.
2. Every field $\mathbf{k}$ is a commutative ring with unity. The **definition of a field** known from linear algebra can also be formulated as follows: A **field** is a commutative ring with unity, such that every $x \in \mathbf{k} \setminus \{0\}$ has an inverse $x^{-1} \in \mathbf{k}$, i.e., with $x \cdot x^{-1} = 1$, exists.
3. If $\mathbf{k}$ is a field, then the set $R := M(n \times n, \mathbf{k})$ of $n \times n$ matrices in $\mathbf{k}$ with $+$ and matrix multiplication, as well as the zero matrix 0, is a **ring with unity**—note, however, that for $n \geq 2$ it is not commutative.
4. If $\mathbf{k}$ is a field, then the polynomial ring $\mathbf{k}[X]$ in one variable is a commutative ring with unity.

The last example can also be generalized: If R is a commutative ring with unity, then so is

$$R[X] := \{\sum_{j=0}^{m} a_j X^j \mid a_j \in R\}$$

with the usual addition and multiplication of polynomials is the **polynomial ring in one variable** over R.

Successively, one can also define the polynomial ring in several variables: For $n \in \mathbb{N}$ we write

$$R[X_1, \ldots, X_n] := (\cdots (R[X_1])[X_2])[X_3]) \cdots)[X_n]$$

for the **polynomial ring over R in n variables**. It is easy to see that each element of it can be represented as

$$f(X_1, \ldots, X_n) = \sum_{i_1, \ldots, i_n}^{N} a_{i_1, \ldots, i_n} X_1^{i_1} \cdots X_n^{i_n} \tag{5.1}$$

with coefficients in R. Addition and multiplication are clear applying the distributive law.

It is also clear that the order in which the variables are "attached" does not matter.

And now?

With the notation (5.1), one should not be surprised that all indices run up to N, this can always be written in this form after "filling up" coefficients that do not appear a priori with zero entries, i.e. with coefficients $a_{i_1, \ldots, i_n} = 0$. It might be better to write it like this:

$$f(X_1, \ldots, X_n) = \sum_{(i_1, \ldots, i_n) \in \mathbb{N}_0^n} a_{i_1, \ldots, i_n} X_1^{i_1} \cdots X_n^{i_n}$$

requiring that for all but finitely many indices $(i_1, \ldots, i_n)$ the coefficient vanishes: $a_{i_1, \ldots, i_n} = 0$.

Also note that the polynomial $f(X_1, \ldots, X_n)$ is uniquely determined by its coefficients $a_{i_1, \ldots, i_n}$ by definition. Such a polynomial also induces a map

$$R \times \cdots \times R \to R, \quad (x_1, \ldots, x_n) \mapsto f(x_1, \ldots, x_n),$$

but it can happen (if the ring R is finite) that two different polynomials

$$f = \sum a_{i_1, \ldots, i_n} X_1^{i_1} \cdots X_n^{i_n}, \ g = \sum b_{i_1, \ldots, i_n} X_1^{i_1} \cdots X_n^{i_n}$$

(i.e. there is at least one index $(i_1, \ldots, i_n)$, such that $a_{i_1, \ldots, i_n} \neq b_{i_1, \ldots, i_n}$) induce the same function.

If one wants to write out the multiplication for the polynomial ring $R[X]$ over a commutative ring with unity in detail, one obtains the following rule:

$$\left(\sum_{i=0}^{n} a_i X^i \right) \cdot \left(\sum_{j=0}^{m} b_j X^j \right) = \sum_{\nu=0}^{\infty} \underbrace{\left(\sum_{\substack{i,j \in \mathbb{N}_0 \\ i+j=\nu}} a_i b_j \right)}_{=: c_\nu} X^\nu,$$

keeping in mind that the coefficient c_v is non-zero only for finite many $v \in \mathbb{N}_0$, hence the first sum (over the index v) is only a finite sum.

> From now on, we always consider commutative rings with unity. So whenever we will write: *Let R be a ring*, it is assumed to be commutative with unity.

5.2   Ring Homomorphisms

> **And now?**
> These are the maps that preserve all the data that makes a ring a ring.

Definition 5.4 A **ring homomorphism** is a map $\varphi : R \to S$ between two rings, such that

1. $\varphi(0) = 0$ and $\varphi(1) = 1$,
2. $\varphi(x + y) = \varphi(x) + \varphi(y)$ for all $x, y \in R$ and
3. $\varphi(xy) = \varphi(x)\varphi(y)$ for all $x, y \in R$.

Note 5.5 If R, S are rings, then $(R, +, 0)$ and $(S, +, 0)$ are (abelian) groups. A ring homomorphism $\varphi : R \to S$ is then also a group homomorphism of these groups—which also fulfills the third condition and $\varphi(1) = 1$.

From the note, it follows:

Defining Lemma 5.6 The **kernel** of a ring homomorphism $\varphi : R \to S$ is the kernel of φ as a group homomorphism:

$$\ker(\varphi) := \{x \in R \mid \varphi(x) = 0\}.$$

We have: φ is injective if and only if $\ker(\varphi) = \{0\}$ holds.

Proof already proven. $\square$

Remark 5.7

1. With this, we have also reintroduced the concept of a **ring isomorphism**, namely one that possesses an inverse ring homomorphism. The following also applies here: A ring homomorphism $\varphi : R \to S$ is an isomorphism if and only if it is bijective.

2. A **field** is indeed a ring K with the additional property that every $x \in K \smallsetminus \{0\}$ has a multiplicative inverse. For the definition of homomorphisms, this does not matter: A **field homomorphism** is a ring homomorphism $\varphi : K \to L$, in which the two rings are fields.
3. An important observation is the following: If $\varphi : K \to R$ is a ring isomorphism and K is a field, then R is also a field.

Proof If $x \in R \smallsetminus \{0\}$, then $\varphi^{-1}(x) \in K \smallsetminus \{0\}$ (I write φ^{-1} for the inverse homomorphism) has an inverse $\lambda \in K$, so with $\lambda \cdot \varphi^{-1}(x) = 1$. Then we deduce

$$\varphi(\lambda) \cdot x = \varphi(\lambda) \cdot \varphi(\varphi^{-1}(x)) = \varphi(\lambda \cdot \varphi^{-1}(x)) = \varphi(1) = 1,$$

so $x \in R$ has an inverse as well, namely $\varphi(\lambda)$.□

5.3   Units and Zero Divisors

And now?
In the rest of this section, we will consider some definitions and considerations that arise directly from the definition of a ring. After that, we will pick up the analogies (constructing factor rings, the fundamental homomorphism theorem, ...) to group theory again in the next section.

Definition 5.8 Let R be a ring[1], then an element $x \in R$ is called a **unit**, if there is an $y \in R$ such that $xy = 1$ is. The subset of units is written as $R^\times$.

Remark 5.9 If R is not necessarily commutative, but still with unity, then one can analogously define **left-** or **right-units**. Since we want to limit ourselves to commutative rings, we will not pursue these notions.

And now?
However, i would like to consider not necessarily commutative rings and briefly discuss two terms for an element in such a ring R:

- a **nilpotent element** is an $x \in R$, such that there is an $n \in \mathbb{N}$ with $x^n = 0$,
- an **idempotent element** is an $x \in R$, such that $x^2 = x$ holds.

These two terms also play a role in commutative rings with unity.

---

[1] thus commutative with unity.

There are some tricks when handling with such elements which might be worth considering in this generality. Let's assume that R comes equipped with a unity.

- If x is nilpotent—let's say with $x^n = 0$-- then the element $1 - x \in R$ is a (left and right) unit, because

$$(1-x) \cdot \sum_{j=0}^{n-1} x^j = \sum_{j=0}^{n-1} x^j - \sum_{j=1}^{n} x^j = 1 - x^n = 1$$

and similarly

$$\sum_{j=0}^{n-1} x^j \cdot (1-x) = 1$$

holds. This is often used in commutative rings as well where we do not have to distinguish between left- and right-units anyway.
- If $x \in R$ is idempotent, so is $1 - x$, furthermore the two elements satisfy the equation

$$x(1-x) = (1-x)x = 0.$$

With this in mind, one obtains a decomposition

$$R = x \cdot R + (1-x) \cdot R,$$

(which holds for any x because $r = xr + (1-x)r$ is always true), but now (with x being idempotent) the decomposition is also a direct one in the following sense: If $0 = xa + (1-x)b$, then $xa = 0$ and $(1-x)b = 0$, because

$$0 = xa + (1-x)b \Rightarrow 0 = x \cdot 0 = x(xa + (1-x)b) = xa \Rightarrow (1-x)b = 0.$$

Lemma 5.10 *If R is a ring, then $R^\times$ (with multiplication as the operation) is an abelian group.*

Proof Actually clear: Associativity is already given in R, the neutral element is one and every $x \in R^\times$ has an inverse by definition of $R^\times$. $\square$

Examples:

1. $\mathbb{Z}^\times = \{\pm 1\}$.
2. A field $\mathbf{k}$ has by definition $\mathbf{k}^\times = \mathbf{k} \smallsetminus \{0\}$.
3. If $\mathbf{k}$ is a field, then

$$\mathbf{k}[X]^\times = \mathbf{k}^\times, \tag{5.2}$$

applies, where we read **k** as a subring in **k**[X] via the injective mapping

$$\mathbf{k} \hookrightarrow \mathbf{k}[X], \ a \mapsto a$$

(to be read on the right side as the polynomials of degree 0).

Check it out!
Can you justify (5.2)? The solution follows now.

In the last example, note that we have the concept of the **degree of a polynomial**, namely (for a given ring R):

$$f(X) = a_n X^n + \ldots + a_0 \in R[X] \text{ with } a_n \neq 0 \text{ has degree } \deg(f) = n.$$

Lemma 5.11 *If R is a ring and $f, g \in R[X]$ are polynomials, we have*

$$\deg(f \cdot g) \leq \deg(f) + \deg(g).$$

If R is a field, then $\deg(f \cdot g) = \deg(f) + \deg(g)$.

Proof The inequality is clear. Since the product of the leading coefficients in a field cannot vanish, the equation for fields also follows easily. □

Example: Let R be the ring $R := \mathbb{Z}/4\mathbb{Z}$. Then, for $f(X) = g(X) = 2X + 1 \in R[X]$, we observe that

$$\deg(f) = \deg(g) = 1 \text{ and } \deg(f \cdot g) = 0 < 2 = \deg(f) + \deg(g),$$

because $f \cdot g = (2X + 1)(2X + 1) = 4X^2 + 4X + 1 = 1 \in \mathbb{Z}/4\mathbb{Z}[X]$.

This also solves the **Check it out!** from before: If **k** is a field and $f \in (\mathbf{k}[X])^\times$ is a polynomial that is a unit in the polynomial ring, then there exists a $g \in \mathbf{k}[X]$ with $f \cdot g = 1$. According to Lemma 5.11 (in the case of a field), then $\deg(f) = \deg(g) = 1$ and $\deg(f \cdot g) = 0$

$$0 = \deg(f \cdot g) = \deg(f) + \deg(g) \Longrightarrow \deg(f) = 0,$$

so $f(X) = a \in \mathbf{k} \smallsetminus \{0\} = \mathbf{k}^\times$.

Check it out!
Find an example of a ring R and a polynomial $f(X) \in (R[X])^\times$ with $\deg(f) \geq 1$!

Because it is so important, let me remind you again that we have polynomial division with remainder when working over a field. If we look more closely to its proof we get the immediate generalization:

Lemma 5.12 (Polynomial division **with remainder when the leading coeffi-cient is a unit**) *Let R be a ring and let $f, g \in R[X]$ be polynomials with $g \neq 0$ and assume that*

$$g(X) = a_n X^n + \ldots + a_0 \in R[X]$$

*has the **leading coefficient** a_n which is a **unit** $a_n \in R^\times$ (this is always the case when R is a field).*

Then there are uniquely determined polynomials $q, r \in R[X]$, such that

$$f(X) = q(X) \cdot g(X) + r(X) \text{ and } \deg(r) < \deg(g).$$

Proof The proof is the same we gave in Sect. 1.2, because it works exactly as in the case of fields. In its procedure, the only division that appears is the one by the leading coefficient of g and this division is possible according to the assumption. $\square$

In the examples on the unit group $R[X]^\times$ we have seen an aspect we would like to analyze in an abstract notion:

Definition 5.13 Let R be a ring.

1. An element $x \in R \setminus \{0\}$ is called a **zero divisor**, if there exists a $y \in R \setminus \{0\}$ such that $xy = 0$ holds.
2. A ring that has no zero divisors is called **integral** or equivalently an **integral domain** or sometimes just a **domain**.

And now?

An important concept in rings (again recalling that we always assume it to be commutative with unity) is that of **divisibility**. Quite analogous to the situation in the special ring $\mathbb{Z}$:

An element $x \in R$ **divides** the element $y \in R$, if there is a $z \in R$ with $x \cdot z = y$. We then write $x|y$.

The units and the zero divisors then each play an interesting role:

1. Zero divisors divide zero (without being zero themselves), this has the following effect: If you encounter an equation $xa = xb$ in a ring (with $x \neq 0$ otherwise it is certainly true for all $a, b \in R$), in general there is no way to divide by x. The best way to handle such an equation is to write in the following way:

$$xa = xb \implies 0 = xa - xb = x(a - b) \iff$$

$$\begin{cases} a = b \text{ or} \\ x \text{ is a zero divisor and } a - b \text{ is a corresponding other zero divisor.} \end{cases}$$

Therefore, it is often desirable to work with rings without zero divisors since then the second line in the last statement cannot occur.

The statement

A product xy is zero if and only if one of the two factors is zero holds **in rings without zero divisors**. If you have zero divisors, this is **false**.

2. If $x \in R^{\times}$ is a unit, then there exists a $y \in R$ such that $xy = 1$ and thus we have:

$$x \in R^{\times} \implies x|a \text{ for every } a \in R,$$

since choosing an inverse element y, we obtain $x(ya) = a$.

Lemma 5.14 *If $f(X) \in R[X]$ is a polynomial of degree $n \geq 1$ over an integral domain R, then f has at most n distinct roots in R.*

Proof If $a_1 \in R$ is a root, then we have a factorization

$$f(X) = (X - a_1) \cdot q(X)$$

with a $q(X) \in R[X]$—to do this, divide f by $X - a_1$ according to Lemma 5.12 and obtain $f = (X - a_1)q + r$ with $\deg(r) < \deg(X - a_1) = 1$, so $r \in R$. If you insert a_1 into both sides, you get $r = 0$.

Successively, one obtains

$$f(X) = (X - a_1)^{\nu_1}(X - a_2)^{\nu_2} \cdots (X - a_r)^{\nu_r} \cdot q(X)$$

with pairwise different $a_j \in R$ and $q \in R[X]$, such that q has no more roots in R. We observe that $r \leq n$ due to Lemma 5.11.

Now if $a \in R \smallsetminus \{a_1, \dots, a_r\}$, then $a - a_j \neq 0$ holds for all $j = 1, \dots, r$ and additionally $q(a) \neq 0$, thus we get:

$$f(a) = (a - a_1)^{\nu_1} \cdots (a - a_r)^{\nu_r} \cdot q(a) \neq 0,$$

because R has no zero divisors. $\square$

Lemma 5.15 *If $\varphi : R \to S$ is a ring homomorphism, then $\varphi(R^{\times}) \subset S^{\times}$, i.e., units are mapped to units.*

Proof We have already seen the argument in Remark 5.7 (3). If $s \in \varphi(R^{\times})$, so $s = \varphi(r)$ for a unit $r \in R^{\times}$, and r has an inverse $\lambda \in R^{\times}$, i.e. $r\lambda = 1$ holds. Then

$$\varphi(\lambda) \cdot s = \varphi(\lambda) \cdot \varphi(r) = \varphi(\lambda \cdot r) = \varphi(1) = 1 \ .$$

$\square$

5.4   Ideals, Factor Rings, and the Fundamental Homomorphism Theorem

And now?
We want to further advance ring theory in analogy to the chapter on groups. The question now is: What substructures do we have that lead to a good equivalence relation $\sim$ on a ring R, such that the quotient set $R/\sim$ is again a ring in a natural way?

Recall that rings are assumed to be commutative with unity.

We have already defined whathe notion of a subring $S \subset R$ —Definition 2.12. Let us now care about quotients with respect to a suitable equivalence relation.

And now?
Given a group G and a subgroup H, we only had one operation $\cdot$ available and introduced the equivalence relation

$$g \sim g' :\Longleftrightarrow \text{there exists an } h \in H \text{ with } g' = g \cdot h.$$

Given a ring, we have $+$ and $\cdot$ to choose from and one can ask oneself which operation should be taken into account for such a definition. Perhaps one can also look at it in such a way that given a ring $(R, +, \cdot, 0, 1)$ (commutative with unity), the triple $(R, +, 0)$ indeed forms an (abelian) group—one simply focuses on this aspect and defines the relation $\sim$ on R as if there were no $\cdot$. Therefore, we define:

Given a subgroup $\mathfrak{a} \subset R$ (of the group $(R, +, 0)$)[2]—possibly putting with further requirements on $\mathfrak{a}$—we define for $x, y \in R$:

$$x \sim y :\Longleftrightarrow \text{there exists an } a \in \mathfrak{a} \text{ with } y = x + a.$$

We then write $R/\mathfrak{a} : = R/\sim$. We observe:

1. Since $\mathfrak{a} \subset R$ is a normal subgroup in $(R, +, 0)$ (they are abelian groups!), addition $+$ induces again the structure of an abelian group on $R/\mathfrak{a}$, namely $(R/\mathfrak{a}, +, 0 = [0])$ and we have the surjective **group homomorphism**

    $$\pi : R \to R/\mathfrak{a}, \ x \mapsto [x].$$

2. We now remember that R also has a multiplication: We have to take care of the well-definedness of the induced multiplication

---

[2] The letter $\mathfrak{a}$ has been used now because it will be the usual notation.

$\cdot_{\text{candidate}} : R/\mathfrak{a} \times R/\mathfrak{a} \to R/\mathfrak{a}, \; ([x], [y]) \mapsto [x] \cdot_{\text{candidate}} [y] := [x \cdot y]$

(i.e. we have to find conditions on $\mathfrak{a}$ so that it is well-defined).

Let $[x] = [x']$ and $[y] = [y']$ be two choices of representatives for the classes. Then there are $a, b \in \mathfrak{a}$ with $x' = x + a$ and $y' = y + b$. We obtain

$$x'y' = (x + a)(y + b) = xy + ay + xb + ab \overset{\text{our wish}}{\sim} xy.$$

So our wish is that

$$ay + xb + ab \in \mathfrak{a} \tag{5.3}$$

applies More precisely, this should hold **for all** $x, y \in R$ **and all** $a, b \in \mathfrak{a}$. In particular, this should hold for $x = 0$, $b = 0$ hence we should require that

$$a \in \mathfrak{a} \text{ and } x \in R \Longrightarrow xa \in \mathfrak{a}$$

holds. This is the solution to the well-definedness—see Definition 5.16.

Definition 5.16 Let R be a ring. A subset $\mathfrak{a} \subset R$ is called an **ideal**, if the following properties are fulfilled:

1. We have $0 \in \mathfrak{a}$,
2. For any $a, b \in \mathfrak{a}$ we have $a + b \in \mathfrak{a}$,
3. For any $a \in \mathfrak{a}$ and any $x \in R$ we have $ax \in \mathfrak{a}$.

The usual notation is $\mathfrak{a} \lhd R$ if $\mathfrak{a}$ is an ideal in R .

And now?
If one has read the **And now?** before the definition, one might wonder why the requirement $a \in \mathfrak{a} \Rightarrow -a \in \mathfrak{a}$ does not appear in Definition 5.16 (which we must have for the additive subgroup $\mathfrak{a} \subset R$). This is because in every ring

$$-a = (-1) \cdot a$$

holds and thus this desirable property already follows from the third requirement—and we always eliminate unnecessary requirements in definitions.

And now?
If one does not proceed as we did above in the **And now?** one could of course have tried to approach the problem with the following requirement:

$a, b \in \mathfrak{a} \Rightarrow ab \in \mathfrak{a}$ which looks very natural to impose. However, this is not sufficient to ensure that the quotient $R/\mathfrak{a}$ inherits the ring structure.

To illustrate this subtlety, take $R = \mathbb{Z}/n\mathbb{Z}$, that is $R = \mathbb{Z}$ and $\mathfrak{a} = n\mathbb{Z}$. Then we have $a, b \in n\mathbb{Z}$, so $a = \alpha n$ and $b = \beta n$ with $\alpha, \beta \in \mathbb{Z}$. This also implies

$$ab = n^2(\alpha\beta) \in n\mathbb{Z}$$

(so far so good) but one can see that this is automatically better than planned: ab is contained in $n^2\mathbb{Z}$. The (additive) subgroup $\mathfrak{a} = n\mathbb{Z}$ is characterized by the fact that for $a = \alpha n \in \mathfrak{a}$ and $x \in \mathbb{Z}$ arbitrary, we also have $xa = nx\alpha \in \mathfrak{a}$ —$\mathfrak{a}$ is an ideal.

Remark 5.17 In non-necessarily commutative rings (with/without unity), we have similar terms, namely **left-** or **right-** or **two-sided ideals**, depending on whether in the third condition of the definition $xa \in \mathfrak{a}$ or $ax \in \mathfrak{a}$ or both are required.

A trivial, but simple consequence of the definition is that

Lemma 5.18 *If $\mathfrak{a} \lhd R$ is an ideal and there exists a unit $x \in \mathfrak{a} \cap R^\times$ in $\mathfrak{a}$, then $\mathfrak{a} = R$.*

Proof Given the unit x, there is an inverse $y \in R$ with $xy = 1$. We deduce (for any $z \in R$):

$$x \in \mathfrak{a} \Longrightarrow 1 = xy \in \mathfrak{a} \Longrightarrow z = z \cdot 1 \in \mathfrak{a}.$$

Corollary 5.19 A field K has exactly the two ideals (0) and $(1) = K$.

The considerations that led to Definition 5.16 now yield:

Definition 5.20 If $\mathfrak{a} \lhd R$ is an ideal, we define the equivalence relation on R by:

$$x \sim y :\Longleftrightarrow \text{there exists an } a \in \mathfrak{a} \text{ such that } y = x + a.$$

The quotient is then written as $R/\mathfrak{a} := R/\!\sim$ and one also writes

$$x \sim y \Leftrightarrow: x \equiv y \bmod \mathfrak{a}$$

and says, the elements x, y are **equivalent modulo $\mathfrak{a}$**.

As we have proven in the **And now?** using Eq. (5.3), we also have:

Defining Lemma 5.21 *If $\mathfrak{a} \lhd R$ is an ideal, then $+, \cdot$ induce operations on $R/\mathfrak{a}$ such that $R/\mathfrak{a}$ becomes again a ring with $[0]$ and $[1]$ as neutral elements and such that*

$$\pi : R \to R/\mathfrak{a}, \ x \mapsto [x]$$

is a surjective ring homomorphism. One calls $R/\mathfrak{a}$ the **factor ring of R modulo** $\mathfrak{a}$.

There is a relatively simple way to generate ideals:

Definition 5.22 Let R be a ring and $a \in R$ be given. Then the ideal

$$(a) := R \cdot a := \{xa \in R \mid x \in R\}$$

generated by a is called the **principal ideal with generator** a

If every ideal $\mathfrak{a} \lhd R$ is a principal ideal, then the ring is called a **principal ideal ring**. A **principal ideal domain** is a principal ideal ring without zero divisors.

Examples:

1. As we have already considered above: $\mathfrak{a} = n\mathbb{Z} \lhd \mathbb{Z}$. Then we obtain the ring $\mathbb{Z}/n\mathbb{Z}$ and in it a representative system

$$\{[0], [1], \ldots, [n-1]\}.$$

Often the square brackets are omitted (so for example one can write $5 \in \mathbb{Z}/7\mathbb{Z}$) or sometimes people write $\overline{5} \in \mathbb{Z}/7\mathbb{Z}$ for the class of 5 .

2. Let K be a field and $f(X) \in K[X]$ be a polynomial. Then we have the principal ideal $(f(X))$ and thus the factor ring $K[X]/(f(X))$.

   Further example inside the example: Let $f(X) = X^2 - d \in \mathbb{Q}[X]$ for a given square-free number $d \in \mathbb{Z}$. Then we have the ring

   $$\mathbb{Q}[X]/(X^2 - d).$$

   Note that the map

   $$\varphi : \mathbb{Q}[X]/(X^2 - d) \to \mathbb{Q}(\sqrt{d}) \,, \; [g(X)] \mapsto g(\sqrt{d}) \tag{5.4}$$

   is well-defined.

Check it out!

Think about what needs to be proven and then prove it!—I will do so in the next few lines as well.

The map φ of (5.4) is also a ring homomorphism and even an **isomorphism of rings**

Check it out!

You guessed what the task will be: Prove this! To show injectivity, compute $\ker(\varphi)$—we know that $X^2 - d$ is the minimal polynomial of $\sqrt{d}$ over $\mathbb{Q}$; for surjectivity, we know that $\mathbb{Q}(\sqrt{d}) = \{a + b\sqrt{d} \mid a, b \in \mathbb{Q}\}$ holds.

So, $\mathbb{Q}[X]/(X^2 - d)$ is a field!

Check it out!
Calculate the inverse of any arbitrary $[g(X)] \in \mathbb{Q}[X]/(X^2 - d)$.

Resolution of the *Check it out!*:

Proof Firstly, φ is well-defined, because if $[g(X)] = [h(X)]$ are two choices of representatives, then there is an $a(X) \in \mathbb{Q}[X]$ with

$$h(X) = g(X) + a(X) \cdot (X^2 - d) \ .$$

If one inserts $\sqrt{d} \in \mathbb{C}$, one obtains $h(\sqrt{d}) = g(\sqrt{d})$ because the second term on the right hand side disappears.

Furthermore, φ is obviously a ring homomorphism. Its kernel is

$$\ker(\varphi) = \{[g(X)] \in \mathbb{Q}[X]/(X^2 - d) \mid g(\sqrt{d}) = 0\} \ .$$

Now, $X^2 - d$ is the minimal polynomial of $\sqrt{d}$ over $\mathbb{Q}$ and thus we know: If $g(\sqrt{d}) = 0$ holds for a given polynomial $g \in \mathbb{Q}[X]$, then g is a multiple of $X^2 - d$ in $\mathbb{Q}[X]$—see Defining Lemma 2.16. Hence g lies inside the principal ideal generated by X^{2-d} : $g \in (X^2 - d)$. But then $[g(X)] = [0] = 0 \in \mathbb{Q}[X]/(X^2 - d)$. Therefore, φ is injective.

Regarding surjectivity: An arbitrary element $\xi \in \mathbb{Q}(\sqrt{d})$ has the form $\xi = a + b\sqrt{d}$ with $a, b \in \mathbb{Q}$. Then $\xi = \varphi([a + bX])$.

Therefore, $\mathbb{Q}[X]/(X^2 - d)$ is a field. Now, if $[g(X)] \neq 0$ an arbitrary element ($\neq 0$), then division with remainder yields

$$g(X) = q(X)(X^2 - d) + r(X) \text{ with } \deg(r) \leq \deg(X^2 - d) = 2,$$

thus $r(X) = a + bX$. We then obtain

$$[g(X)] = [r(X)] \in \mathbb{Q}[X]/(X^2 - d) \ .$$

But then $\varphi([g(X)]) = a + b\sqrt{d}$. As we have seen in Chap. 2, the latter element in $\mathbb{Q}(\sqrt{d})$ has the inverse

$$\frac{1}{a + b\sqrt{d}} = \frac{a}{a^2 - b^2 d} + \frac{-b}{a^2 - b^2 d}\sqrt{d}$$

and this is the image with respect to φ of the element

$$\left[\frac{a}{a^2 - b^2 d} + \frac{-b}{a^2 - b^2 d}X\right] \in \mathbb{Q}[X]/(X^2 - d) \ .$$

In fact, we observe:

$$g(X) \cdot (\frac{a}{a^2 - b^2 d} + \frac{-b}{a^2 - b^2 d} X) \equiv (a + bX)(\frac{a}{a^2 - b^2 d} + \frac{-b}{a^2 - b^2 d} X) =$$

$$\frac{a^2}{a^2 - b^2 d} - \frac{b^2}{a^2 - b^2 d} X^2 = 1 + \frac{b^2 d}{a^2 - b^2 d} - \frac{b^2}{a^2 - b^2 d} X^2 =$$

$$1 - \frac{b^2}{a^2 - b^2 d} \cdot (X^2 - d) \equiv 1 \bmod (X^2 - d).$$

Lemma 5.23 *The ring $\mathbb{Z}$ is a principal ideal domain.*

Proof Let $\mathfrak{a} \neq \{0\}$ (the zero ideal is a principal ideal) be an ideal and consider the element

$$a := \min(\mathfrak{a} \cap \mathbb{N}), \tag{5.5}$$

i.e. the smallest positive element in $\mathfrak{a}$. Then $\mathfrak{a} = (a)$, because given any $b \in \mathfrak{a}$ one obtains by division with remainder

$$b = qa + r \text{ with } 0 \leq r < a . \tag{5.6}$$

Then $r = b - qa \in \mathfrak{a}$ holds and since a was the smallest positive element in $\mathfrak{a}$, we deduce that $r = 0$ and hence $b \in (a)$. $\square$

One also proves analogously:

Lemma 5.24 *If K is a field, then $K[X]$ is a principal ideal domain.*

Proof The same proof applies, with the only difference that the elements in the ring are not compared via the $<$-relation, but via the degree $\deg(f)$. Thus, in (5.5) we put

$$a := \text{an element} \neq 0 \text{ in } \mathfrak{a} \text{ with minimal degree}$$

i.e. we choose a polynomial with *smallest* degree inside the ideal and instead of (5.6) we use the division with remainder

$$b = qa + r \text{ with } \deg(r) < \deg(a).$$

We then conclude in the same way as before. $\square$

The construction of a principal ideal can also be generalized:

Definition 5.25 Let R be a ring and $a_1, \ldots, a_m \in R$ be elements. Then

$$(a_1, \ldots, a_m) := Ra_1 + \ldots + Ra_m := \{x_1 a_1 + \ldots + x_m a_m \mid x_j \in R\} \lhd R$$

is called the **ideal generated by** $a_1, \ldots, a_m$.

Example: Let K be a field and $R = K[X, Y]$ be the polynomial ring in two variables. Then consider the ideal

$$(X, Y) := \{X \cdot g(X, Y) + Y \cdot h(X, Y) \mid g, h \in K[X, Y]\} \lhd K[X, Y].$$

Note that $(X, Y) \neq K[X, Y]$, because $1 \notin (X, Y)$.

The ideal (X, Y) is **not** a principal ideal, because if it were, i.e. $(X, Y) = (f(X, Y))$ for some $f \in K[X, Y]$, then in particular $X \in (f(X, Y))$ must hold, so that $X = f(X, Y) \cdot a(X, Y)$ for some $a \in K[X, Y]$. Since $K[Y]$ is free of zero divisors, we deduce $\deg_X(f) + \deg_X(a) = \deg_X(X) = 1$, where $\deg_X(f(X, Y))$ denotes the degree of f as an element in $(K[Y])[X]$. Then f is of the form

$$f(X, Y) = f_0(Y) + f_1(Y)X$$

with $f_0, f_1 \in K[Y]$. Since $f(X, Y) \mid Y$, we get $Y = f(X, Y)b(X, Y)$ for some $b \in K[X, Y]$ but then even $f_1(Y) = 0$ must hold. Therefore, $f(X, Y) = f_0(Y)$ and since $f(X, Y) \mid X$, we obtain that $f_0(Y) \in K^\times$ is a constant polynomial, in contradiction to $(X, Y) \neq K[X, Y]$.

Let us summarize this:

Lemma 5.26 *Let K be a field. For $n \geq 2$ the polynomial ring $K[X_1, \ldots, X_n]$ is not a principal ideal ring.*

Back to general commutative rings with unity. The Fundamental Homomorphism Theorem now follows quite easily. First a preliminary remark (analogous to Lemma 4.19):

Lemma 5.27 *If $\varphi : R \to S$ is a ring homomorphism, then $\ker(\varphi)$ is an ideal in R.*

Proof Easy to verify. □

Theorem 5.28 (Fundamental Homomorphism Theorem for rings) If $\varphi : R \to S$ is a ring homomorphism, then φ induces a well-defined injective ring homomorphism

$$\overline{\varphi} : R/\ker(\varphi) \to S , \quad [x] \mapsto \varphi(x) .$$

Proof There is actually nothing more to prove here, well-definedness and injectivity can already be seen at the level of the underlying groups. The fact that $\overline{\varphi}$ is again a ring homomorphism is very easy to see. □

Using Lemma 5.27 and Corollary 5.19 we deduce:

Lemma 5.29 *If $\varphi : K \to S$ is a ring homomorphism and K is a field, then φ is automatically injective.*

Proof Because $\varphi(1) = 1$, $\ker(\varphi) \neq K$ is not equal to K. But $\ker(\varphi) \lhd K$ and thus $\ker(\varphi) = (0)$ must be the zero ideal. □

5.5 Prime Ideals and Maximal Ideals

And now?

So, we have now seen that given a ring R and an ideal $a \lhd R$ we obtain a new ring R/a.

The following natural **question** arises: What quality does R/a have and how do we read this from R or from a?

What does the **quality** of a ring mean? Note: This is not a mathematical term, but a term we will use in the blocks called *And now?*. We have already seen that rings can have different properties some of which simplify their understanding and could therefore be considered as different levels of *good-behaviour*. For example, think of the two qualities:

- being an integral domain (i.e. having no zero divisors),
- being a field.

There are many other *qualities* that we will discuss partly in the next chapters. For now, focus on these two qualities. Note that being a *field* is a stronger quality than being *integral*, because a field automatically has no zero divisors.

How do we proceed? We examine the desired quality of R/a, try to understand what this implies for R and a, and then possibly give the resulting requirements a nice name.

Defining Proposition 5.30 Let $a \lhd R$ be an ideal, $a \neq R$. Then R/a is integral if and only if a satisfies the following condition:

$$xy \in a \text{ for } x, y \in R \Leftrightarrow x \in a \text{ or } y \in a. \tag{5.7}$$

An ideal with this property is called a **prime ideal**.

Proof For $x \in R$ we have:

$$[x] = 0 \in R/a \Longleftrightarrow x \in a.$$

Therefore $0 \neq [x] \in R/a$ is a zero divisor if and only if there is a $[y] \in R/a$ with $[y] \neq 0$ such that $0 = [xy]$ holds, i.e. such that $xy \in a$. We have obtained:

R/a has zero divisors $\Longleftrightarrow$ there exist $x, y \in R$ with $x \notin a$ and $y \notin a$, but $xy \in a$.

This is equivalent to the assertion. $\square$

Remark 5.31 The ring itself as an ideal $a = R$ is not referred to as a prime ideal. This is the same convention that declares that $1 \in \mathbb{Z}$ is not considered a prime number by definition.

Examples:

1. Inside $\mathbb{Z}$ every ideal is a principal ideal, and since $(a) = (-a)$ holds, every ideal is of the form $\mathfrak{a} = n\mathbb{Z}$ for a $n \in \mathbb{N}_0$.
   The ideal $n\mathbb{Z}$ is a prime ideal if and only if n is a prime number.
   This reflects the characterization of prime numbers p by the following property: A number $n \in \mathbb{N}$, $n \neq 1$ is a prime number if and only if the following holds: If n divides a product $nlxy$, then n divides one of the two factors nlx or nly—this characterization relies on the unique prime factor decomposition in $\mathbb{Z}$.
2. The ideal $(X^2 - d) \in \mathbb{Q}[X]$ with square-free $d \in \mathbb{Z}$ is a prime ideal, because $\mathbb{Q}[X]/(X^2 - d)$ is a field and in particular integral.

Defining Theorem 5.32 Let $\mathfrak{a} \lhd R$ be an ideal, $\mathfrak{a} \neq R$. Then $R/\mathfrak{a}$ is a field if and only if $\mathfrak{a}$ satisfies the following property:

$$\text{If } \mathfrak{a} \subset \mathfrak{b} \subset R \text{ is an ideal with } \mathfrak{b} \neq R, \text{ then } \mathfrak{b} = \mathfrak{a}. \tag{5.8}$$

An ideal $\mathfrak{a} \neq R$ with this property is called a **maximal ideal**.

Remark 5.32 For a (presumably maximal) ideal $\mathfrak{a} \subsetneq R$ the condition (5.8) is equivalent to the condition

$$\text{If } \mathfrak{a} \subset \mathfrak{b} \subset R \text{ is an ideal with } \mathfrak{a} \subsetneq \mathfrak{b}, \text{ then } \mathfrak{b} = R. \tag{5.9}$$

Keeping this in mind, we proceed to the proof of the assertion.

Proof (of 5.32) For $0 \neq [x] \in R/\mathfrak{a}$, thus $x \notin \mathfrak{a}$, one has a multiplicative inverse $[y] \in R/\mathfrak{a}$ if and only if $1 = [x][y] = [xy]$ holds, which in turn is equivalent to $1 - xy \in \mathfrak{a}$.
   If we assume that $R/\mathfrak{a}$ is a field and $\mathfrak{b} \lhd R$ is another ideal with

$$\mathfrak{a} \subsetneq \mathfrak{b} \subset R,$$

then there exists an $x \in \mathfrak{b} \setminus \mathfrak{a}$. The class $[x] \in R/\mathfrak{a}$ has an inverse, so there exists a $y \in R$ with $1 - xy \in \mathfrak{a}$. We deduce

$$1 = \underbrace{1 - xy}_{\in \mathfrak{a} \subset \mathfrak{b}} + \underbrace{xy}_{\in \mathfrak{b}} \in \mathfrak{b} \Longrightarrow \mathfrak{b} = R.$$

Conversely, if $\mathfrak{m} \lhd R$ is a maximal ideal (we usually write $\mathfrak{m}$ to emphasize this property), then consider any $0 \neq [x] \in R/\mathfrak{m}$.

Check it out!

Take a moment to think about how you would proceed in the proof. As a hint: Now you need to apply the assumption that $\mathfrak{m}$ *is a maximal ideal*. To this end, you might want to construct a situation

$$\mathfrak{m} \subset \mathfrak{b} \subset R$$

> with a suitable ideal $\mathfrak{b}$ (which you have to find/produce—a creative act required from your side)! Which $\mathfrak{b} \lhd R$ can you construct using the obejcts that already are on the table: It should come with $\mathfrak{m} \subset \mathfrak{b}$ and have something to do with the additionally given element $x \in R$, $x \notin \mathfrak{m}$.

The idea is, that there is an obvious way to construct an ideal $\mathfrak{b} \lhd R$, which contains the ideal $\mathfrak{m}$ and the element x, namely

$$\mathfrak{b} := \mathfrak{m} + R \cdot x = \{m + rx \mid m \in \mathfrak{m} \text{ und } r \in R\} \lhd R.$$

It satisfies

$$\mathfrak{m} \subsetneq \mathfrak{b} \subset R$$

and since $\mathfrak{m}$ is maximal, it follows that $\mathfrak{b} = R$, in particular $1 \in \mathfrak{b}$. Thus, there are elements $m \in \mathfrak{m}$ and $r \in R$ with

$$1 = m + rx .$$

But then $[r] \in R/\mathfrak{m}$ is an inverse to $[x]$ and we are done. $\square$

Lemma 5.34 *Every maximal ideal $\mathfrak{a} \lhd R$ is also a prime ideal.*

Proof This is clear since every field is integral. $\square$

Lemma 5.35 *A ring R is integral if and only if $(0) \lhd R$ is a prime ideal.*

Proof This is clear since $R/(0) = R$ holds. $\square$

Examples:

1. For the integers $\mathbb{Z}$ we have: An ideal $n\mathbb{Z}$ is maximal if and only if n is a prime number (and thus again if and only if $n\mathbb{Z}$ is a prime ideal).

Proof If p is a prime number, then $\mathbb{Z}/p\mathbb{Z}$ is a field, because for any $0 \neq [x] \in \mathbb{Z}/p\mathbb{Z}$ we get that $p \nmid x$. Then $\gcd(x,p) = 1$ and according to the Euclidean algorithm, there is a representation $1 = ax + bp$, the so-called Bézout representation. We will discuss this in more detail in a more general context later in Lemma 6.4. Thus, $[a] \in \mathbb{Z}/p\mathbb{Z}$ is an inverse to $[x]$.

If n is not a prime number, then $\mathbb{Z}/n\mathbb{Z}$ is not even integral let alone a field. $\square$

2. In $\mathbb{Q}[X]$ the ideal $(X^2 - d)$ for square-free d is a maximal ideal.
3. Let K be a field. The ideal $(X) \lhd K[X,Y]$ is a prime ideal but not a maximal ideal because

$$(0) \subsetneq (X) \subsetneq (X,Y) \subsetneq K[X,Y]$$

holds.

And now?
It is no coincidence that the first example of a ring in which there are prime ideals that are not maximal is constructed as a polynomial ring with more than one variable. In fact, the maximum length of chains

$$\mathfrak{p}_0 \subsetneq \mathfrak{p}_1 \subsetneq \mathfrak{p}_2 \subsetneq \dots \subsetneq \mathfrak{p}_n$$

of prime ideals in R is related to the concept of **dimension** and $K[X, Y]$ should indeed have dimension 2.

Rings without zero divisors in which every prime ideal is already a maximal ideal, are accordingly also called 1-dimensional integral domains. The rings $\mathbb{Z}$ and $K[X]$ for a field K are 1-dimensional.

Theorem 5.36 If $\mathfrak{a} \subsetneq R$ is an arbitrary ideal, then there exists a maximal ideal $\mathfrak{m} \lhd R$ with $\mathfrak{a} \subset \mathfrak{m} \subsetneq R$.

Proof For the proof, we will use Zorn's Lemma (see below). Consider the set

$$B := \{\mathfrak{b} \lhd R \mid \mathfrak{a} \subset \mathfrak{b} \subsetneq R\}.$$

On this set, a partial order is defined by $\mathfrak{b} \leq \mathfrak{c} :\Leftrightarrow \mathfrak{b} \subset \mathfrak{c}$. Now, if $K \subset B$ is a totally ordered subset (i.e., for any two $\mathfrak{b}, \mathfrak{c} \in K$ we always have $\mathfrak{b} \leq \mathfrak{c}$ or $\mathfrak{c} \leq \mathfrak{b}$), then this set K admits an upper bound in B , namely

$$\mathfrak{c} := \bigcup_{\mathfrak{b} \in K} \mathfrak{b} \in B,$$

where we still have to show,

1. that $\mathfrak{c} \subset R$ is an ideal with $\mathfrak{a} \subset \mathfrak{c}$ and
2. $\mathfrak{c} \neq R$ holds,

(so that we can ensure that $\mathfrak{c} \in B$).

to (1): Obviously, $\mathfrak{a} \subset \mathfrak{c}$ since $\mathfrak{a} \subset \mathfrak{b}$ for all $\mathfrak{b} \in K$. For $x, y \in \mathfrak{c}$ we know that there exists a $\mathfrak{b}_x \in K$ with $x \in \mathfrak{b}_x$ and a $\mathfrak{b}_y \in K$ with $y \in \mathfrak{b}_y$. Now we have $\mathfrak{b}_x \leq \mathfrak{b}_y$ or $\mathfrak{b}_y \leq \mathfrak{b}_x$, because we have considered a totally ordered subset, let's say WLOG (otherwise we rename the elements) $\mathfrak{b}_x \leq \mathfrak{b}_y$. Therefore, $\mathfrak{b}_x \subset \mathfrak{b}_y$ and thus $x + y \in \mathfrak{b}_y \subset \mathfrak{c}$. Similarly, for $r \in R$ also $rx \in \mathfrak{b}_x \subset \mathfrak{c}$. With this, (1) is proven.

to (2): Since $\mathfrak{b} \neq R$ for all $\mathfrak{b} \in K$, it follows that $1 \notin \mathfrak{b}$ for all $\mathfrak{b} \in K$. But this also implies $1 \notin \mathfrak{c}$.

Zorn's Lemma states that a partially ordered set (here B with $\leq$), such that every totally ordered subset K has an upper bound in B, has a maximal element. Such an element is then a maximal ideal $\mathfrak{m}$ with $\mathfrak{a} \subset \mathfrak{m}$. $\square$

> **What now?**
> If you want to do mathematics without Zorn's Lemma—e.g. in constructive mathematics, then this theorem will not convince you and you must, if you need such a statement for a ring R, individually consider your given ring and verify that the statement of the theorem is constructively valid. However, we want to accept Zorn's Lemma as valid in this book.

5.6   The Chinese Remainder Theorem

An important general theorem in ring theory is still missing, which we will now prove. First, we see that we can calculate with ideals as follows: Let $\mathfrak{a}, \mathfrak{b} \triangleleft R$ be ideals in R, then one defines

> i) $\mathfrak{a} + \mathfrak{b} := \{a + b \in R \mid a \in \mathfrak{a}, b \in \mathfrak{b}\} \triangleleft R$
> ii) $\mathfrak{a} \cdot \mathfrak{b} := \{\sum_{j=1}^{m} a_j b_j \mid m \in \mathbb{N}, a_j \in \mathfrak{a}, b_j \in \mathfrak{b}\} \triangleleft R.$

Both are ideals in R. Specifically, for $\mathfrak{a} \triangleleft R$ we know its powers, e.g. its square:

$$\mathfrak{a}^2 = \mathfrak{a} \cdot \mathfrak{a} = \{\sum_{j=1}^{m} a_j a_j' \mid m \in \mathbb{N}, a_j, a_j' \in \mathfrak{a}\} \triangleleft R.$$

> **Check it out!**
> Justify why $\mathfrak{a}^2 \subset \mathfrak{a}$ holds!

Definition 5.37   Two ideals $\mathfrak{a}, \mathfrak{b} \triangleleft R$ are called **coprime**, if the equation $\mathfrak{a} + \mathfrak{b} = R$ holds.

> **And now?**
> Why does the word *coprime* fit? It does by analogy to the integers $\mathbb{Z}$. For the integers, the ideals are principal ideals, so $\mathfrak{a} = (a)$ and $\mathfrak{b} = (b)$. Then the following applies
>
> $$a, b \text{ coprime (in the usual sense)} \iff (a) + (b) = \mathbb{Z}.$$

Proof   If $a, b \in \mathbb{Z}$ are coprime, then there exist $x, y \in \mathbb{Z}$ with $xa + yb = 1$ (Bézout). Then $1 \in (a) + (b)$, so $(a) + (b) = \mathbb{Z}$. If, conversely, $(a) + (b) = \mathbb{Z}$, then $1 = xa + yb$ with $x, y \in \mathbb{Z}$. But then a, b are coprime, because a common divisor would also be a divisor of 1. $\square$

Note beforehand:

- Given ideals $\mathfrak{a}_1, \ldots, \mathfrak{a}_n \lhd R$, the intersection $\bigcap_{j=1}^{n} \mathfrak{a}_j \lhd R$ is also an ideal.
- Given rings $R_1, \ldots, R_n$, the product $R_1 \times R_2 \times \ldots \times R_n$ is naturally a commutative ring with one, where $+$ and $\cdot$ are defined component-wise:

$$(r_1, \ldots, r_n) + (s_1, \ldots, s_n) := (r_1 + s_1, \ldots, r_n + s_n)$$
$$(r_1, \ldots, r_n) \cdot (s_1, \ldots, s_n) := (r_1 s_1, \ldots, r_n s_n)$$

and $0 = (0, \ldots, 0)$ as well as $1 = (1, \ldots, 1)$.

Theorem 5.38 (Chinese Remainder Theorem—elegant version) If $\mathfrak{a}_1, \ldots, \mathfrak{a}_n \lhd R$ are pairwise coprime ideals in R, the well-defined ring homomorphism

$$\Psi : R / \bigcap_{j=1}^{n} \mathfrak{a}_j \to R/\mathfrak{a}_1 \times R/\mathfrak{a}_2 \times \ldots \times R/\mathfrak{a}_n$$

$$x \bmod \bigcap_{j=1}^{n} \mathfrak{a}_j \mapsto (x \bmod \mathfrak{a}_1, x \bmod \mathfrak{a}_2, \ldots, x \bmod \mathfrak{a}_n)$$

is a ring isomorphism.

Note: the map is often written like this (where the symbol $[x]$ occurs several times, each time with a different meaning, namely the class modulo $\bigcap$ or the individual $\mathfrak{a}_j$):

$$\Psi : R / \bigcap_{j=1}^{n} \mathfrak{a}_j \to R/\mathfrak{a}_1 \times R/\mathfrak{a}_2 \times \ldots \times R/\mathfrak{a}_n$$

$$[x] \mapsto ([x], [x], \ldots, [x]).$$

Proof Consider the mapping

$$\psi : R \to R/\mathfrak{a}_1 \times \ldots R/\mathfrak{a}_n, \quad x \mapsto (x \bmod \mathfrak{a}_1, \ldots, x \bmod \mathfrak{a}_n).$$

This is a ring homomorphism and the following holds:

$$\ker(\psi) = \{x \in R \mid x \equiv 0 \bmod \mathfrak{a}_j \text{ for all } j = 1, \ldots, n\} = \bigcap_{j=1}^{n} \mathfrak{a}_j,$$

therefore Ψ is a well-defined injective ring homomorphism and it remains to be shown that ψ (and therefore also Ψ) is surjective. This is the classical version of the Chinese remainder theorem—Theorem 5.39. $\square$

Theorem 5.39 (Chinese Remainder Theorem—classical version) If $\mathfrak{a}_1, \ldots, \mathfrak{a}_n \lhd R$ are pairwise coprime ideals in R and $x_1, \ldots, x_n \in R$ are arbitrary elements. Then there exists an $x \in R$, such that

$$x \equiv x_1 \bmod \mathfrak{a}_1$$
$$x \equiv x_2 \bmod \mathfrak{a}_2$$
$$\vdots \qquad\qquad (5.10)$$
$$x \equiv x_n \bmod \mathfrak{a}_n.$$

Proof The idea of the proof is to first show the following:

Claim 1 There exist elements $s_1, \ldots, s_n \in R$, such that

$$s_j \equiv 0 \bmod \mathfrak{a}_i \text{ für } i \neq j$$
$$s_j \equiv 1 \bmod \mathfrak{a}_j$$

If this claim is established, we are done, because then $x := x_1 s_1 + \ldots + x_n s_n \in R$ proves the assertion of the theorem. To prove the claim, one must of course use the pairwise coprimality. Let $j \in \{1, \ldots, n\}$ be fixed and let $i \neq j$. The coprimality of $\mathfrak{a}_j$ and $\mathfrak{a}_i$ ensures that there are elements

$$a_{ij} \in \mathfrak{a}_i \text{ and } b_{ij} \in \mathfrak{a}_j \text{ show that } a_{ij} + b_{ij} = 1. \qquad (5.11)$$

Then

$$s_j := \prod_{\substack{i=1,\ldots,n \\ i \neq j}} a_{ij} \in R$$

solves the claim, because then we deduce:

- For $i \neq j$ we have $a_{ij} \in \mathfrak{a}_i$ and thus $s_j = a_{ij} \cdot$ (the other factors) $\in \mathfrak{a}_i$, so $s_j \equiv 0 \bmod \mathfrak{a}_i$.
- Furthermore, for all $i \neq j$ we have the equality $a_{ij} = 1 - b_{ij}$ with $b_{ij} \in \mathfrak{a}_j$ (so $b_{ij} \equiv 0 \bmod \mathfrak{a}_j$) and thus we obtain

$$s_j = \prod_{\substack{i=1,\ldots,n \\ i \neq j}} (1 - b_{ij}) \equiv \prod_{\substack{i=1,\ldots,n \\ i \neq j}} 1 = 1 \bmod \mathfrak{a}_j.$$

Therefore the claim is true and hence the theorem is proven. $\square$

And now?

The proof also shows how to calculate such a solution $x \in R$ if one can find the elements in (5.11). We choose elements a_{ij} and b_{ij} as in the proof and then define the elements s_j accordingly. Finally, we put $x = x_1 s_1 + \ldots + x_n s_n$.

Note that this procedure has the advantage that one only needs to determine the s_j once to solve the **simultaneous equivalences** (5.10) for any $x_1, \ldots, x_n$ without further ado. Once you have computed the s_j, you can use it for all choices of $x_1, \ldots, x_n$.

And now?
The theorem is called the *Chinese Remainder Theorem*, because it already appeared in a calculation book of the Chinese author Sun Zi in the 3rd century AD. There, of course, the case $R = \mathbb{Z}$ was considered and we want to briefly emphasize this special case.

Theorem 5.40 (Chinese Remainder Theorem—specifically for $\mathbb{Z}$) Let $a_1, \ldots, a_n \in \mathbb{Z}$ are pairwise coprime numbers and $x_1, \ldots, x_n \in \mathbb{Z}$ arbitrary. Then there exists a number $x \in \mathbb{Z}$, such that

$$x \equiv x_1 \bmod a_1$$
$$x \equiv x_2 \bmod a_2$$
$$\vdots$$
$$x \equiv x_n \bmod a_n$$

holds. If x is such a solution, then the set of all solutions is the set

$$x + a_1 \cdots a_n \cdot \mathbb{Z}. \tag{5.12}$$

Proof Note that

$$\bigcap_{i=1}^{n} (a_i) = (a_1 \cdots a_n) = a_1 \cdots a_n \mathbb{Z}$$

holds (because these are pairwise coprime—easy to seee using unique prime factor decomposition for $\mathbb{Z}$). The elegant version of the Chinese remainder theorem gives that

$$\mathbb{Z}/a_1 \cdots a_n \mathbb{Z} \xrightarrow{\cong} \mathbb{Z}/a_1 \mathbb{Z} \times \ldots \times \mathbb{Z}/a_n \mathbb{Z}, \ [x] \mapsto ([x], \ldots, [x])$$

is an isomorphism, so $([x_1], \ldots, [x_n])$ is in the image and there is a unique preimage $[x] \in \mathbb{Z}/a_1 \cdots a_n \mathbb{Z}$. This class is then the set of all solutions and it equals the set (5.12) as claimed. $\square$

5.7 Examples of Rings in Quadratic Number Fields

Definition 5.41 A **quadratic number field** is a field extension $K|\mathbb{Q}$ of degree $[K : \mathbb{Q}] = 2$.

These are exactly the examples we have already considered above:

Lemma 5.42 *If $K|\mathbb{Q}$ is a quadratic number field, then there exists a square-free $d \in \mathbb{Z}$, such that K is isomorphic to the field $\mathbb{Q}(\sqrt{d}) \subset \mathbb{C}$.*

Proof Let $\alpha \in K$ be an element $\alpha \notin \mathbb{Q}$. Then α has a minimal polynomial $f(X) \in \mathbb{Q}[X]$ and since $\mathbb{Q} \subsetneq \mathbb{Q}(\alpha) \subset K$, we obtain by the degree formula that $\mathbb{Q}(\alpha) = K$. Let us write $f(X) = X^2 + bX + c \in \mathbb{Q}[X]$. Then this equation has a solution

$$\tilde{\alpha} := \frac{-b + \sqrt{b^2 - 4c}}{2} \in \mathbb{C}$$

(where the root is any chosen root of the radicand in $\mathbb{C}$). Writing the fractions as follows: $b = \frac{x}{y}$ and $c = \frac{r}{s}$, we get

$$b^2 - 4c = \frac{sx^2 - 4ry^2}{y^2 s} = \frac{s^2 x^2 - 4rsy^2}{y^2 s^2},$$

and

$$\tilde{\alpha} = \frac{-b \pm \frac{\sqrt{s^2 x^2 - 4rsy^2}}{ys}}{2}.$$

Again writing $\tilde{d} := s^2 x^2 - 4rsy^2 \in \mathbb{Z}$ in the form $\tilde{d} = q^2 \cdot d$ with square-free $d \in \mathbb{Z}$, one can see that

$$\tilde{\alpha} \in \mathbb{Q}(\sqrt{d})$$

holds.

The $\mathbb{Q}$-vector space isomorphism $\varphi : K = \mathbb{Q}(\alpha) \to \mathbb{Q}(\sqrt{d})$, $1 \mapsto 1$ and $\alpha \mapsto \tilde{\alpha}$ is a field isomorphism—for this, one must verify that it is a ring homomorphism (we already know it is bijective). Since φ is linear and $1 \mapsto 1$ holds, one only needs to see that

$$\varphi((s + t\alpha) \cdot (u + v\alpha)) = \varphi(s + t\alpha) \cdot \varphi(u + v\alpha)$$

holds. For the left side, note that

$$(s + t\alpha)(u + v\alpha) = su + (tu + sv)\alpha + tv\alpha^2 =$$
$$su + (tu + sv)\alpha + tv(-b\alpha - c) = (su - ctv) + (tu + sv - btv)\alpha \qquad (5.13)$$

(due to the minimal polynomial $f(X) = X^2 + bX + c$ of α over $\mathbb{Q}$). Therefore,

$$\varphi((s + t\alpha)(u + v\alpha)) = (su - ctv) + (tu - sv - btv)\tilde{\alpha}.$$

The same calculation as in (5.13) (with $\tilde{\alpha}$ instead of α) shows that

$$\varphi(s + t\alpha) \cdot \varphi(u + v\alpha) = (s + t\tilde{\alpha})(u + v\tilde{\alpha}) = (su - ctv) + (tu + sv - btv)\tilde{\alpha}$$

holds, since $\tilde{\alpha} \in \mathbb{C}$ has the same minimal polynomial $f(X) \in \mathbb{Q}[X]$. $\square$

And now?
The observation from the lemma will be an essential starting point for the entire Galois theory. Here it presented itself as follows: Given two field extensions $K|\mathbb{Q}$ and $\Omega|\mathbb{Q}$ and elements $\alpha \in K$ and $\widetilde{\alpha} \in \Omega$ which share the same minimal polynomial $f(X) = X^2 + bX + c \in \mathbb{Q}[X]$ over $\mathbb{Q}$, the isomorphism of $\mathbb{Q}$-vector spaces

$$\varphi : \mathbb{Q}(\alpha) \to \mathbb{Q}(\widetilde{\alpha}), \; s + t\alpha \mapsto s + t\widetilde{\alpha}$$

is indeed an **isomorphism of fields**.
  One might wonder whether the fact that $\deg(f) = 2$ plays a role here. The answer is: No, but then the proof is not so easy to write down. We will find an elegant proof in Chap. 10.

Now, let us consider the field $\mathbb{Q}(\sqrt{d})$. We will find the following subring:

$$\mathbb{Z}[\sqrt{d}] := \{a + b\sqrt{d} \in \mathbb{Q}(\sqrt{d}) \mid a, b \in \mathbb{Z}\} .$$

It is certainly without zero divisors (because it is a subring in a field) hence a domain.
  If $d \equiv 1 \bmod 4$, we can also look at the subring

$$\mathbb{Z}[\frac{1 + \sqrt{d}}{2}] := \{a + b\frac{1 + \sqrt{d}}{2} \mid a, b \in \mathbb{Z}\}.$$

Why should we take this particular subring into account? Let's determine the minimal polynomial $f(X) \in \mathbb{Q}[X]$ of the element $\omega := \frac{1+\sqrt{d}}{2}$ over $\mathbb{Q}$. We deduce:

$$2\omega = 1 + \sqrt{d} \Rightarrow 2\omega - 1 = \sqrt{d} \Rightarrow (2\omega - 1)^2 = d \Rightarrow 4\omega^2 - 4\omega + 1 - d = 0$$

therefore

$$f(X) = X^2 - X + \frac{1 - d}{4} \in \mathbb{Q}[X]$$

the minimal polynomial of ω over $\mathbb{Q}$. Now under the assumption $d \equiv 1 \bmod 4$, the constant coefficient of this polynomial is $\frac{1-d}{4} \in \mathbb{Z}$, an integer, thus

$$d \equiv 1 \bmod 4 \Longrightarrow f(X) \in \mathbb{Z}[X].$$

This is something special, the element $\frac{1+\sqrt{d}}{2} \in \mathbb{Q}(\sqrt{d})$ is called **integral over** $\mathbb{Z}$ - note that there is some risk of confusion: a ring is called **integral** if it has no zero divisors, an element $\omega \in S$ for some ring extension $R \subset S$ (in the obvious sense) is called **integral** if there is a monic polynomial $f(X) \in R[X]$, $f \neq 0$, such that $f(\omega) = 0$. These two notions of being integral are not related (in some other languages than English -- e.g. in German -- different adjectives are used).

In summary, we can define:

Definition 5.43 Let $d \in \mathbb{Z}$ be square-free and $K = \mathbb{Q}(\sqrt{d})$. Then, **the ring of integers in** $\mathbb{Q}(\sqrt{d})$, written $\mathcal{O}_K$, is defined[3] in the following way:

$$d \equiv 2, 3 \bmod 4 \Longrightarrow \mathcal{O}_K := \mathbb{Z}[\sqrt{d}]$$

$$d \equiv 1 \bmod 4 \Longrightarrow \mathcal{O}_K := \mathbb{Z}[\frac{1 + \sqrt{d}}{2}]$$

Remark 5.44 In the case of $d \equiv 1 \bmod 4$, note that the elements can also be described as follows

$$\mathbb{Z}[\frac{1 + \sqrt{d}}{2}] = \{\frac{1}{2}(u + v\sqrt{d}) \mid u, v \in \mathbb{Z} \text{ and } u \equiv v \bmod 2\}.$$

Proof Firstly, with $a, b \in \mathbb{Z}$:

$$a + b\frac{1 + \sqrt{d}}{2} = \frac{2a + b + b\sqrt{d}}{2} = \frac{1}{2}((2a + b) + b\sqrt{d})$$

so that it indeed admits the form on the right hand side of the assertion.

If, conversely, u, v are given with $u \equiv v \bmod 2$, we deduce that $u - v = 2a$ for some $a \in \mathbb{Z}$ and thus

$$\frac{1}{2}(u + v\sqrt{d}) = \frac{u - v + v + v\sqrt{d}}{2} = \frac{2a + v + v\sqrt{d}}{2} = a + v\frac{1 + \sqrt{d}}{2},$$

as claimed. $\square$

Red Thread to the Previous Chapter
In this chapter, we have seen

- what a commutative ring with untiy,
- what the unit group $R^\times$ in such a R is,
- which elements we callzero divisors,
- how a ring homomorphism $\varphi : R \to S$ is defined,
- which subrings are called ideals, written as $\mathfrak{a} \lhd R$,
- how the factor ring $R/\mathfrak{a}$ for an ideal is defined and that the fundamental homomorphism theorem holds in an analogous sense,
- that an ideal $\mathfrak{p} \lhd R$ with $\mathfrak{p} \neq R$ is a prime ideal, if $xy \in \mathfrak{p} \Longrightarrow x \in \mathfrak{p}$ or $y \in \mathfrak{p}$ holds, and that this is the case if and only if $R/\mathfrak{p}$ is a domain,
- that an ideal $\mathfrak{m} \lhd R$ with $\mathfrak{m} \neq R$ is a maximal ideal if:

$$\mathfrak{m} \subsetneq \mathfrak{a} \subset R \Longleftrightarrow \mathfrak{m} = \mathfrak{a},$$

---

[3] actually, one defines the *ring of integers in a finite field extension* $K|\mathbb{Q}$ more generally, namely to be the elements which are integral over. Then our definition is a lemma for the special case $K = \mathbb{Q}(\sqrt{d})$.

and that this is the case if and only if the ring $R/\mathfrak{m}$ is a field,
- that $\mathbb{Z}$ and $K[X]$ (for a field K) are principal ideal domains,
- that given pairwise coprime ideals $\mathfrak{a}_1, \ldots, \mathfrak{a}_n \lhd R$ one has the natural isomorphism $R/\bigcap_{i=1}^n \mathfrak{a}_i \to R/\mathfrak{a}_1 \times \ldots \times R/\mathfrak{a}_n$ which maps $[x] \mapsto ([x], \ldots, [x])$ (Chinese remainder theorem).

Exercises:

5.1 Consider the factor ring $R := \mathbb{Q}[X]/(f(X))$ for $f(X) := X^2 - 5X + 6$. Determine all $a \in \mathbb{Q}$, such that $a + X \bmod (f(X))$ is a unit.

5.2 Let $p \in \mathbb{N}$ be a prime number. Consider the subset

$$\mathbb{Z}_{(p)} := \{\frac{a}{b} \in \mathbb{Q} \text{ in reduced form } \mid a, b \in \mathbb{Z} \text{ and } p \not\mid b\} \subset \mathbb{Q}.$$

(a) Apparently, $\mathbb{Z} \subset \mathbb{Z}_{(p)}$. Show: If $\mathfrak{a} \lhd \mathbb{Z}_{(p)}$ is an ideal in $\mathbb{Z}_{(p)}$, then $\mathfrak{a} \cap \mathbb{Z}$ is an ideal in $\mathbb{Z}$.
(b) Determine all prime ideals of $\mathbb{Z}_{(p)}$.

5.3 We denote by $\sqrt{-5} \in \mathbb{C}$ any root of -5 (e.g., $\sqrt{-5} = i\sqrt{5}$). Consider the ring

$$\mathbb{Z}[\sqrt{-5}] := \{a + b\sqrt{-5} \mid a, b \in \mathbb{Z}\} \subset \mathbb{C}$$

and the ideal generated by 2 and $1 + \sqrt{-5}$ in this ring:

$$\mathfrak{p} := (2, 1 + \sqrt{-5}) = \{x \cdot 2 + y \cdot (1 + \sqrt{-5}) \mid x, y \in \mathbb{Z}[\sqrt{-5}]\} \lhd \mathbb{Z}[\sqrt{-5}].$$

Prove the following:

(a) We have $\mathfrak{p} = \{a + b\sqrt{-5} \mid a \equiv b \bmod 2\}$.
(b) The ideal $\mathfrak{p}$ is a prime ideal.

5.4 Let K be a field. Show that for $a \in K$ the ideal $(X - a) \lhd K[X, Y]$ is a prime ideal but not a maximal ideal, and that for $a, b \in K$ the ideal $(X - a, Y - b) \lhd K[X, Y]$ is a maximal ideal.

5.5 Number of roots: Find an example of a ring R (necessarily with zero divisors) and a polynomial $f(X) \in R[X]$ of a degree $n \geq 1$, which has more than n roots in R.

5.6 A ring R is called a **local ring** if it has exactly one maximal ideal. Show that R is a local ring if and only if the complement of the units $\mathfrak{a} := R \setminus R^\times$ is an ideal in R.

5.7 A farmer wants to present all his cows at a country fair. However, if he lines them up in rows of 3, then 2 cows are left over. If he lines them up in rows of 4, one cow is left over. Only with rows of 7, no cow is left over. How many cows does the farmer have at least?

5.8 Consider the ring $R := \mathbb{Z}[\omega]$ with $\omega = \frac{1+\sqrt{-d}}{2}$ for a square-free $d \in \mathbb{N}$ as a subset $R \subset \mathbb{C}$, by considering $\sqrt{-d} := i\sqrt{d}$. Show the following:

(a) We have: $\mathbb{Z}[\omega] = \{a + b\omega \mid a, b \in \mathbb{Z}\}$.
(b) The   map $N : \mathbb{Z}[\omega] \to \mathbb{Z}$, $z \mapsto |z|^2 = z \cdot \bar{z}$ is   multiplicative,   i.e.
    $N(zw) = N(z)N(w)$.
(c) An element $z \in \mathbb{Z}[\omega]$ is a unit if and only if $N(z) = \pm 1$.

Euclidean Rings, Principal Ideal Rings, Noetherian Rings

6

6.1 Euclidean Rings

Euclidian rings are rings with a division with remainder. More precisely:

Definition 6.1 An integral domain R is called **euclidean**, if there is a map $d : R \smallsetminus \{0\} \to \mathbb{N}$ such that for any two elements $a, b \in R \smallsetminus \{0\}$ there are elements $q, r \in R$ such that

$$a = qb + r \text{ with } r = 0 \text{ or } d(r) < d(b) \tag{6.1}$$

holds.

In a Euclidean ring, we therefore have *division with remainder* as indicated in (6.1), where the smallness of the remainder is induced by a map $d : R \smallsetminus \{0\} \to \mathbb{N}$ which we have to find if we want to test for the Euclidian property. Often, we additionally put $d(0) = 0$ and thus consider $d : R \to \mathbb{N}_0$.

Examples
1. $R = \mathbb{Z}$ with $d = |.| : \mathbb{Z} \smallsetminus \{0\} \to \mathbb{N}$, $x \mapsto |x|$ is Euclidean.
2. $R = K[X]$ for a field K and $d = \deg$ is Euclidean.
3. $R = \mathbb{Z}[i] = \{a + bi \in \mathbb{C} \mid a, b \in \mathbb{Z}\}$ is a Euclidean ring with the map

$$N : \mathbb{Z}[i] \smallsetminus \{0\} \to \mathbb{N} , \ a + bi \mapsto a^2 + b^2 .$$

(This map is often denoted N because it is also called the *norm map of this ring*—see next *And now?*)

Proof We need to show that we have a division with remainder (6.1) in $\mathbb{Z}[i]$ with respect to the map N. Assume we have two elements $a + bi, c + di \in \mathbb{Z}[i] \smallsetminus \{0\}$. We want to write $a + bi = (c + di)q + r$ with suitable $q, r \in \mathbb{Z}[i]$.

© The Author(s), under exclusive license to Springer-Verlag GmbH, DE, part of Springer Nature 2024
M. Hien, *Abstract Algebra*, Mathematics Study Resources 7,
https://doi.org/10.1007/978-3-662-67974-6_6

The map N is obviously the restriction of the square of the absolute value in the complex numbers $|.|^2 : \mathbb{C} \to \mathbb{R}_{\geq 0}, z \mapsto |z|^2$ restricted to $\mathbb{Z}[i]$. We take advantage of this. We can first take the quotient in $\mathbb{C}$: Let

$$\omega := \frac{a+bi}{c+di} \in \mathbb{C}.$$

Now we choose $x, y \in \mathbb{Z}$, so that the point $x + iy \in \mathbb{Z}[i]$ is as close as possible (in terms of the distance given by $|.|$) to ω —see Fig. 6.1. Then

$$|x - \mathrm{Re}(\omega)| \leq \frac{1}{2} \text{ and } |y - \mathrm{Im}(\omega)| \leq \frac{1}{2},$$

so the distance satsfies

$$|(x+iy) - \omega|^2 \leq \frac{1}{4} + \frac{1}{4} = \frac{1}{2} < 1$$

(the maximal possible distance is obtained if ω lies exactly in the middle of a mesh for the grid determined by integer linear combinations of 1 and i , in which case every surrounding grid point is equally far away). Now if we set

$$\eta := \omega - (x+iy) = \frac{a+ib}{c+id} - (x+iy) \in \mathbb{C},$$

we   deduce   that $|\eta|^2 < 1.$ If   we   finally   put
$r + is := a + ib - (x+iy)(c+id) \in \mathbb{Z}[i]$, we get

$$a + ib = (x+iy)(c+id) + (r+is)$$

with $r + is = \eta \cdot (c+id)$ and thus

$$N(r+is) = |\eta \cdot (c+id)|^2 = |\eta|^2 \cdot N(c+id) < N(c+id),$$

as was to be shown. $\square$

Fig. 6.1 For the example of $\mathbb{Z}[i]$. The crosses are the elements in $\mathbb{Z}[i]$ (the grid), the element $x + iy \in \mathbb{Z}[i]$ is a neighbor to $\omega \in \mathbb{C}$

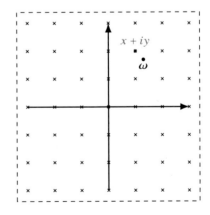

And now?

The mapping N from the last example exists for all *rings of integers in number fields* (which I do not want to define now). We know the case of quadratic number fields from Sect. 5.7.

Given a square-free $d \in \mathbb{Z}$ consider the field $K := \mathbb{Q}(\sqrt{d})$ and define the following **norm map**

$$N := N_{K|\mathbb{Q}} : \mathbb{Q}(\sqrt{d}) \to \mathbb{Q}, \ s + t\sqrt{d} \mapsto s^2 - dt^2 = (s + t\sqrt{d})(s - t\sqrt{d}) \tag{6.2}$$

It is obviously multiplicative

$$N\big((s + t\sqrt{d})(u + v\sqrt{d})\big) = N(s + t\sqrt{d}) \cdot N(u + v\sqrt{d}).$$

Computing the minimal polynomial of $\omega := s + t\sqrt{d}$ (with $t \neq 0$) over $\mathbb{Q}$,

one obtains (by an analogous calculation as in Sect. 5.7):

$$\omega = s + t\sqrt{d} \Rightarrow (\omega - s)^2 = t^2 d \Rightarrow \omega^2 - 2s\omega + s^2 - t^2 d = 0,$$

so the minimal polynomial is $p(X) = X^2 - 2sX + N(\omega) \in \mathbb{Q}[X]$ and hence $N(\omega) \in \mathbb{Q}$ turns out to be the constant term in the minimal polynomial.

- If $d < 0$, we call $\mathbb{Q}(\sqrt{d}) \subset \mathbb{C}$ an **imaginary quadratic number field**. Then

$$N_{K|\mathbb{Q}} = |.|^2 : \mathbb{Q}(\sqrt{d}) \to \mathbb{Q}, \ s + t\sqrt{d} \mapsto s^2 - t^2 d = s^2 + t^2|d| = |s + ti\sqrt{|d|}|^2$$

  is again the square of the complex absolute value.
- If $d > 0$, we call $\mathbb{Q}(\sqrt{d})$ a **real quadratic number field**. Then $N(s + t\sqrt{d})$ also takes on negative values.

One can now restrict N to the ring of integers $\mathcal{O}_K$ in K and obtains:

$$N : \mathcal{O}_K \to \mathbb{Z}, \ \xi \mapsto N(\xi)$$

being a map with integer values—*verify this!!! This is true in all cases, regardless of whether $d \equiv 2, 3 \bmod 4$ or $d \equiv 1 \bmod 4$! In the case $d \equiv 1 \bmod 4$ note that the equation 5.44 should be applied.*

And now?

Some more comments on the norm map:

Observation 1 Since $N_{K|\mathbb{Q}}$ is multiplicative, furthermore $N_{K|\mathbb{Q}}(1) = 1$ and $N_{K|\mathbb{Q}}|_{\mathcal{O}_K} : \mathcal{O}_K \to \mathbb{Z}$ takes values in $\mathbb{Z}$, we deduce:

$$\xi \in \mathcal{O}_K \text{ is a unit} \iff N_{K|\mathbb{Q}}(\xi) = \pm 1$$

Proof If ξ is a unit, then there exists $\eta \in \mathcal{O}_K$ with $\xi\eta = 1$. Applying the map $N_{K|\mathbb{Q}}$ and using its multiplicativity, we observe that

$$1 = N_{K|\mathbb{Q}}(1) = N_{K|\mathbb{Q}}(\xi\eta) = N_{K|\mathbb{Q}}(\xi) \cdot N_{K|\mathbb{Q}}(\eta) \Rightarrow N_{K|\mathbb{Q}}(\xi) = \pm 1 \,.$$

If conversely $N_{K|\mathbb{Q}}(\xi) = \pm 1$, we briefly distinguish the cases depending on d mod 4.

- If $d \equiv 2, 3 \bmod 4$, we write $\xi = a + b\sqrt{d}$ with $a, b \in \mathbb{Z}$. Then

$$\pm 1 = N_{K|\mathbb{Q}}(\xi) = (a + b\sqrt{d})(a - b\sqrt{d}) = \xi \cdot (a - b\sqrt{d}) \,,$$

  so we see that ξ is a unit with inverse $\pm(a - b\sqrt{d}) \in \mathcal{O}_K$.
- If $d \equiv 1 \bmod 4$, the same applies. Then write $\xi = \frac{1}{2}(u + v\sqrt{d})$ with $u, v \in \mathbb{Z}$ and $u \equiv v \bmod 2$ – see Remark 5.44. Then again

$$\pm 1 = N_{K|\mathbb{Q}}(\xi) = \frac{1}{2}(u + v\sqrt{d}) \cdot \frac{1}{2}(u - v\sqrt{d}) = \xi \cdot \frac{1}{2}(u - v\sqrt{d})$$

  and ξ is a unit with inverse $\pm\frac{1}{2}(u - v\sqrt{d}) \in \mathcal{O}_K$.□

The following theorem justifies why the term *principal ideal ring* is included in the title of this chapter:

Theorem 6.2 A Euclidean ring is a principal ideal domain.

Proof Let $\mathfrak{a} \lhd R$ be an ideal and R be a Euclidian ring with respect to the map $d : R \to \mathbb{N}_0$. The set of values $d(\mathfrak{a} \setminus \{0\}) \subset \mathbb{N}$ has a smallest element. So let $a \in \mathfrak{a}$ be an element in the ideal with minimal $d(a)$. Then $\mathfrak{a} = (a)$ holds, because if $x \in \mathfrak{a} \setminus \{0\}$ is an arbitrary element of the ideal, we obtain

$$x = qa + r \text{ with } q, \ r \in R \text{ with } d(r) < d(a).$$

But then $r = x - qa \in \mathfrak{a}$ lies in the ideal as well and due to the minimality of $d(a)$, we deduce that $r = 0$ must hold, hence $x \in (a)$. □

6.2   The Euclidean Algorithm

Why are Euclidean rings named after Euclid? Because given a Euclidian ring, the **Euclidean algorithm** works for it. Let R be a Euclidean ring with respect to the map $d : R \to \mathbb{N}_0$. Let $x, y \in R \setminus \{0\}$. We perform division with remainder successively as in Fig. 6.2 – always with $d(r_1) > d(r_2) > d(r_3) > \ldots$, all of these numbers lying in $\mathbb{N}_0$, which is why the algorithm will terminate with $r_{n+1} = 0$ after a finite number of steps.

Then one can substitute backwards the second but final non-vanishing remainder r_{n-1} into the penultimate equation and obersves that r_{n-2} is a multiple of r_n. Proceeding successively in this way, one observes that $r_n | r_j$ for all $j \le k$. In particular, $r_n | y$ (from the second line) and $r_n | x$ (first line) also hold. Thus, we see that we have found a **common divisor** r_n of x and y.

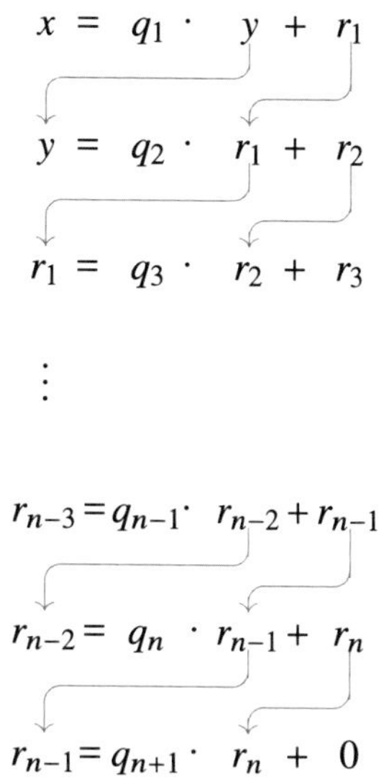

Fig. 6.2 Euclidean
Algorithm

$$x = q_1 \cdot y + r_1$$

$$y = q_2 \cdot r_1 + r_2$$

$$r_1 = q_3 \cdot r_2 + r_3$$

$$\vdots$$

$$r_{n-3} = q_{n-1} \cdot r_{n-2} + r_{n-1}$$

$$r_{n-2} = q_n \cdot r_{n-1} + r_n$$

$$r_{n-1} = q_{n+1} \cdot r_n + 0$$

However, one can achieve even more if one observes that from the penultimate line we can deduce that r_n is an R-linear combination of r_{n-1} and r_{n-2}: $r_n \in Rr_{n-2} + Rr_{n-1}$—and the coefficients are easily determined from the corresponding line in the algorithm.

From the next line above, we obtain that $r_{n-1} \in Rr_{n-2} + Rr_{n-3}$ (with explicitly given coefficients) and thus we get $r_n \in Rr_{n-2} + Rr_{n-3}$ (with coefficients in R easily computable).

Proceeding successively in this way, we solves all equations backwards up to the last one and finally obtain: $r_n \in Rx + Ry$, so we have obtained the common divisor r_n as a linear combination of the initial elements $x, y \in R$. We will see in the defining Lemma 6.3 that this common divisor is in a certain sense the **greatest common divisor** of x and y. In the Euclidean ring $\mathbb{Z}$ it is clear what we mean by "greatest", in a general ring there is no $<$-relation, but we still have the divisibility relation we can use as a (partial) order.

Defining Lemma 6.3 Let R be an integral domain and $x, y \in R \smallsetminus \{0\}$. An element $d \in R \smallsetminus \{0\}$ is called a **greatest common divisor**, written as $d = \gcd(x, y)$, if the following two properties are fulfilled:

1. d divides x and d divides y.
2. If $s \in R$ is another element such that s divides x and s divides y, then s also divides d.

If $x, y \in R$ admit a greatest common divisor (gcd), then it is unique up to multiplication with a unit $\varepsilon \in R^\times$.

Proof Let $d, e \in R$ be the greatest common divisors of x and y. It follows that $d|e$ (because condition 1) applies to $e \in R$ and then 2) applies for the element d). Conversely, we get $e|d$ interchanging the roles. Thus, there exist $a, b \in R$ with $e = ad$ and $d = be$. We deduce that

$$d = be = abd \Rightarrow 0 = d - abd = d(1 - ab) \stackrel{R \text{ is a domain}}{\Longrightarrow} ab = 1 \Rightarrow a \in R^\times$$

as claimed. $\square$

Lemma 6.4 *In a Euclidean ring R there exists a gcd for any elements $x, y \in R \setminus \{0\}$ and it has a representation* (**Bézout's identity**)

$$\gcd(x, y) = ax + by$$

for suitable elements $a, b \in R$.

Proof Applying the Euclidean algorithm, one obtains a common divisor $r \in R$ for our given $x, y \in R \setminus \{0\}$. This gcd can be written (passing through the Euclidean algorithm in backwards direction) as $r = ax + by$ with $a, b \in R$.

This element r is indeed even a greatest common divisor (gcd) of x and y, because if $s \in R$ is any divisor of x and y, then consider the Euclidean algorithm Fig. 6.2 (from top to bottom):

$$s|x, y \Longrightarrow s|r_1$$
$$s|y, r_1 \Longrightarrow s|r_2$$
$$s|r_1, r_2 \Longrightarrow s|r_3$$
$$\vdots$$
$$s|r_{n-1}, r_{n-2} \Longrightarrow s|r_n = r.$$

This had to be verified to finsh the proof. $\square$

6.3   Noetherian Rings

Definition 6.5 A ring R is called **Noetherian**, if every ideal $\mathfrak{a}$ is generated by a finite number of elements: $\mathfrak{a} = (a_1, \ldots, a_m)$ for some $a_1, \ldots a_m \in R$.

We only want to make two observations:

Lemma 6.6 *A ring is Noetherian if and only if every ascending chain*

$$\mathfrak{a}_1 \subset \mathfrak{a}_2 \subset \mathfrak{a}_3 \subset \ldots$$

of ideals in R becomes stationary (i.e., there exists an $N \in \mathbb{N}$ such that $\mathfrak{a}_m = \mathfrak{a}_N$ for all $m \geq N$).

Proof If R is Noetherian and a chain of ideals is given, then $\mathfrak{a} := \bigcup_{j=1}^{\infty} \mathfrak{a}_j \subset R$ is also an ideal *(think about this briefly!)*. By assumption, $\mathfrak{a} = (a_1, \ldots, a_m)$ for finitely many elements. But then there exists an $N \in \mathbb{N}$, such that $a_1, \ldots, a_m \in \mathfrak{a}_N$ holds. This N will do the job.

Conversely, if the condition concerning ascending chains of ideals is met and if $\mathfrak{a} \subset R$ is an ideal, we proceed as follows. Suppose $\mathfrak{a}$ is not finitely generated. Then we could find an infinite sequence $a_1, a_2, \ldots$ of elements in $\mathfrak{a}$, such that for all $m \in \mathbb{N}$ we have $(a_1, \ldots, a_m) \subsetneq \mathfrak{a}$ and $a_{m+1} \notin (a_1, \ldots, a_m)$. But then

$$(a_1) \subsetneq (a_1, a_2) \subsetneq (a_1, a_2, a_3) \subsetneq \ldots$$

is a chain that becomes non-stationary which contradicts our assumption. $\square$

Theorem 6.7 (Hilbert's Basis Theorem) If R is a Noetherian ring, then $R[X]$ is also Noetherian.

Proof Assume that there exists a non-finitely generated ideal $\mathfrak{a} \lhd R[X]$. We obtain a sequence of polynomials by successively choosing

$$f_1 \in \mathfrak{a} \smallsetminus \{0\} \text{ of minimal degree,}$$
$$f_2 \in \mathfrak{a} \smallsetminus (f_1) \text{ of minimal degree,}$$
$$f_3 \in \mathfrak{a} \smallsetminus (f_1, f_2) \text{ of minimal degree,}$$
$$\vdots$$

Define $n_k := \deg(f_k)$ and write the polynomials in the form $f_k = a_k X^{n_k} + \text{smaller powers of} X$. Then $n_1 \leq n_2 \leq n_3 \leq \ldots$ *(justify this!)*, and for all k we deduce that

$$(a_1, \ldots, a_k) \subsetneq (a_1, \ldots, a_k, a_{k+1}), \qquad (6.3)$$

due to the following reasoning: Suppose this was not the case. Then for some $k \in \mathbb{N}$ we would have $a_{k+1} \in (a_1, \ldots, a_k)$, let's say

$$a_{k+1} = r_1 a_1 + \ldots + r_k a_k \text{ with suitable } r_1, \ldots, r_k \in R$$

.

Then consider the polynomial

$$g(X) := \sum_{j=1}^{k} r_j \cdot X^{n_{k+1}-n_j} \cdot f_j \in (f_1, \ldots, f_k).$$

It has the form

$$g(X) = \sum_{j=1}^{k} r_j \cdot a_j X^{n_{k+1}} + \text{smaller powers} = a_{k+1} X^{n_{k+1}} + \text{smaller powers.}$$

But then $\deg(f_{k+1} - g) < n_{k+1} = \deg(f_{k+1})$ holds as well as

$$f_{k+1} - g \in \mathfrak{a} \smallsetminus (f_1, \ldots, f_k)$$

(since $g \in (f_1, \ldots, f_k)$ and $f_{k+1} = g + (f_{k+1} - g) \notin (f_1, \ldots, f_k)$).

This is a contradiction to the minimality of $\deg(f_{k+1})$ among all elements of $\mathfrak{a} \smallsetminus (f_1, \ldots, f_k)$.

Hence, (6.3) is shown and we have constructed a non-stationary ascending chain of ideals in R, contradicting the assumption. $\square$

Corollary 6.8 If K is a field and $n \in \mathbb{N}$, the polynomial ring $K[X_1, \ldots, X_n]$ is Noetherian.

Red Thread to the Previous Chapter
We have seen in this chapter:

- how to define division with remainder in an abstract way. An integral domain R together with a mapping $d : R \smallsetminus \{0\} \to \mathbb{N}$, for which a division with remainder can be formulated, is called Euclidean,
- that Euclidean rings are principal ideal domains,
- that Noetherian rings are by definition those whose ideals are all finitely generated,
- that given a Noetherian R the polynomial ring $R[X]$ is again Noetherian.

Exercises:
6.1 Let R be an integral domain. Show that $R[X]$ is a principal ideal domain if and only if R is a field.

6.2 Consider the ring $R : = \mathbb{Z}[\frac{1+\sqrt{-7}}{2}] \subset \mathbb{Q}(\sqrt{-7})$ of the integers in $\mathbb{Q}(\sqrt{-7})$, see Definition 5.43. Show that R is a Euclidean ring with respect to the norm map— cp. Exercise 7.2.

6.3 Let R be Euclidean with respect to $d : R \smallsetminus \{0\} \to \mathbb{N}$ and $a \in R \smallsetminus (R^\times \cup \{0\})$ be an element with minimal $d(a)$.

(a) Show that $R^\times \cup \{0\} \to R/(a)$, $u \mapsto [u]$ defines a surjective map between sets.
(b) Assume that $R^\times = \{\pm 1\}$ holds. Conclude that for an element a as in (a), it follows that $R/(a) \cong \mathbb{Z}/2\mathbb{Z}$ or $R/(a) \cong \mathbb{Z}/3\mathbb{Z}$.

6.4 Let $R : = \mathbb{Z}[\frac{1+\sqrt{-19}}{2}] \subset \mathbb{Q}(\sqrt{-19})$ be the ring of integers in $\mathbb{Q}(\sqrt{-19})$. Prove the following:

(a) The units of this ring are $R^\times = \{\pm 1\}$.
(b) The element $\omega : = \frac{1+\sqrt{-19}}{2}$ satisfies the equation $\omega^2 - \omega + 5 = 0$.

(c) For each of the two rings $\mathbb{Z}/2\mathbb{Z}$ and $\mathbb{Z}/3\mathbb{Z}$ we have that there is no element x that solves the equation $x^2 - x + 5 = 0$.

(d) There is no map d such that the ring R could be Euclidean with respect to d. (Hint: Exercise 6.3).

Note: This example is interesting insofar as R is a principal ideal domain (but this is not so easy to show) but not Euclidean (with respect to any possible map d).

Unique Factorization Domains

7

7.1   Prime Elements and Irreducible Elements, Unique Factorization Domains

And now?

In $\mathbb{Z}$ we have a unique prime factor decomposition. One could of course ask oneself if this is a general property of any ring, for example considering the ring $\mathbb{Z}[\zeta_n] := \{f(\zeta_n) \in \mathbb{C} \mid f(X) \in \mathbb{Z}[X]\}$ for $\zeta_n := \exp(2\pi i/n)$. (We will define the abstract notion of prime factorization in Definition 7.6).

- The ring $\mathbb{Z}[\zeta_n]$ has a unique prime factorization only for

$$n \in \{1, \ldots, 22, 24, 25, 26, 27, 28, 30, 32, 33, 34, 35, 36, 38,$$
$$40, 42, 44, 45, 48, 50, 54, 60, 66, 70, 84, 90\}$$

  For all other n, it does not.[1]
- If $\mathbb{Z}[\zeta_n]$ has a unique prime factorization for some $n \geq 3$, the **Fermat's Problem** for this n:

  There is no solution $x, y, z \in \mathbb{Q}^\times$ to the equation $x^n + y^n = z^n$

  has been known since at least the mid-nineteenth century (Eduard Kummer even generalized the approach).
  The general statement for any $n \geq 3$ was proven by A. Wiles in 1994 (using much more advanced methods from algebraic geometry).

In the following, let R always be an integral domain.

---

[1] See Chap. 11 in *Washington, L.C.* (1997): Introduction to Cyclotomic Fields, Graduate Texts in Mathematics, vol 83. Springer, New York, NY.

© The Author(s), under exclusive license to Springer-Verlag GmbH, DE, part of
Springer Nature 2024
M. Hien, *Abstract Algebra*, Mathematics Study Resources 7,
https://doi.org/10.1007/978-3-662-67974-6_7

And now?

The first question when one tries to define the unique prime factorization in a ring R is the following: What should the analogue to a prime number be in R? One has two characterizations to choose from: A number $p \in \mathbb{N} \smallsetminus \{1\}$ is a prime number if and only if it fulfills one of the following equivalent conditions:

1. Given a porduct of integer numbers $xy \in \mathbb{Z}$, such that $p|xy$, then $p|x$ or $p|y$.
2. p admits only two positive divisors, namely itself and 1.

The second point should be rephrased so that the adjective *positive* does not appear -- there is no notion of being positive in an arbitrary ring. The sign enters the play here in $\mathbb{Z}$ because the group of units is $\mathbb{Z}^\times = \{\pm 1\}$. When asking about divisors, however, units do not play a role anyway. Therefore, it is better to rephrase the characterization in 2. as follows:

2. $\Leftrightarrow$ if p decomposes into a product $p = xy$, then $x \in \mathbb{Z}^\times$ or $y \in \mathbb{Z}^\times$

Definition 7.1 Let R be an integral domain. An element $p \in R$ is called

- a **prime element**, if $p \notin R^\times \cup \{0\}$ and the following statement holds:

$$p|xy \text{ for } x, y \in R \Longleftrightarrow p|x \text{ or } p|y$$

- **irreducible**, if $p \notin R^\times \cup \{0\}$ and if the following holds:

$$\text{If } p = xy \text{ with } x, y \in R, \text{ then } x \in R^\times \text{or } y \in R^\times.$$

Note 7.2 Let us make the following remarks:

- In general, these terms are not equivalent. We will see examples in the exercises.
- Apparently, $r \in R$ is a prime element if and only if $r \neq 0$ and $(r) \lhd R$ is a prime ideal.

Lemma 7.3 *We have the implication: r is a prime element $\Longrightarrow r$ is irreducible*

Proof Let r be a prime element and $r = xy$ with $x, y \in R$. Then obviously $r|xy$, so it follows from the assumption that $r|x$ or $r|y$. Let's say $r|x$ (otherwise we rename x and y). Then there is an element $a \in R$ with $x = ra$. Thus,

$$r = xy = ray \Longrightarrow r(1 - ay) = 0 \Longrightarrow 1 = ay \Longrightarrow y \in R^\times,$$

where the middle $\Longrightarrow$ follows, because $r \neq 0$ and R has no zero divisors. $\square$

At least in principal ideal domains, we can read off irreducible from a generator of the principal ideal:

Lemma 7.4 *If R is a principal ideal domain and $\pi \in R$, Then we have the following equivalence:*

$$\pi \text{ is irreducible } \Longleftrightarrow (\pi) \lhd R \text{ is a maximal ideal.}$$

Proof We show both directions individually:

$\Rightarrow$: Let $(\pi) \subsetneq \mathfrak{a} = (a) \subset R$ be a strictly larger ideal—a principal ideal by assumption. Since $\pi \in (\pi) \subset (a)$ there exists an element $b \in R$ with $\pi = ab$. Since π is irreducible, it follows that $a \in R^\times$ or $b \in R^\times$. If $b \in R^\times$, then $a = \pi b^{-1} \in (\pi)$, which is not the case. Therefore, we deduce $a \in R^\times$.

$\Leftarrow$: So let $\pi = ab$ be a decomposition into a product and let $a \notin R^\times$. Then consider

$$(\pi) \subset (a) \subset R .$$

Since $a \notin R^\times$, $(a) \subsetneq R$ holds. Now, (π) being a maximal ideal, it follows that $(\pi) = (a)$, so $a \in (\pi)$ and thus $a = \pi c$ for some $c \in R$. Therefore,

$$\pi = ab = \pi cb \Longrightarrow \pi(1 - cb) = 0 \Longrightarrow cb = 1 \Longrightarrow b \in R^\times,$$

which was to be shown. $\square$

An important application of this will be the special case:

Corollary 7.5 *If K is a field and $f(X) \in K[X]$ be a polyomial then*

$$(f(X)) \lhd K[X] \text{ is a maximal ideal} \Leftrightarrow$$
$$K[X]/(f(X)) \text{ is a field} \Leftrightarrow f(X) \in K[X] \text{ is irreducible}$$

Definition 7.6 A domain R is called **unique factorization domain(=UFD)**, if the following holds:

Every $r \in R \setminus (\{0\} \cup R^\times)$ has a decomposition which is unique up to reordering and multiplication by units

$$r = \pi_1 \cdots \pi_m$$

with irreducible factors $\pi_1, \ldots, \pi_m \in R$.

Remark 7.7 The uniqueness is to be understood as follows: First, the factors can be reordered arbitrarily. Moreover, one easily observes the following: If

$$r = \pi_1 \cdots \pi_m$$

is such a decomposition, one can choose arbitrary units $\varepsilon_1, \ldots, \varepsilon_{m-1} \in R^\times$ and put $\varepsilon_m := (\varepsilon_1 \cdots \varepsilon_{m-1})^{-1} \in R^\times$. Then:

$$r = \pi_1 \cdot \pi_2 \cdots \pi_m = (\varepsilon_1 \pi_1) \cdot (\varepsilon_2 \pi_2) \cdots (\varepsilon_m \pi_m).$$

Since the elements π_j are irreducible, so are $\varepsilon_j \pi_j$. Thus, a different decomposition into irreducible factors has been constructed.

The uniqueness in the definition hence has to be understood in the following sense: If

$$r = \pi_1 \cdots \pi_m = \omega_1 \cdots \omega_n$$

are decompositions into irreducible elements, then $m = n$ and the ω_j arise from the π_j as indicated above—via reordering and/or multiplying with units.

The following related notion is commonly used:

Definition 7.8 Let R be a domain and $x, y \in R$. Then x, y are called **associated to each other**, if there is a unit $\varepsilon \in R^\times$ such that $y = \varepsilon x$. We then write $x \sim y$.

7.2   Properties

And now?
The **uniqueness** of the decomposition into irreducible factors (except for unavoidable changes by reordering or by units) is essential for the theory. We can observe this also in most applications in $\mathbb{Z}$ already: It is not enough that every number a has a prime factor decomposition $a = \pm p_1 \ldots p_m$, in the usual application we almost always argue with uniqueness. For example, considering fractions $\frac{a}{b} \in \mathbb{Q}$ one often asks for the fraction to be of reduced form, that is, with co-prime $a, b \in \mathbb{Z}$ In doing so, one uses the **uniqueness** of the prime factor decomposition: Decomposing a and b into prime factors we ask that these are mutually disjoint—otherwise we can eliminate common prime factors. But if the decomposition were not unique, this argument would collapse since then there could be different prime factorizations of a and b, some with common factors some without.

And now?
One might now ask why the term **irreducible** was used in the definition and not the notion of **prime elements**. A possible answer to this question comes from Lemma 7.3:

$$\text{prime element} \implies \text{irreducible},$$

hence **irreducible** is a priori the weaker condition. So, one only needs to find a unique decomposition into irreducibles, which appears easier a priori than a decomposition into prime elements—due to Lemma 7.3 there are presumably more irreducibles than prime elements in general domains.

This is all admittedly a bit nitpicky, the next theorem shows that one could also demand the stronger condition using *prime elements* for the factors, then one doesn't even have to demand uniqueness additionally, it will come for free—roughly speaking one demands stronger prerequisites for the factors and gets the uniqueness as a gift.

The whole discussion even appears really nitpicky a posteriori, since we will see—as a part of the next theorem—that in unique factorization domains the two terms *irreducible* and *prime element* indeed coincide. But this is not the case in more general domains.

The question on the unique factorization property in a domain is highly interesting and important in number theory because the rings of integers $\mathcal{O}_K$ in an algebraic number field are very often (mostly) not UFDs. The same applies to many rings that occur in algebraic geometry.

Theorem 7.9 Let R be a domain.

1. If every $r \in R \setminus (\{0\} \cup R^{\times})$ admits a decomposition into a product

$$r = p_1 \ldots p_m$$

   of prime elements, then R is a UFD.
2. In a UFD R, every irreducible element is a prime element.

Proof We first demonstrate the claim that compares the two versions of decompositions -- one into irreducible and the other into prime elements.
Claim 1 If $\pi_1 \cdots \pi_n \sim p_1 \cdots p_m$ with irreducible π_j and prime elements p_i, then $m = n$ and up to renumbering, $\pi_j \sim p_j$ for all $j = 1, \ldots, m = n$. $\square$

Proof Since p_1 is a prime element and $p_1 | r = \pi_1 \cdots \pi_n$, it follows by definition that there is a j such that $p_1 | \pi_j$. WLOG (relabeling) $p_1 | \pi_1$. Therefore, $\pi_1 = u_1 p_1$ and, since π_1 is irreducible and $p_1 \notin R^{\times}$, we deduce that $u_1 \in R^{\times}$ holds, hence $\pi_1 \sim p_1$. So we have

$$u_1 \pi_1 p_2 \cdots p_m \sim \pi_1 \pi_2 \cdots \pi_n \Leftrightarrow p_2 \cdots p_m \sim \pi_2 \cdots \pi_n$$

and we can successively continue in the same way—note that $\sim$ always means *left side = unit · right side* and the units do not matter in the arguments. $\square$

From this the assertion (1) easily follows, because if $r = p_1 \cdots p_m$ is a decomposition into prime elements, each p_j is also a decomposition into irreducibles. The uniqueness then follows from the claim we just proved. Given any other decomposition $r = \pi_1 \cdots \pi_n$ into irreducible elements, one can apply the claim directly.

To (2): Let $\pi \in R$ be irreducible and $\pi | xy$, so $\pi a = xy$ for an $a \in R$. Since R is a UFD, all involved elements have a unique decomposition into irreducibles:

$$a = q_1 \cdots q_m, \ x = p_1 \cdots p_n, \ y = \rho_1 \cdots \rho_k.$$

It follows that

$$\pi \cdot q_1 \cdots q_m = p_1 \cdots p_n \cdot \rho_1 \cdots \rho_k.$$

Due to the uniqueness of the decompositions, $\pi \sim p_j$ must hold for some j or $\pi \sim \rho_i$ holds for an i. In the first case, we obtain $\pi | x$, in the second case $\pi | y$. $\square$

Theorem 7.10 Every principal ideal domain is a UFD.

> **Check it out!**
> How do you prove that every number $n \in \mathbb{N}$ has a unique prime factorization? This is exactly how we will prove the theorem.

Proof Let $a \in R \setminus (R^\times \cup \{0\})$. We show that a has an irreducible divisor. Suppose this is not the case: Then a is not irreducible (otherwise it would be its own irreducible divisor), so there are $x, y \in R \setminus R^\times$ with $a = xy$. Both factors are not irreducible, because otherwise this would be an irreducible divisor of a. Let's write $a_1 := x$. Then (because $y \notin R^\times$)

$$(a) \subsetneq (a_1)$$

Continuing successively in this way (a_1 has no irreducible divisor, so it is not irreducible, thus it decomposes as a product, ...) one obtains an infinite ascending sequence.

$$(a) \subsetneq (a_1) \subsetneq (a_2) \subsetneq (a_3) \subsetneq \cdots$$

of principal ideals. Then $\mathfrak{a} := \bigcup_{n=1}^{\infty}(a_n) \lhd R$ is an ideal. Since R is a principal ideal domain, $\mathfrak{a}$ is therefore of the form $\mathfrak{a} = (\alpha)$ for an $\alpha \in R$. Since $\alpha \in \bigcup_{n=1}^{\infty}(a_n)$, there is an $N \in \mathbb{N}$ with $\alpha \in (a_N)$, but then

$$(\alpha) \subset (a_N) \subsetneq (a_{N+1}) \subset (\alpha),$$

is a contradiction.

Thus, a also has a prime divisor, because if $q \in R$ is irreducible, then (q) is a maximal ideal (because R is a principal ideal ring), so (q) is also a prime ideal and thus q is a prime element.

Then we now have $a = q_1 \cdot b_1$ with a prime element $q_1 \in R$ and $b_1 \in R$. If $b \in R^\times$ we are done. If not, we can continue with b_1 in the same way: There is a prime divisor $q_2 | b_1$, so that $a = q_1 q_2 b_2$ with a $b_2 \in R$:

$$(b_1) \subsetneq (b_2)$$

Continuing successively, one obtains

$$a = q_1 q_2 \cdots q_r b_r \tag{7.1}$$

with prime elements q_j and a b_r, such that

$$(b_1) \subsetneq (b_2) \subsetneq (b_3) \subsetneq \ldots \subsetneq (b_r)$$

holds—as long as $b_1, b_2, \ldots, b_r \notin R^\times$. As above, there cannot be an infinite sequence like this, so at some point $b_r \in R^\times$ must hold and thus (7.1) is a prime factor decomposition. With Theorem 7.9 the assertion follows. $\square$

Theorem 7.11 If R is a UFD, then so is $R[X]$.

Proof here without proof, see the exercises. $\square$

Corollary 7.12 If K is a field, then $K[X_1, \ldots, X_n]$ is a UDF.

A small remark that we will need in the next chapter.

Lemma 7.13 *If R is a UFD and $x, y \in R \setminus \{0\}$, there exists a greatest common divisor, cp. Defining Lemma 6.3.*

Proof If one of the two elements is a unit, then $\gcd(x, y) = 1$. Note that the greatest common divisor is only well-defined up to units, so by writing $\gcd(x, y) = 1$, we state that $\gcd(x, y) \in R^\times$.

If both elements are not units, they have unique decompositions into irreducibles

$$x = p_1^{v_1} \cdots p_m^{v_m} \text{ and } y = p_1^{\mu_1} \cdots p_m^{\mu_m}$$

with $v_j, \mu_j \in \mathbb{N}_0$ (since we allow 0 as a power, we can take the same irreducible factors $p_1, \ldots, p_m$ for both). Then

$$d := p_1^{\min\{v_1, \mu_1\}} \cdots p_1^{\min\{v_m, \mu_m\}} \in R$$

is a greatest common divisor of x, y —which can be easily proven. $\square$

Red Thread to the Previous Chapter
In this chapter we have seen:

- how the concept of prime number can be generalized for arbitrary domains and that two natural versions arise: prime elements or irreducible elements,
- that in a principal ideal domain R an element $q \in R \setminus (R^\times \cup \{0\})$ is irreducible if and only if $(q) \lhd R$ is a maximal ideal,
- that in particular for a field K and a $f(X) \in K[X]$ with $\deg(f) \geq 1$, $K[X]/(f(X))$ is a field if and only if $f(X) \in K[X]$ is irreducible,
- how to define a **unique factorization domain (UFD)** and that
- principal ideal domains are UFDs,
- if R is a UFD then so is $R[X]$,
- that in particular for a field K, the polynomial ring $K[X_1, \ldots, X_n]$ is a UFD,
- that in a UFD , given two elements $x, y \in R \setminus \{0\}$, there is a $\gcd(x, y)$.

Exercises:

7.1 Let R be a UFD. Show: Given $a, b \in R$ such that $a^3 | b^2$, then already $a | b$.

7.2 Two "similar" rings with different qualities:

We consider the ring $\mathbb{Z}[\omega] \subset \mathbb{C}$ with $\omega = \frac{1+\sqrt{-d}}{2}$. For $d \equiv 3 \bmod 4$ this is the ring of integers in $\mathbb{Q}(\sqrt{d})$, see definition 5.43.

(a) The case $d = 7$. Show that $\mathbb{Z}[\omega]$ together with the norm mapping $N : a + b\omega \mapsto |a + b\omega|^2$ is a Euclidean ring, and in particular, therefore, is a principal ideal domain and thus also a UFD—see also Exercise 6.2.

(b) The case $d = 15$. Show that ω is an irreducible element but not a prime element. Conclude that $\mathbb{Z}[\omega]$ is not a UFD.

Hint: Use the multiplicative norm map as in (6.2).

7.3 (The proof of Theorem 7.11: with step-by-step instructions): For this, one uses the quotient field of R and Gauss's lemma, which will be discussed in the following chapters. Therefore, the proof will be postponed and put as an exercise later: Exercise 9.2.

7.4 Let K be a field. Show that the ring

$$R := K[X, Y]/(Y^2 - X^3)$$

is not a UFD.

7.5 Give an integral domain R which contains elements $a, b \in R$ with $a^3 | b^2$ but $a \nmid b$.

Quotient Fields for Domains

8

Defining Lemma 8.1 Let R be a domain. We define the following equivalence relation on the set $R \times (R \setminus \{0\})$

$$(a, b) \sim (c, d) :\Longleftrightarrow ad = bc.$$

This is indeed an equivalence relation and we write

$$\frac{a}{b} := [(a, b)]$$

for the class of $(a, b) \in R \times (R \setminus \{0\})$ and $\mathrm{Quot}(R) := \big(R \times (R \setminus \{0\})\big)/\sim$ for the quotient according to this equivalence relation.

Proof The only assertion we have to prove are the axioms of an equivalence relation: Obviously, $\frac{a}{b} = \frac{a}{b}$ and symmetry is also clear. Now, if

$$\frac{a}{b} = \frac{c}{d} \Longleftrightarrow ad = bc \text{ and}$$
$$\frac{c}{d} = \frac{r}{s} \Longleftrightarrow cs = rd,$$

we can use the following trick:

$$(as - br) \cdot d = asd - brd = asd - bcs = (ad - bc) \cdot s = 0,$$

and since $s \neq 0$ and R has no zero divisors, we obtain $as = br$, so $\frac{a}{b} = \frac{r}{s}$. $\square$

Theorem 8.2 Together with the operations (the usual calculus with fractions)

$$\frac{a}{b} + \frac{c}{d} := \frac{ad + bc}{bd} \text{ and } \frac{a}{b} \cdot \frac{c}{d} := \frac{ac}{bd},$$

as well as $0 := \frac{0}{1}$ and $1 := \frac{1}{1}$, the ring $\mathrm{Quot}(R)$ is a field, the **quotient field of R**.

© The Author(s), under exclusive license to Springer-Verlag GmbH, DE, part of
Springer Nature 2024
M. Hien, *Abstract Algebra*, Mathematics Study Resources 7,
https://doi.org/10.1007/978-3-662-67974-6_8

Note that we have an injective ring homomorphism

$$R \hookrightarrow \mathrm{Quot}(R), \ r \mapsto \frac{r}{1}$$

and we will simply write $r := \frac{r}{1} \in \mathrm{Quot}(R)$ for the elements in the image.

Proof The well-definedness of the operations is a straightforward computation. But this is just the usual calculus of fractions known from high-school. All other claims follow trivially. $\square$

Red Thread to the Previous Chapter
In this chapter, we have essentially just repeated the calculus of fractions. Every domain R gives rise to its quotient field $\mathrm{Quot}(R)$, in which R is again embedded as a subring.

Exercises:
8.1 Let R be a domain. A subset $S \subset R$ is called a **multiplicative subset**, if $1 \in S$ and if $a, b \in S \Rightarrow ab \in S$ holds. We define the **localization of R with respect to S**, written as $S^{-1}R$ as the set of equivalence classes of pairs $(r, s) \in R \times S$, where

$$(r_1, s_1) \sim (r_2, s_2) :\Longleftrightarrow r_1 s_2 = r_2 s_1. \tag{8.1}$$

Show:

(a) In a natural way, $S^{-1}R$ is again a commutative ring with one.
(b) There is a canonical, injective ring homomorphism $S^{-1}R \hookrightarrow \mathrm{Quot}(R)$.

8.2 Let R be a commutative ring with one and $S \subset R$ a multiplicative subset as in Exercise 8.1. We now define the following variant of the relation (8.1):

$$(r_1, s_1) \sim (r_2, s_2) :\Longleftrightarrow \text{ there exists a } t \in S \text{ with } t(r_1 s_2 - r_2 s_1) = 0. \tag{8.2}$$

Show that this defines an equivalence relation on $R \times S$ and that $S^{-1}R$ formed as above is a commutative ring with unity. What goes wrong if R has zero divisors and one wanted to use the relation (8.1) instead of (8.2)?

8.3 Let K be a field. We consider two versions of the polynomial ring over K, namely $K[t]$ in the variable t and $K[u]$ in the variable u. Let $K(t) := \mathrm{Quot}(k[t])$ and $K(u) := \mathrm{Quot}(k[u])$ be the respective quotient fields. Let $d \in \mathbb{N}$. Show that

$$K(t) \hookrightarrow K(u), \ \frac{f(t)}{g(t)} \mapsto \frac{f(u^d)}{g(u^d)}$$

is a well-defined field homomorphism with respect to which $K(u)|K(t)$ is a field extension of degree d.

Irreducible Polynomials in UFDs

9

9.1 Content of Polynomials

We now want to consider the following situation:

- Z is a UFD and
- $Q := \mathrm{Quot}(Z)$ its field of quotients.

Examples are:

- $Z = \mathbb{Z}$ and $Q = \mathbb{Q}$ (and one can always also think of this example in the following—this also explains the letters we chose for the objects),
- $Z = K[X]$ and $Q = K(X) := \mathrm{Quot}(K[X])$ for a field K.

> **And now?**
> As we have seen in Chap. 2 , given a field extension $L|K$ and an algebraic $\alpha \in L$ over K , it is important to determine the *minimal polynomial* $p(X) \in K[X]$ of α over K, since then we have a better understanding of the extension $K(\alpha)|K$. We will examine this even more closely in the next chapter. After all, we already know that, for example,
>
> $$[K(\alpha) : K] = \deg(p)$$
>
> holds. Obviously, we have the following observation:

Lemma 9.1 *Let $L|K$ be a field extension and $\alpha \in L$ be algebraic over K . Then, for a monic polynomial $f(X) \in K[X]$ with $\deg(f) \geq 1$ we have:*

M. Hien, *Abstract Algebra*, Mathematics Study Resources 7,
https://doi.org/10.1007/978-3-662-67974-6_9

f is the minimal polynomial of α over K $\Longleftrightarrow$

$$f(\alpha) = 0 \text{ and } f(X) \in K[X] \text{ is irreducible.}$$

Proof Let $p(X) \in K[X]$ be the minimal polynomial of α over K. Then $p(X)$ is irreducible, because otherwise it would be a product $p(X) = a(X)b(X)$ with $a, b \in K[X]$ and $\deg(a) < \deg(p)$ and $\deg(b) < \deg(p)$. But then $a(\alpha) = 0$ or $b(\alpha) = 0$, in contradiction to the minimality of p.

If $f(X)$ is a polynomial with $f(\alpha) = 0$, then $f(X)$ is a multiple of $p(X)$—cp. Defining Lemma 2.16. Thus

$$f(X) = p(X)g(X)$$

with a $g(X) \in K[X]$. If f is therefore irreducible in $K[X]$, we obtain $g(X) \in (K[X])^{\times} = K^{\times}$ and since both f and p are monic, $g(X) = 1$. $\square$

And now?

In general, however, it is usually very difficult to determine the minimal polynomial. Even if one has found a polynomial $f(X) \in K[X]$ that satisfies $f(\alpha) = 0$ and one suspects that it is the minimal polynomial, one still has to prove the irreducibility of f. This is not easy.

The aim of this chapter is to provide techniques in the more specific situation as stated at the beginning of the section.

If $f(X) \in Q[X]$ is of the form

$$f(X) = X^n + q_{n-1}X^{n-1} + \ldots + q_0 \in Q[X],$$

then all coefficients $q_j = \frac{a_j}{b_j} \in Q$ are fractions and after multiplication with a common denominator (e.g., with $\prod_{j=0}^{n-1} b_j$), one can use the following polynomial instead of $f(X)$:

$$\tilde{f}(X) := \prod_{j=0}^{n-1} b_j \cdot f(X) \in Z[X]$$

which still satisfies $\tilde{f}(\alpha) = 0$. So we start by first examining polynomials in $Z[X]$.

Definition 9.2 Let $f(X) = a_nX^n + \ldots + a_0 \in Z[X]$. Then we define the **content of** f as

$$\text{cont}(f) := \gcd(a_i \mid i = 0, \ldots, n, \ a_i \neq 0) \in Z$$

The polynomial is called **primitive,** if $\text{cont}(f) = 1$ holds.

Note 9.3

1. So far, we have defined the gcd of two elements. The definition of gcd for a finite number of elements is completely analogous. In particular, such a gcd always exists since Z is factorial.
2. The gcd is uniquely determined up to units only, and therefore the same holds for the content of f. In particular, the statement $\text{cont}(f) = 1$ more precisely means that $\text{cont}(f) \in Z^\times$.
3. If a coefficient $a_j \in Z^\times$ is a unit, then f is primitive. In particular, every monic f is primitive.

And now?

What is the purpose for us if we consider $f(X) \in Z[X]$, but are actually interested in the question of whether $f(X) \in Q[X]$ is irreducible? Regarding this question, we will observe two aspects that become interesting/important:

1. In $Z[X]$ we have more techniques available than we have in $Q[X]$—for example, the reduction modulo primes will enter here, see the subsection below.
2. For this to be useful, we need to find a relationship for $f(X) \in Z[X]$

$$\left(f(X) \text{ irreducible in } Z[X]\right) \quad \text{compared to} \quad \left(f(X) \text{ irreducible in } Q[X]\right).$$

These are the next goals in this chapter.

9.2   Reduction Modulo Prime Elements

Let $p \in Z$ be a prime element. Let us consider the principal ideal $(p) = pZ$ and the following definition:

Definition 9.4 For a prime element $p \in Z$ we call the ring homomorphism

$$\pi : Z[X] \to Z/pZ[X]$$

$$f(X) = a_n X^n + \ldots + a_0 \mapsto \overline{f}(X) := \overline{a_n} X^n + \ldots + \overline{a_0},$$

where we write $\overline{a} := [a] = a + pZ \in Z/pZ$ for the residue class of a, the **reduction modulo** p.

Examples Let's take $Z = \mathbb{Z}$ and $p = 3$. Then

$$\pi : \mathbb{Z}[X] \to \mathbb{Z}/3\mathbb{Z}[X], \, a_n X^n + \ldots + a_0 \mapsto \overline{a_n} X^n + \ldots + \overline{a_0},$$

is the reduction modulo 3 and for example

$$\pi(X^5 + 4X^4 - 3X^2 + 5X - 2) = X^5 + X^4 + 2X + 1 \in \mathbb{Z}/3\mathbb{Z}[X],$$

where we did not write $2 \in \mathbb{Z}/3\mathbb{Z}$ as $\overline{2}$.

From this example, we can now also see the benefit: The ring Z/pZ is generally often much easier to understand than Z itself. In the example, we obtain

- a field $\mathbb{Z}/3\mathbb{Z} = \mathbb{F}_3$ (and this is correct for every prime number p: $\mathbb{Z}/p\mathbb{Z} = \mathbb{F}_p$ is then a field) and
- this field even has only a finite number of elements!

If you want to test a polynomial in $\mathbb{F}_p[X]$ for irreducibility, you can simply do this by trying out the finite number of possibilities how the polynomial could possibly decompose into a product.

Example The polynomial $X^3 + 2X + 1 \in \mathbb{F}_3[X]$ is irreducible, because if it were reducible, it would have to have a factor of degree 1 for reasons of degree, but this would have a root (because $aX + b$ with $a \neq 0$ in the field has the zero $-ba^{-1}$). But $X^3 + 2X^2 + 1$ has no zero in $\mathbb{F}_3$, as one can see by inserting the three(!) elements into $\mathbb{F}_3$ immediately:

$$0^3 + 2 \cdot 0 + 1 = 1 \neq 0, \ 1^3 + 2 \cdot 1 + 1 = 4 = 1 \neq 0, \ 2^3 + 2 \cdot 2 + 1 = 13 = 1 \neq 0.$$

Note 9.5 If $f(X) \in Z[X]$ is a non-primitive polynomial and $\mathrm{cont}(f) = d$ has the prime element $p \in Z$ as a divisor, then it follows that $p|a_j$ for all coefficients a_j of Z and thus

$$\bar{f}(X) = 0 \in Z/pZ[X].$$

If conversely $\bar{f}(X) = 0 \in Z/pZ[X]$, then p obviously divides all coefficients of f and thus we have $p|\mathrm{cont}(f)$ -- in particular f is non-primitive.

9.3   The Gauss Lemma

And now?
When you briefly think about irreducibility in $Z[X]$, it is immediately clear that the concept of content plays a role. If $d \in Z \setminus Z^\times$ is a common divisor of all coefficients $a_j = db_j$ of $f(X) \in Z[X]$, then one has the product decomposition

$$f(X) = a_n X^n + \ldots + a_0 = d(b_n X^n + \ldots + b_0) \tag{9.1}$$

If $n = \deg(f) \geq 1$, then neither of the two factors is a unit (we had assumed this for $d \in Z$).

This is the difference to the field case, where a polynomial $g(X) \in Q[X]$ is a unit exactly when the degree $\deg(g) = 0$ vanishes. In $\mathbb{Z}[X]$ the constant degree-zero polynomial $2 \in \mathbb{Z}[X]$ is **not** a unit.

Back to (9.1): A common divisor $d \in Z \setminus Z^{\times}$ of the coefficients of $f(X) \in Z[X]$ thus leads to the conclusion that $f(X)$ in $Z[X]$ is not irreducible, even though it could be irreducible in $Q[X]$. We can control this phenomenon with the content of f. If f is primitive, we cannot factor out such a simple factor as in (9.1).

Interesting and very useful are the following considerations, which in principle show that the phenomenon related to the content we just discussed is the only issue that can happen for the relation of the two notions of irreducibility.

Lemma 9.6 (by Gauss) *If $f, g \in Z[X]$ are polynomials, we have:*

$$\mathrm{cont}(f \cdot g) = \mathrm{cont}(f) \cdot \mathrm{cont}(g)$$

In particular: If f, g are primitive, then so is $f \cdot g$.

(Reminder: The content is only unique up to units. One could/should perhaps better write $\mathrm{cont}(fg) \sim \mathrm{cont}(f) \cdot \mathrm{cont}(g)$ where $\sim$ stands for being associated to each other).

And now?
Even though the formula in the Gauss lemma seems so inconspicuous, it contains the key argument for the entire chapter. And the formula is anything but trivial: If we write

$$f(X) = a_n X^n + \ldots + a_0 \text{ and } g(X) = b_m X^m + \ldots + b_0$$

with $d = \mathrm{cont}(f) = \gcd(a_i \mid a_i \neq 0)$ and $e = \mathrm{cont}(g) = \gcd(b_i \mid b_i \neq 0)$, then fg has the form

$$fg(X) = \sum_{\ell=0}^{n+m} \left(\sum_{i+j=\ell} a_i b_j \right) X^{\ell}$$

and it is not clear why the greatest common divisor of these coefficients is exactly de.

Proof (Gauss's Lemma.) We first show that the claim starting with *In particular* holds: Let $f, g \in Z[X]$ be primitive. Suppose fg is not primitive. Then $d := \mathrm{cont}(fg)$ is not a unit and thus has a prime divisor $p \mid \mathrm{cont}(fg)$. Considering reduction modulo p (see Remark 9.5), we obtain $\overline{fg}(X) = 0 \in Z/pZ[X]$. Since the reduction is a ring homomorphism, we deduce

$$\bar{f} \cdot \bar{g} = 0 \in Z/pZ[X].$$

But $Z/pZ[X]$ is a domain (since Z/pZ is a domain, because pZ is a prime ideal), so $\bar{f} = 0$ or $\bar{g} = 0$ must hold. But this means that f or g is non-primitive (Remark 9.5), a contradiction to the assumption.

This also shows the general assertion. Let $d := \mathrm{cont}(f)$ and $e := \mathrm{cont}(g)$. Then consider the elements

$$\varphi := \frac{1}{d}f(X), \gamma := \frac{1}{e}g(X) \in Q[X].$$

Their coefficients are again in Z (because, for example, d divides all coefficients of f), so $\varphi, \gamma \in Z[X]$. Furthermore, φ, γ are primitive and due to what has already been shown, their product $\varphi \cdot \gamma \in Z[X]$ is primitive as well. But

$$f \cdot g = (d \cdot \varphi) \cdot (e \cdot \gamma) = de \cdot (\varphi\gamma)$$

and thus it is clear that $\mathrm{cont}(fg) = de$. $\square$

Check it out!
Given two polynomials $f, g \in Z[X]$, which of the two following implications is easy to see?

$$g \mid f \ \text{in} \ Q[X] \left\{ \begin{array}{c} \overset{?}{\Rightarrow} \\ \overset{?}{\Leftarrow} \end{array} \right\} g \mid f \ \text{in} \ Z[X]?$$

As an application of Gauss's lemma, we obtain:

Theorem 9.7 (Gauss's Theorem) Let Z be a UFD and $Q := \mathrm{Quot}(Z)$ its field of quotients. Given two polynomials $f, g \in Z[X]$, the following holds:

$$\left. \begin{array}{l} g \ \text{is primitive and} \\ g \mid f \ \text{in} \ Q[X] \end{array} \right\} \Longrightarrow g \mid f \ \text{in} \ Z[X]$$

Proof By assumption, there is a polynomial $q(X) \in Q[X]$ with $f = gq$ in $Q[X]$. The coefficients of $q(X)$ are in Q and when multiplied by the common denominator (the lcm of the coefficients—similar to the gcd, this can be determined from the prime factorizations), it becomes a polynomial in $Z[X]$, i.e., there exists an $b \in Z \setminus \{0\}$, such that

$$b \cdot q(X) \in Z[X].$$

Thus, the content of bq exists in Z and from the Gauss Lemma we deduce:

$$bf(X) = g(X) \cdot bq(X) \Rightarrow \mathrm{cont}(bf(X)) = \mathrm{cont}(g(X)) \cdot \mathrm{cont}(bq(X)). \quad (9.2)$$

According to the assumption, $\text{cont}(g(X)) = 1$. Since $f(X) \in Z[X]$ has coefficients in Z, one can also consider the content $\text{cont}(f(X)) \in Z$ and obviously we have

$$\text{cont}(bf(X)) = b \cdot \text{cont}(f(X)).$$

Therefore, it follows from (9.2) that

$$\text{cont}(bq(X)) = b \cdot \text{cont}(f(X)) \in Z$$

and hence in particular, that

$$b \mid \text{cont}(bq(X)) \tag{9.3}$$

holds. If we write $bq(X) = \alpha_m X^m + \ldots + \alpha_0$ with $\alpha_j \in Z$, then it follows that $b \mid \alpha_j$ holds for all j. Therefore, there exist $\beta_j \in Z$, such that $\alpha_j = b\beta_j$ and thus we have

$$bq(X) = b\beta_m X^m + \ldots + b\beta_0 \implies q(X) = \beta_m X^m + \ldots + \beta_0 \in Z[X],$$

and this is what we wanted to show. $\square$

Actually, we have shown a stronger version of the theorem:

Theorem 9.8 (Statement reformulated) Let Z be a factorial ring and $Q := \text{Quot}(Z)$ its field of quotients. Let $f, g \in Z[X]$ be two polynomials such that g is primitive. If $f(X) = g(X) \cdot q(X)$ with a $q(X) \in Q[X]$, it follows that $q(X) \in Z[X]$.

As a consequence of Gauss's theorem, we obtain:

Theorem 9.9 (also often called Gauss's Theorem) Let Z be a factorial ring and $Q := \text{Quot}(Z)$ its field of quotients. Then for a **primitive** polynomial $f \in Z[X]$:

$$f \text{ is irreducible in } Q[X] \iff f \text{ is irreducible in } Z[X].$$

Proof We show both directions.

"$\Rightarrow$":   So let f be irreducible in $Q[X]$ and consider a decomposition

$$f = gh \text{ with } g, h \in Z[X].$$

It must be shown that g or h is a unit in $Z[X]$. By assumption, one of the factors is a unit in $Q[X]$, let's say w.l.o.g. $g(X) \in (Q[X])^\times = Q^\times$, but then also $g(X) \in Z[X]$, so $g(X) = a \in Z$ is a constant polynomial with $a \neq 0$. This implies that

$$a \mid \text{cont}(gh) = \text{cont}(f) = 1$$

since by assumption f is primitive. Therefore, $a \in Z^\times$ must be a unit.

"$\Leftarrow$":   Let's assume that f is irreducible in $Z[X]$. If $f = gh$ with $g, h \in Q[X]$ holds, then we have again, after multiplication with the respective common denominators of the coefficients (the latter in maximally reduced form—we have a unique prime factor decomposition in Z):

$$b \cdot g(X) \in Z[X] \text{ with cont } (b \cdot g(X)) = 1,$$
$$c \cdot h(X) \in Z[X] \text{ with cont } (c \cdot h(X)) = 1.$$

We deduce the following equality in $Q[X]$:

$$f(X) = \underbrace{(b \cdot g(X))}_{\in Z[X] \text{ and primitive}} \cdot \underbrace{\left(\frac{1}{b} \cdot h(X) \right)}_{\in Q[X]} \in Q[X].$$

According to theorem 9.8, it follows that $\frac{1}{b} \cdot h(X) \in Z[X]$ holds and thus the decomposition happens in fact already in $Z[X]$:

$$f(X) = \underbrace{(b \cdot g(X))}_{\in Z[X] \text{ and primitive}} \cdot \underbrace{\left(\frac{1}{b} \cdot h(X) \right)}_{\in Z[X]} \in Z[X].$$

By assumption, $f(X)$ is irreducible in $Z[X]$, so one of the factors lies in $(Z[X])^{\times} = Z^{\times}$ and thus is of degree 0 and hence also a unit in $Q[X]$. $\square$

And now?

Why is Theorem 9.9 so useful? In most applications:

1. We have a given $f(X) \in Q[X]$ and want to understand if f is irreducible in $Q[X]$ or not.
2. If moreover, let's say coincidentally, $f(X) \in Z[X]$ and is monic, then it is also primitive.
3. Theorem 9.9 states that we now need to investigate whether

f is irreducible in $Z[X]$

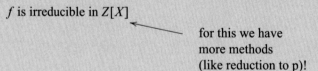

for this we have
more methods
(like reduction to p)!

9.4   Application of the Reduction Mod p

Now we consider the additional methods that we have in $Z[X]$ compared to $Q[X]$. As above, Z is a UFD.

Theorem 9.10 (Reduction Criterion) Let Z be a UFD and $p \in Z$ a prime element. Furthermore, let $f(X) \in Z[X]$ be monic and let us assume that the reduction $\bar{f}(X) \in Z/pZ[X]$ is irreducible. Then $f(X) \in Z[X]$ is also irreducible.

Proof Assume $f(X) = g(X)h(X) \in Z[X]$ is a factorization $g(X), h(X) \notin (Z[X])^{\times}$. Then, $\deg(g), \deg(h) \geq 1$, because if, for example, $\deg(g) = 0$, then $g(X) = a \in Z$ is a constant polynomial and $a \mid \mathrm{cont}(f) = 1$, so $a \in Z^{\times} = (Z[X])^{\times}$.

Let's write $g(X) = b_r X^r + \ldots + b_0$ and $h(X) = c_s X^s + \ldots + c_0$ with $b_r \neq 0$ and $c_s \neq 0$. Then $b_r \cdot c_s = 1$ (because f is monic), thus $b_r, c_s \in Z^{\times}$.

Then, however, the leading coefficients of

$$\bar{g}(X) = \bar{b}_r X^r + \bar{b}_{r-1} X^{r-1} + \ldots + \bar{b}_0,$$
$$\bar{h}(X) = \bar{c}_s X^s + \bar{c}_{s-1} X^{s-1} + \ldots + \bar{c}_0 \in Z/pZ[X]$$

are units and the polynomials are of degree ≥ 1, so $\bar{f}(X) = \bar{g}(X) \cdot \bar{h}(X)$ is reducible, contradicting the assumption. $\square$

Remark 9.11 Note that this is a criterion for *irreducibility*. An analogous statement for *reducibility* does not hold.

If one knows that $\bar{f}(X) = \bar{g}(X) \cdot \bar{h}(X)$ decomposes with $\deg(\bar{g}), \deg(\bar{h}) \geq 1$, one cannot **deduce** that f must decompose as well. Example: The monic polynomial

$$f(X) := X^3 + 3X + 2 \in \mathbb{Z}[X]$$

decomposes modulo 3 into

$$\bar{f}(X) = X^3 + 2 = (X - 1)(X^2 + X + 1) \in \mathbb{Z}/3\mathbb{Z}[X],$$

but $f(X)$ itself is irreducible in $\mathbb{Z}[X]$, see the following lemma.

Let's introduce two more lemmas that sometimes help with the question of irreducibility.

Lemma 9.12 (Zero Criterion) *Let R be a domain and $f(X) \in R[X]$ be a monic polynomial of degree $2 \leq \deg(f) \leq 3$. Then the following holds:*

$$f \text{ is irreducible in } R[X] \iff f \text{ has no zero in } R.$$

Proof Both directions individually:

"$\Rightarrow$:"   If f has a zero $\alpha \in R$, one can factor out the linear term $X - \alpha \in R[X]$, thus

$$f(X) = (X - \alpha) \cdot g(X)$$

with a polynomial $g(X) \in R[X]$ of degree $\deg(g) = \deg(f) - 1$. Therefore, f is reducible in $R[X]$.

"$\Leftarrow$:"   If $f(X)$ were reducible in $R[X]$, then there would be a decomposition

$$f(X) = g(X)h(X) \in R[X]$$

into two non-units $g, h \in R[X]$. As in the proof of Theorem 9.10, the leading coefficients of g and h are units in R. In particular, $\deg(g), \deg(h) \geq 1$ must hold (otherwise it would be a unit). For reasons of degree, $\deg(g) = 1$ and $\deg(h) \in \{1, 2\}$ or vice versa. Let's assume the former holds. Then $g(X)$ has the form $g(X) = aX + b$ with a unit $a \in R^{\times}$. This factor then has the root $-b \cdot a^{-1} \in R$, which is also a root of f.

$\square$

Lemma 9.13 *Let Z be a UFD and $Q := \mathrm{Quot}(Z)$ its quotient field . Let $f(X) = a_n X^n + \ldots + a_0 \in Z[X]$ be a polynomial of degree n, which has a root $\alpha = \frac{a}{b} \in Q$ with coprime $a, b \in Z$. Then the following holds:*

$$b \mid a_n \text{ and } a \mid a_0.$$

In particular, if $f(X)$ is monic, then every root $\alpha \in Q$ of $f(X)$ in Q already lies in Z and satisfies $\alpha \mid a_0$.

Proof By multiplying with b^n one obtains:

$$0 = f(\alpha) = f\left(\frac{a}{b}\right) = a_n \frac{a^n}{b^n} + a_{n-1} \frac{a^{n-1}}{b^{n-1}} + \ldots + a_0 \implies$$
$$0 = a_n a^n + a_{n-1} b a^{n-1} + \ldots + a_1 a b^{n-1} + a_0 b^n \implies$$
$$-a_n a^n = b \cdot \left(a_{n-1} a^{n-1} + \ldots + a_1 a b^{n-2} + a_0 b^{n-1}\right)$$

Since a and b are coprime, it follows that $b \mid a_n$ (every prime power $p^k \mid b$ must also satisfy $p^k \mid a_n$ because $p \nmid a$).

If $a_n = b \cdot r$ for an $r \in R$, then it further follows that

$$-b r a^n = b \cdot (a^{n-1} a^{n-1} + \ldots + a_1 a b^{n-2} + a_0 b^{n-1}) \overset{Z \text{ a domain}}{\implies}$$
$$-r a^n = a_{n-1} a^{n-1} + \ldots + a_1 a b^{n-2} + a_0 b^{n-1} \implies$$
$$a_0 b^{n-1} = -r a^n - a_{n-1} a^{n-1} - \ldots - a_1 a b^{n-2} = a \cdot (-r a^{n-1} - a_{n-1} a^{n-2} - \ldots - a_1 b^{n-2})$$

and again, since a and b are coprime, we deduce $a \mid a_0$. $\square$

And now?

The lemma above is often useful to determine whether a polynomial has a root. Example: Determine all $a \in \mathbb{Z}$ such that the polynomial

$$f(X) = X^4 + aX + 3 \in \mathbb{Z}[X]$$

has a root in $\mathbb{Q}$.

According to the lemma, for any root $\alpha = \frac{r}{s} \in \mathbb{Q}$ (in reduced form), we know that $s \mid 1$ (and thus the roots must be integers) and $r \mid 3$, so the only possibilities are:$\alpha \in \{1, -1, 3, -3\}$. Inserting these four candidates into the polynomial gives:

$$0 = f(1) = a + 4 \Rightarrow a = -4 \qquad 0 = f(-1) = -a + 4 \Rightarrow a = 4$$
$$0 = f(3) = 3a + 84 \Rightarrow a = -28 \quad 0 = f(-3) = -2a + 84 \Rightarrow a = 28.$$

A very special, but often useful criterion in exercises is the following

Theorem 9.12 (Eisenstein Criterion) Let Z be a UFD. Let $p \in Z$ be a prime element and assume that the coefficients of the polynomial

$$f(X) = a_n X^n + a_{n-1} X^{n-1} + \ldots + a_0 \in Z[X]$$

of degree $n \geq 2$ satisfy the condition[1]:

$$p \nmid a_n, \; p \mid a_{n-1}, \; \ldots, \; p \mid a_0 \text{ and } p^2 \nmid a_0$$

Then $f(X)$ is irreducible in $Q[X]$.

Proof The assertion that $f(X)$ is irreducible in $Q[X]$ remains the same if we multiply $f(X)$ with any unit $q \in Q^\times$, in particular with $q = \frac{1}{\text{cont}(f)}$. According to the assumptions, $p \nmid \text{cont}(f)$ and thus these assumptions remain true even after dividing by $\text{cont}(f)$. So we can assume without loss of generality that $f(X) \in Z[X]$ is primitive. We show that $f(X)$ is then irreducible in $Z[X]$, and by Gauss also in $Q[X]$.

We use reduction modulo p. According to the assumptions, the reduction $\bar{f}(X) \in Z/pZ[X]$ has the form

$$\bar{f}(X) = \bar{a}_n X^n \in Z/pZ[X],$$

with $\bar{a}_n \neq 0 \in Z/pZ$. Now suppose $f(X)$ was reducible in $Z[X]$, hence $f(X) = g(X) \cdot h(X)$ with

$$g(X) := b_r X^r + \ldots + b_0,$$
$$h(X) := c_s X^s + \ldots + c_0 \in Z[X]$$

of degree r or s _respectively_. Since f is primitive (w.l.o.g.), $\deg(g), \deg(h) \geq 1$ must hold (we argue as in the proof of Theorem 9.10). Note that

$$b_0 \cdot c_0 = a_0 \tag{9.4}$$

holds. We deduce that the reduction of f modulo p reads as

$$\bar{f}(X) = \bar{a}_n X^n = \underbrace{(\bar{b}_r X^r + \ldots + \bar{b}_0)}_{=\bar{g}(X)} \cdot \underbrace{(\bar{c}_s X^s + \ldots + \bar{c}_0)}_{=\bar{h}(X)}.$$

Since $0 \neq \bar{a}_n = \bar{b}_r \cdot \bar{c}_s$, both factors are $\neq 0$.

---

[1] Note that $p \mid a_j$ also holds for $a_j = 0$.

Let us define the numbers $k = \min\{j \mid \overline{b}_j \neq 0\} \leq r$ and $\ell = \min\{j \mid \overline{c}_j \neq 0\} \leq s$. Then

$$\overline{g}(X) = \overline{b}_r X^r + \ldots + \overline{b}_k X^k,$$
$$\overline{h}(X) = \overline{c}_s X^s + \ldots + \overline{c}_\ell X^\ell \in Z[X]$$

and thus

$$\overline{a}_n X^n = (\overline{b}_r X^r + \ldots + \overline{b}_k X^k) \cdot (\overline{c}_s X^s + \ldots + \overline{c}_\ell X^\ell) = \overline{b}_r \overline{c}_s X^n + \ldots + \underbrace{\overline{b}_k \overline{c}_\ell}_{\neq 0} X^{k+\ell}$$

from which we deduce that $k = r$ and $s = \ell$. In particular, $\overline{b}_0 = 0 = \overline{c}_0$, so $p \mid b_0$ and $p \mid c_0$, from which

$$p^2 \mid b_0 c_0 = a_0$$

follows, in contradiction to the assumption. $\square$

Corollary 9.15 Let Z be a UFD. Let $f(X) \in Z[X]$ be a **primitive** polynomial that fulfills the conditions of the Eisenstein criterion. Then $f(X)$ is irreducible in $Z[X]$.

Proof We proceed exactly as in the proof above, however, we do not need to divide the coefficients of f with $\mathrm{cont}(f)$. But then the proof above directly verfies the irreducibility of $f(X)$ in $Z[X]$ for a primitive f. $\square$

And now?
Note that one usually has to be creative and find the appropriate prime element p when trying to apply the Eisenstein criterion.

Finally, let us comment on a common trick:

Lemma 9.16 *Let R be any ring, $a \in R$ an element and $f(X) \in R[X]$. Then the following holds:*

$$f(X) \text{ is irreducible in } R[X] \iff f(X + a) \text{ is irreducible in } R[X].$$

Proof The map

$$\varphi : R[X] \to R[X], \ f(X) \mapsto f(X + a)$$

is a ring isomorphism (with inverse mapping $g(X) \mapsto g(X - a)$). $\square$

This lemma is often useful because, for example, it might happen that we cannot find a prime element for which a given polynomial $f(X) \in Z[X]$ fulfills the

Eisenstein criterion, but for instance $f(X + 1)$ could be Eisenstein for some prime element p.

A prominent example is the following:

Lemma 9.17 *If $p \in \mathbb{Z}$ is a prime number, the polynomial*

$$\Phi_p(X) := X^{p-1} + X^{p-2} + \ldots + X + 1 \in \mathbb{Q}[X]$$

is irreducible.

Proof For $p = 2$ it is clear. Therefore let $p \geq 3$ be an odd prime number.

We observe that $\Phi_p(X) \cdot (X - 1) = X^p - 1$. If we consider reduction modulo p, it follows that

$$\overline{\Phi_p(X)} \cdot \overline{(X - 1)} = \overline{X^p - 1}.$$

Consider also the polynomial $(X - 1)^p$. Applying the binomial expansion, we obtain

$$(X - 1)^p = X^p + \sum_{j=1}^{p-1} \binom{p}{j} (-1)^{p-j} X^j - 1.$$

Now, each of the binomial coefficients is divisible by p:

$$p \mid \binom{p}{j} \quad \text{für } j = 1, \ldots, p - 1.$$

Check it out!
Prove this!

We   deduce   that $\overline{(X - 1)^p} = \overline{X^p - 1}$ and   thus $\overline{\Phi_p(X)} = \overline{(X - 1)^{p-1}}$. Consider   the   polynomial $\Phi_p(X + 1) \in \mathbb{Z}[X]$.   Let's   write $\Phi_p(X + 1) =: X^{p-1} + c_{p-2}X^{p-2} + \ldots + c_0$ with $c_j \in \mathbb{Z}$. Then we obtain that

$$\overline{\Phi_p(X + 1)} = \overline{X^{p-1}}$$

and from that we deduce that $p \mid c_j$ for all $j = 0, \ldots, p - 2$. Furthermore, $c_0 = p$ and thus $p^2 \nmid c_0$. The polynomial $\Phi_p(X + 1) \in \mathbb{Z}[X]$ therefore fulfills the Eisenstein criterion and since it is monic, $\Phi_p(X + 1)$ is irreducible in $\mathbb{Q}[X]$. With Lemma 9.16, $\Phi_p(X)$ is also irreducible in $\mathbb{Q}[X]$. $\square$

Red Thread to the Previous Chapter
In this chapter, we have considered a UFD Z and its quotient field Q and we have seen

- that for $f(X) \in Z[X]$ the ring element $\text{cont}(f) = \gcd(\text{coefficients})$ is an important invariant for the polynomial,
- that $\text{cont}(fg) = \text{cont}(f) \cdot \text{cont}(g)$ holds (Gauss's Lemma),
- that a primitive polynomial $f(X) \in Z[X]$ is irreducible in $Z[X]$ if and only if it is irreducible in $Q[X]$ (Gauss's Theorem),
- that one can use the reduction modulo a prime element $\pi : Z[X] \to Z/pZ[X], f \mapsto \bar{f}$,
- that a monic polynomial $f \in Z[X]$, such that $\bar{f} \in Z/pZ[X]$ is irreducible, is itself also irreducible in $Z[X]$,
- that one has a very explicit criterion for irreducibility in $Z[X]$—Eisenstein's Criterion.

Exercises:

9.1 Apply the Eisenstein criterion to prove that

(a) the polynomial $X^3 - 3X + 1 \in \mathbb{Q}[X]$ is irreducible.
(b) the polynomial $Y^2 - X^3 - X \in \mathbb{Q}[X, Y]$ is irreducible. (Hint: It may be helpful to consider $\mathbb{Q}[X, Y] = (\mathbb{Q}[Y])[X]$.)

9.2 (The proof of Theorem 7.11 with step-by-step instructions): Let R be a UFD. Show the following:

(a) If $\pi \in R$ is a prime element in R, then it is also a prime element in $R[X]$.
(b) Considering the quotient field $Q := \text{Quot}(R)$ of R, the following holds: If $f \in R[X]$ with $\deg(f) \geq 1$ is a prime element, then f is irreducible in $Q[X]$ and thus also a prime element in the principal ideal domain $Q[X]$. (Note: This is one of the reasons why this exercise was put at the end of this chapter since one needs some results we have only seen now.)
(c) If $f \in R[X]$ is irreducible, then f is already a prime element in $R[X]$.
(d) Every $f \in R[X] \smallsetminus (R^\times \cup 0)$ has a product decomposition $f = q_1 \cdots q_n$ with irreducible factors $q_i \in R[X]$. (Hint: Induction by the degree of f.)
(e) The polynomial ring $R[X]$ is also a UFD. (Hint: Theorem 7.9.)

9.3 (as preparation for Exercise **9.4**): Let $p \geq 3$ be a prime number.

(a) Show that there are as many squares as non-squares in $\mathbb{Z}/p\mathbb{Z} \smallsetminus \{0\}$.
(b) For $a \in \mathbb{Z}$ we define the **Legendre symbol** $\left(\frac{a}{p}\right)$ by

$$\left(\frac{a}{p}\right) := \begin{cases} 0 & \text{if } p|a, \\ 1 & \text{if } p \nmid a \text{ and the equation } x^2 \equiv a \bmod p \text{ has a solution,} \\ -1 & \text{if } p \nmid a \text{ and the equation } x^2 \equiv a \bmod p \text{ does not have a solution.} \end{cases}$$

Show that for $a, b \in \mathbb{Z}$ the two formulas

$$\left(\frac{a}{p}\right) \cdot \left(\frac{b}{p}\right) = \left(\frac{ab}{p}\right) \quad \text{and} \quad \left(\frac{a^2 b}{p}\right) = \left(\frac{b}{p}\right)$$

hold. In particular, we have

$$\left(\frac{2}{p}\right) \cdot \left(\frac{-2}{p}\right) = \left(\frac{-1}{p}\right) \tag{9.5}$$

9.4 Consider the polynomial $f(X) = X^4 + 1 \in \mathbb{Z}[X]$. Let $p \in \mathbb{N}$ be a prime number. Show the following:

(a) If there is an $a \in \mathbb{Z}$ with $a^2 \equiv -1 \bmod p$, then the reduction of $X^4 + 1$ modulo p is reducible and $X^4 + 1 = (X^2 - a)(X^2 + a) \in \mathbb{Z}/p\mathbb{Z}[X]$.
(b) If there is a $b \in \mathbb{Z}$ with $b^2 \equiv 2 \bmod p$ or a $c \in \mathbb{Z}$ with $c^2 \equiv -2 \bmod p$, then the reduction of $X^4 + 1$ modulo p is reducible and

$$X^4 + 1 = (X^2 + bX + 1)(X^2 - bX + 1) \in \mathbb{Z}/p\mathbb{Z}[X]$$

or

$$X^4 + 1 = (X^2 + cX - 1)(X^2 - cX - 1) \in \mathbb{Z}/p\mathbb{Z}[X].$$

(c) For every prime number p, the reduction of $X^4 + 1$ modulo p is reducible. (Hint: Use (a) and (b) together with (9.5) from Exercise 9.3.)
(d) The polynomial $X^4 + 1 \in \mathbb{Z}[X]$ is irreducible. (Note: This can now be solved directly by an Ansatz for a decomposition into products and comparing coefficients, involving some cumbersome computations. One can also postpone this problem and wait for later chapters where it will follow easily from Theorem 15.18 and the equality (15.3).)

Galois Theory (I)—Theorem A and Its Variant A′

10

10.1 The Miraculous Creation of some Field

Let K be a field and $f(X) \in K[X]$ an irreducible polynomial. We want to examine the equation $f(X) = 0$.

> **And now?**
> Demanding irreducibility is not a significant restriction. Since $K[X]$ is a UFD, every $f(X)$ decomposes into irreducible factors $f(X) = f_1(X) \cdots f_m(X)$, so we can examine the individual equations $f_j(X) = 0$.

In this situation, the question arises whether there always is a (preferably small) field extension $L|K$ such that $f(X) \in K[X] \subset L[X]$ has a root in L.

If one starts with $K = \mathbb{Q}$, we know that $\mathbb{C}$ is an algebraically closed field, so every polynomial $f(X) \in \mathbb{C}[X]$ of degree ≥ 1 has a root. Then one can consider

$$K = \mathbb{Q} \subset L := \mathbb{Q}(\alpha) \subset \mathbb{C}$$

and $L|\mathbb{Q}$ is a finite field extension (thus small) with this property.

If there is no natural candidate for a field Ω with $K \subset \Omega$ in sight, in which a root $\alpha \in \Omega$ of $f(X)$ is known to exist, the question is more delicate of course.

Examples: Let's consider the field $\mathbb{F}_5 = \mathbb{Z}/5\mathbb{Z}$ and the polynomial $f(X) = X^2 - 2 \in K[X]$. Since it has degree 2 and no root in $\mathbb{F}_5$, it is irreducible. So, we want to have a field $L|\mathbb{F}_5$ in which $X^2 - 2$ has a root, i.e. in which an element $\alpha \in L$ exists that we then could call a root of two: $\alpha =: \sqrt{2}$.

© The Author(s), under exclusive license to Springer-Verlag GmbH, DE, part of Springer Nature 2024
M. Hien, *Abstract Algebra*, Mathematics Study Resources 7,
https://doi.org/10.1007/978-3-662-67974-6_10

The next theorem shows that there always is such a field extension, and the construction is quite simple since we have studied the polyomial ring in detail in previous chapters and now understand the polynomial ring $K[X]$ very well:

Theorem 10.1 Let K be a field and $f(X) \in K[X]$ be an irreducible polynomial of degree ≥ 1. Then there exists a finite field extension $L|K$, such that there is a root $\alpha \in L$ of $f(X)$ in L.

Proof Let

$$f(X) = a_n X^n + \ldots + a_0. \tag{10.1}$$

Let's call the indeterminate T instead of X (to avoid confusion later). So we consider $f(T) = a_n T^n + \ldots + a_0 \in K[T]$.

Since $f(T) \in K[T]$ is irreducible and $K[T]$ is a principal ideal ring, the ideal $(f(T)) \lhd K[T]$ is a maximal ideal. Therefore, the factor ring

$$L := K[T]/(f(T))$$

is a field. We have the injective ring homomorphism

$$K \hookrightarrow K[T] \twoheadrightarrow K[T]/(f(T)), \ a \mapsto [a]$$

(mapping the element a to the class of the constant polynomial $a \in K$), so $L|K$ is a field extension.

We define $\alpha := [T] \in L := K[T]/(f(T))$.

We now consider the polynomial ring in one variable (now denoted by X) over K and L:

$$K[X] \hookrightarrow L[X], \ c_r X^r + \ldots + c_0 \mapsto [c_r]X^r + \ldots + [c_0].$$

We want to insert elements $\beta \in L$ into the polynomial $f(X) \in K[X]$ with coefficients as in (10.1). If $\beta = [g(T)] = [b_m T^m + \ldots + b_0] \in L$, then

$$f(\beta) = [a_n]\beta^n + [a_{n-1}]\beta^{n-1} + \ldots + [a_0] =$$
$$[a_n]([b_m T^m + \ldots + b_0])^n + [a_{n-1}]([b_m T^m + \ldots + b_0])^{n-1} + \ldots + [a_0] =$$
$$\left[a_n(b_m T^m + \ldots + b_0)^n + a_{n-1}(b_m T^m + \ldots + b_0)^{n-1} + \ldots + a_0\right].$$

In the latter equality, recall that the operations (addition and multiplication) in $L = K[T]/(f(T))$ are defined by choosing representatives of the classes, adding/multiplying these and then taking the class of the result.

In particular, for $\beta := \alpha = [T]$:

$$f(\alpha) = \left[a_n T^n + a_{n-1} T^{n-1} + \ldots + a_0\right] = [f(T)] = [0] = 0 \in L,$$

hence $\alpha \in L$ is a root of $f(X)$ in L . $\square$

Continuation of the example from above: We proceed as the theorem tells us to, defining

$$L := \mathbb{F}_5[T]/(T^2 - 2)$$

and $\alpha := [T] \in L$.

And now?

The field in this example can also be described differently: As we know, $[L : \mathbb{F}_5] = 2$ with the base $1, [T]$, i.e. we have

$$L = \{a + b[T] \mid a, b \in \mathbb{F}_5\}.$$

Let's consider the $\mathbb{F}_5$-vector space isomorphism (basis isomorphism)

$$\psi : \mathbb{F}_5^2 \xrightarrow{\cong} L, \; e_1 = \begin{pmatrix} 1 \\ 0 \end{pmatrix} \mapsto 1, \; e_2 = \begin{pmatrix} 0 \\ 1 \end{pmatrix} \mapsto [T],$$

then we can transfer the operations on $L' := \mathbb{F}_5^2$ from L to L':

$$v + w := \psi^{-1}(\psi(v) + \psi(w)) \text{ and } v \cdot w := \psi^{-1}(\psi(v) \cdot \psi(w)).$$

We obtain the field $L' = \mathbb{F}_5^2$ (which is isomorphic to L). If one writes out the operations explicitly, one obtains

$$\begin{pmatrix} a \\ b \end{pmatrix} + \begin{pmatrix} c \\ d \end{pmatrix} = \begin{pmatrix} a + c \\ b + d \end{pmatrix} \text{ and } \begin{pmatrix} a \\ b \end{pmatrix} \cdot \begin{pmatrix} c \\ d \end{pmatrix} = \begin{pmatrix} ac + 2bd \\ ad + bc \end{pmatrix} \quad (10.2)$$

The latter holds since

$$\psi(\begin{pmatrix} a \\ b \end{pmatrix}) \cdot \psi(\begin{pmatrix} c \\ d \end{pmatrix}) = (a + b[T])(c + d[T]) =$$

$$ac + (ad + bc)[T] + bd[T]^2 = ac + 2bd + (ad + bc)[T].$$

The element $\alpha = [T] \in L$ corresponds to the vector $\begin{pmatrix} 0 \\ 1 \end{pmatrix}$. Indeed, using (10.2) we obtain

$$\alpha^2 = \begin{pmatrix} 0 \\ 1 \end{pmatrix} \cdot \begin{pmatrix} 0 \\ 1 \end{pmatrix} = \begin{pmatrix} 2 \\ 0 \end{pmatrix}$$

(in slight abuse of notation, the most correct way to write the left hand side would be $\psi(\alpha)^2$, but since ψ is an isomorphism, ...) so $\alpha^2 = 2$ holds and we can call this element a $\sqrt{2} := \alpha \in L$ in a meaningful way. But note: This is just a notation which could be misunderstood and hence is a little bit dangerous: This element $\sqrt{2}$ has nothing to do with the real number $\sqrt{2} \in \mathbb{R}$ -- **really nothing at all!**

Of course, $-\alpha \in L$ is also a root of 2, because $(-\alpha)^2 = \alpha^2 = 2$. We have $-\alpha = -[T] = [4T]$.

Also note here: There are further square roots of certain elements in K in L (and thus in L'). For example:

$$(2[T])^2 = 4[T]^2 = 8 = 3 \in \mathbb{F}_5, \; (3[T])^2 = 9[T]^2 = 18 = 3 \in \mathbb{F}_5,$$

so $2[T] \in L$ (corresponds to $\begin{pmatrix} 0 \\ 2 \end{pmatrix} \in L'$) and $3[T] = -2[T]$ are the two roots of $3 \in \mathbb{F}_5$ in L.

The remaining elements $1, 4 \in \mathbb{F}_5$ already have a root there, because $1^2 = 1$ and $2^2 = 4$. If one wants to continue with the sloppy notation

$\sqrt{2} = [T] \in L$ one can also write $i := 2 \in F_5$, because indeed $2^2 = -1$ holds—but again with caution: This element $i \in \mathbb{F}_5$ has nothing to do with the complex number $i \in \mathbb{C}$ -- **nothing at all!**

If one already has a field $\Omega|K$ with a root $\alpha \in \Omega$ of a given polynomial, one has obtained nothing new (up to isomorphism):

Theorem 10.2 Let $f(X) \in K[X]$ be irreducible and $\Omega|K$ a field extension, such that a root $\alpha \in \Omega$ of $f(X)$ exists. Then we have the intermediate field $K[\alpha] \subset \Omega$ and the morphism given by insertion of the element α:

$$\psi : K[X]/(f(X)) \to K[\alpha], \; [g(X)] \mapsto g(\alpha)$$

is an isomorphism.

Proof Well-definedness follows, because choosing two representatives for a given class $[g(X)] = [h(X)] \in K[X]/(f(X))$ there is a polynomial $a(X) \in K[X]$ such that $h(X) = g(X) + a(X)f(X)$. But then $h(\alpha) = g(\alpha)$ since $f(\alpha) = 0$.

Injectivity follows looking at $\ker(\psi) = \{[g(X)] \mid g(\alpha) = 0\}$. Since $f(X)$ is irreducible, $f(X)$ is the minimal polynomial of α (up to multiplication by a unit). Therefore, $f(X)$ divides every $g(X)$ with $[g(X)] \in \ker(\psi)$ (Defining Lemma 2.16). Thus, $g(X) \in (f(X))$ and consequently $[g(X)] = 0$.

Surjectivity is clear. $\square$

And now?

One must be careful here: Whenever one wants to write $K[\alpha]$, one already needs to have a field extension $L|K$ with an element $\alpha \in L$ available/given. Unfortunately, the usual notation $K[\alpha]$ does not reflect this fact -- the field L is not included into this notation. One can easily forget that one needs to have L a priori.

If one considers an irreducible polynomial $f(X) \in K[X]$ and only the field K is given/known (as it was the case in the example with $K = \mathbb{F}_5$), then $K[\alpha]$ makes no sense if one has not yet constructed a field extension $L|K$ with an element $\alpha \in L$.

10.2   The Splitting Field

Again, let us consider an irreducible polynomial $f(X) \in K[X]$ of degree $n \geq 2$. Then we want to build a **tower of fields** so that we will finally end up with a field containing all possible roots of f (in an appropriate sense):

1.   **Step:** Let $L_1|K$ be a finite field extension, such that L_1 contains a root $\alpha_1 \in L_1$ of $f(X)$—this is possible according to Theorem 10.1. We then take the smallest intermediate field that contains α_1, and call it $K_1 := K[\alpha_1] \subset L_1$.

And now?

Let me make two remarks on this:

1. If we define $L_1 := K[X]/(f(X))$ as in the theorem, then the field L_1 already coincides with $K[\alpha_1]$ inside itself, so that $K_1 = L_1$.
2. We proceeded in a way compatible with the discussion in the previous **And now?**: First, we constructed an extension $L_1|K$ with a root $\alpha_1 \in L_1$, then we can consider $K_1 := K[\alpha_1] \subset L_1$ as a potentially smaller intermediate extension containing this root.

So we have

$$K_1 = K[\alpha_1] \lhook\joinrel\longrightarrow L_1$$

$$n \Big| \diagup \qquad\qquad (10.3)$$

$$K$$

where we can now forget about the field L_1 again, or rather, L_1 no longer plays an important role in the following.

According to Corollary 2.19, $[K[\alpha_1] : K] = \deg(f) = n$, which we included into the diagram by putting n next to the line denoting this field extension (10.3).

2. **Step:** Now consider the polynomial $f(X) \in K[X] \subset K_1[X]$ as a polynomial with coefficients in K_1. Then it splits off the linear factor $(X - \alpha_1) \in K_1[X]$:

$$f(X) = (X - \alpha_1)^{\nu_1} \cdot g_1(X) \in K_1[X].$$

Let $\nu_1 \geq 1$ be the exact multiplicity of the root, i.e. such that the other factor satisfies $g_1(\alpha_1) \neq 0$. We will later mainly consider the cases where $\nu_1 = 1$ applies— and for $K = \mathbb{Q}$ this is anyway the general case for irreducible polynomials—see Chap. 13.

If $\deg(g_1(X)) \geq 1$, we can proceed with $g_1(X) \in K_1[X]$ in the same manner, in principal, with a caution: **There is no general reason why $g_1(X)$ should be irreducible.** But nevertheless, $g_1(X)$ in any case has an irreducible factor in the factorial ring $K_1[X]$, let's call it $h_1(X) \in K_1[X]$ and let's write $g_1(X) = h_1(X) \cdot \ell_1(X) \in K_1[X]$. Then

$$f(X) = (X - \alpha_1)^{\nu_1} \cdot \underbrace{h_1(X)}_{\text{irreducible in} K_1[X]} \cdot \ell_1(X) \qquad\qquad (10.4)$$

We construct a field $L_2|K_1$, which contains a root $\alpha_2 \in L_2$ of $h_1(X)$ and again set $K_2 := K_1[\alpha_2] \subset L_2$ (if one has constructed L_2 directly as $L_2 = K_1[X]/(h_1(X))$, already $K_2 = L_2$).

Note that

$$h_1(X) \mid f(X) \in K_1[X]$$

and thus $\alpha_2 \in K_2$ is another root of $f(X)$ (note that $\alpha_2 \neq \alpha_1$). Note that the largest field at this point is L_2 and one could wonder if the intermediate field $K[\alpha_1] \subset L_2$ will play some role, but this field is obviously identical to the intermediate field $K[\alpha_1] \subset L_1$ (**Think about it!**)and had been included into the play already in the first step. As a consequence, we can completely forget the "auxiliary field" L_1 which does not play any role any more.

Furthermore, we have

$$K_1[\alpha_2] = K[\alpha_1, \alpha_2] \subset L_2.$$

We now have obtained the following field extensions:

$$
\begin{array}{c}
K_2 = K_1[\alpha_2] = K[\alpha_1, \alpha_2] \lhook\joinrel\longrightarrow L_2 \\[2mm]
\scriptstyle{\leq n - v_1} \Big| \\[2mm]
K_1 = K[\alpha_1] \\[2mm]
\scriptstyle{n} \Big| \\[2mm]
K
\end{array}
$$

Let us emphasize that

- the degree of $g_1(X)$ is exactly $\deg(g_1) = \deg(f) - v_1 = n - v_1$,
- $h_1(X)$ is the minimal polynomial of α_2 over K_1 (w.l.o.g. we can choose $h_1(X)$ to be monic),
- and that $\deg(h_1) \leq \deg(g_1) = n - v_1$ (since $h_1 \mid g_1$ in $K_1[X]$).

Therefore, the degree of the last extension satisfies $\big[K[\alpha_1, \alpha_2] : K[\alpha_1]\big] \leq n - v_1$.

Let's continue in the same way   We repeat the arguments of the 2nd step accordingly. Consider $f(X) \in K[X] \subset K_2[X]$. In this ring, it decomposes as follows:

$$f(X) = (X - \alpha_1)^{v_1} \cdot g_1(X) = (X - \alpha_1)^{v_1} \cdot h_1(X) \cdot \ell_1(X) =$$
$$(X - \alpha_1)^{v_1} \cdot (X - \alpha_2)^{v_2} \cdot g_2(X) \in K_2[X]$$

with a polynomial $g_2(X) \in K_2[X]$. We have $\deg(g_2) \leq n - v_1 - v_2$, in particular $\deg(g_2) \leq n - 2$ (we don't want to take the multiplicities too seriously anyway).

Now we are looking for an irreducible factor $h_2(X) \mid g_2(X)$ in $K_2[X]$ and proceed exactly as above.

After a finite number of, let's say m, steps (for reasons regarding the degrees we have $m \leq n$) we obtain (less important) field $L_m \mid K$ and inside of it the tower of field extensions:

$$K[\alpha_1, \alpha_2, \ldots, \alpha_m] \hookrightarrow L_m$$

$$K[\alpha_1, \alpha_2, \alpha_3]$$

$\leq n - \nu_1 - \nu_2 \leq n-2$

$$K[\alpha_1, \alpha_2] \tag{10.5}$$

$\leq n - \nu_1 \leq n-1$

$$K[\alpha_1]$$

n

$$K$$

Over the field $Z := K[\alpha_1, \ldots, \alpha_m]$ finally, $f(X)$ finally decomposes into linear factors:

$$f(X) = \prod_{j=1}^{m} (X - \alpha_j)^{\nu_j} \in Z[X]$$

and we observe that $[Z : K] \leq n!$.

We have constructed a field Z that plays an important role and therefore deserves its own name:

Definition 10.3 Let $f(X) \in K[X]$ be a polynomial of degree $\deg(f) \geq 1$ (not necessarily irreducible). Then a field $L|K$ is called a **splitting field of** f, if the following applies:

1. In $L[X]$ the polynomial $f(X) \in K[X] \subset L[X]$ decomposes into linear factors

$$f(X) = \prod_{j=1}^{m} (X - \alpha_j)^{\nu_j} \in L[X]$$

(and thus $\alpha_1, \ldots, \alpha_m \in L$ are exactly the roots of $f(X)$ in L),
2. additionally, L is generated by these roots, in the following sense: If $K[\alpha_1, \ldots, \alpha_m] \subset L$ is the intermediate extension inside L generated by these roots, then automatically

$$K[\alpha_1, \ldots, \alpha_m] = L.$$

In particular, a splitting field is always a finite extension over K.

Remark 10.4 If there is some field $\Omega|K$ available in which $f(X)$ decomposes into linear factors (e.g. if $K = \mathbb{Q}$ and $\Omega = \mathbb{C}$), the constructions of the "auxiliary" fields L_j can be omitted and in each individual step, we can consider $K[\alpha_1, \ldots, \alpha_j] \subset \Omega$ inside Ω. Then the splitting field **inside** Ω is the intermediate field

$$K[\alpha_1, \ldots, \alpha_m] \subset \Omega.$$

But even if there is no such large field Ω in sight, we have seen above that we can always construct a splitting field.

For the record, let us state the last remark as a theorem:

Theorem 10.5 If $f(X) \in K[X]$ is a polynomial of degree $\deg(f) \geq 1$, there exists a splitting field $Z|K$ for f.

Proof Perhaps it should be noted that we have considered the construction above (the tower of field extensions) for an irreducible f. If f is not irreducible, we consider an irreducible factor $f_1 \mid f$ in the first step. Then the construction works as above. Arriving at a splitting field Z_1 of f_1, one continues with $f_2 \in Z_1[X]$, decomposes this into irreducible factors and carries out the same construction again, and continues in this way. This is a finite process. $\square$

And now?

I would like to briefly emphasize that a major technical difficulty arises in the process of building the tower of extensions starting from the 2nd step. Let's go to the beginning of the 2nd step and say we are working over $\mathbb{Q}$. So we start with $f(X) \in \mathbb{Q}[X]$ and we know, for example if $f(X) \in \mathbb{Z}[X]$ is a monic Eisenstein polynomial, that $f(X)$ is irreducible in $\mathbb{Q}[X]$. Inside the complex numbers $\mathbb{C}$, let α_1 be as root of $f(x)$:

$$\mathbb{Q}[\alpha_1] \qquad f(X) = (X - \alpha_1) \cdot g_1(X) \in \mathbb{Q}[\alpha_1][X]$$

$$\deg(f) \Big|$$

$$\mathbb{Q} \qquad\qquad f(X) \text{ irreducible in } \mathbb{Q}[X].$$

Now we have to examine $g_1(X) \in \mathbb{Q}[\alpha_1][X]$ for irreducibility. In general, this is difficult! In the first step, working in $\mathbb{Q}$ with its UFD subring $\mathbb{Z} \subset \mathbb{Q}$ there were several methods that could help to prove irreducibility of f (Gauss, reduction, Eisenstein). In $\mathbb{Q}[\alpha_1]$ this is usually not possible. Although one still finds

$$\mathbb{Z}[\alpha_1] = \{h(\alpha_1) \in \mathbb{Q}[\alpha_1] \mid h(X) \in \mathbb{Z}[X]\} \subset \mathbb{Q}[\alpha_1]$$

as a subring, and again $\mathbb{Q}[\alpha_1] = \mathrm{Quot}(\mathbb{Z}[\alpha_1])$ *(think about it briefly!)*, but **in general $\mathbb{Z}[\alpha_1]$ is no longer a UFD!**

And now?

What use is the tower of extensions (10.5) to us? It contains the information about the roots of $f(X)$. For example, the degrees of the extensions in the tower tell us something about the **relations** between the roots.

Let's illustrate this with an easy example (where we can directly compute the roots): Consider the polynomial $f(X) = X^4 + 2X^2 + 2 \in Q[X]$. Inside the complex numbers $\Omega = \mathbb{C}$, it has the roots (where $\sqrt{-1+i} \in \mathbb{C}$ denotes one of the two complex numbers ω with $\omega^2 = -1 + i$):

$$\alpha_1 = \sqrt{-1+i} \quad \alpha_2 = -\alpha_1$$
$$\alpha_3 = \sqrt{-1-i} \quad \alpha_4 = -\alpha_3.$$

Hence, we have two obvious relations between these, namely $\alpha_2 = -\alpha_1$ and $\alpha_4 = -\alpha_3$.

If we build the tower of field extensions as above, we obtain the following degrees in this case:

$$\mathbb{Q}[\alpha_1, \alpha_2, \alpha_3, \alpha_4] \lhook\joinrel\longrightarrow \mathbb{C}$$
$$1 \Big\| $$
$$\mathbb{Q}[\alpha_1, \alpha_2, \alpha_3]$$
$$2 \Big| \text{ see the subsequent task}$$
$$\mathbb{Q}[\alpha_1, \alpha_2] \hspace{3cm} (10.6)$$
$$1 \Big\|$$
$$\mathbb{Q}[\alpha_1]$$
$$4 \Big|$$
$$\mathbb{Q}$$

The relation $\alpha_2 = -\alpha_1$ is reflected by $[K[\alpha_1, \alpha_2] : K[\alpha_1]] = 1$, since $\alpha_2 \in K[\alpha_1]$. In the 2nd step of the construction of the tower, we are on top of the first layer:

$$\mathbb{Q}[\alpha_1]$$
$$4 \Big|$$
$$\mathbb{Q}$$

and $f(X)$ decomposes inside $\mathbb{Q}[\alpha_1][X]$ into the factors

$$f(X) = (X - \alpha_1) \cdot g_1(X) \overset{\text{here}}{=} (X - \alpha_1) \cdot \underbrace{(X + \alpha_1) \cdot \tilde{g}_1(X)}_{=: h_1(X)}.$$

We now take the irreducible factor $h_1 \mid g_1$ with which we perform the 2nd step. But since $h_1 \in \mathbb{Q}[\alpha_1][X]$ (coincidentally) holds, we don't have to do anything but can choose $K_2 = K_1 = \mathbb{Q}[\alpha_1]$.

The fact that $[Z : \mathbb{Q}] = 8 < 4!$ reflects the fact (considering that the roots are all pairwise different), **that there must be relations between the roots!**

In this example, you can see the relations between the roots

1. directly from the roots—which is possible in this case since we can directly compute them,
2. quite directly from the irreducible polyomial $f(X)$, because it has a special form (only quadratic powers of X occur),
3. indirectly, because the degree of a splitting field $[Z : \mathbb{Q}]$ is not the maximum possible, namely 4!, but only 8.

One of the **fundamental ideas of Galois theory** is the following: If one wants to investigate whether there are relations between the roots of a polynomial $f(X) \in K[X]$ (in a splitting field), and one cannot calculate the roots (which rules out (1)) and the polynomial f doesn't have an obvious peculiarity at first glance (which rules out (2)), one can use method (3) to obtain some information on these possible relations - or e.g. prove that there are no relations at all!

And it won't stop at the degree of the field extensions alone, we will extract even finer information from the tower of fields.

Check it out!

Justify the entry 2 as the degree of the field in (10.6)! This is not so trivial and therefore also put as an exercise at the end of the chapter.

Extreme cases—with examples:

1. It may happen that we already arrived at the splitting field after the first step. For example: Given an $n \in \mathbb{N}$ let $\zeta := \exp(\frac{2\pi i}{n}) \in \mathbb{C}$. Then ζ is a root of $X^n - 1$. We easily see

$$X^n - 1 = (X - 1) \cdot (X^{n-1} + X^{n-2} + \ldots + X + 1).$$

Now, if $n = p$ is a prime number, $\Phi_p(X) := X^{p-1} + X^{p-2} + \ldots + X + 1 \in \mathbb{Z}[X]$ is irreducible, see Lemma 9.17.

If we carry out the first step in the construction (since we have $\mathbb{C}$ with all the roots available, we do not need to construct L_1), we obtain the extension $\mathbb{Q}[\zeta] \subset \mathbb{C}$.

The roots of Φ_p in $\mathbb{C}$ are precisely the pairwise different complex numbers

$$\alpha_1 = \zeta, \ \alpha_2 = \zeta^2, \ \alpha^3 = \zeta^3, \ \ldots, \ \alpha_{p-1} = \zeta^{p-1}.$$

and these all already lie in $\mathbb{Q}[\zeta]$. Thus, $\mathbb{Q}[\zeta]|\mathbb{Q}$ is the splitting field of $\Phi_p(X)$ in $\mathbb{C}$ with degree $[\mathbb{Q}[\zeta] : \mathbb{Q}] = p - 1$.

2. In the next very explicit example we will have to accept some results we will prove in later chapters. The example will be one of the first applications of Galois' idea along with some arguments coming from group theory to prove the statement below—here, we want to prepare ourselves for this application and also get a feeling for what we are supposed to develop in the next chapters. The results we have not developped yet will be written in *italics*.

Consider the polynomial $f(X) = X^5 - 777X + 7 \in \mathbb{Q}[X]$ (irreducible by Eisenstein). According to Theorem 10.5, f has a splitting field Z. The latter contains all five *pairwise different* (we will prove this in Chap. 13) roots $\alpha_1, \ldots, \alpha_5 \in Z$ and we have the tower of fields

$$
\begin{array}{c}
\mathbb{Q}[\alpha_1, \alpha_2, \alpha_3, \alpha_4, \alpha_5] =\!=\!=\!= Z \\
1 \,\| \\
\mathbb{Q}[\alpha_1, \alpha_2, \alpha_3, \alpha_4] \\
\leq 2 \,| \\
\mathbb{Q}[\alpha_1, \alpha_2, \alpha_3] \\
\leq 3 \,| \\
\mathbb{Q}[\alpha_1, \alpha_2] \\
\leq 4 \,| \\
\mathbb{Q}[\alpha_1] \\
5 \,| \\
\mathbb{Q}
\end{array}
$$

At first glance, I see no idea how to further investigate the roots and possible relations. Of course we can also apply some elementary calculus methods and observe that the polynomial (with real, even rational, coefficients) $f(X)$ has exactly three roots in the real numbers, so WLOG $\alpha_1, \alpha_2, \alpha_3 \in \mathbb{R}$ and $\alpha_4, \alpha_5 \in \mathbb{C} \setminus \mathbb{R}$. If $\overline{w} \in \mathbb{C}$ denotes the complex conjugate of $w \in \mathbb{C}$, it follows that

$$0 = f(\alpha_4) = \overline{f(\alpha_4)} = f(\overline{\alpha_4})$$

hence $\overline{\alpha_4} \in \mathbb{C}$ is also a root and therefore $\alpha_5 = \overline{\alpha_4}$. Note that this not an **algebraic** relation between these two roots in the sense of algebraic relations, we are looking for. *With the help of Galois theory, we will later see that* $[Z : \mathbb{Q}] = 5!$ holds, so the maximum degree is *assumed*. We can therefore deduce (once we filled in the gaps) that there is no algebraic relation between the roots of f(X)

10.3 Theorem A and A'

And now?
How can we grasp the finer structure that is embedded in the tower of the field extensions (10.5)? We start with an observation and subsequent question:

1. A weakness of our approach is that we have put the roots $\alpha_1, \ldots, \alpha_n$ in some order and started with the first chosen one α_1. One might ask whether something significantly different happened if we had started— let's say—with α_2 instead of α_1. Let's look at the first step of the construction. We obtain two first floors in the towers above the common ground floor K (again we take an irreducible $f(X) \in K[X]$ of degree $n \geq 2$ as the polynomial of our concern):

$$
\begin{array}{cc}
K[\alpha_1] & K[\alpha_2] \\
n \,| & n \,| \\
K =\!\!=\!\!= & K
\end{array}
\tag{10.7}
$$

Do these two fields have any relation to each other? What could we wish for? Hopefully, there should be an isomorphism between them! Preferably even one that maps one chosen root to the other $\alpha_1 \mapsto \alpha_2$. This is indeed the case (see Theorem A below), but it also leads to the following question:

2. When considering (10.7), one might generally ask the following: Given two elements $\alpha, \beta \in L$ in a field extension $L|K$, such that $[K[\alpha] : K] = [K[\beta] : K]$ holds (otherwise the answer will definitely be no!), can we find an isomorphism

$$
\begin{array}{ccc}
K[\alpha] & \xrightarrow[\;???\;]{\;\cong\;} & K[\beta] \\
n \,| & & n \,| \\
K & =\!\!=\!\!= & K
\end{array}
$$

with $\alpha \mapsto \beta$.

An affirmative answer would be bad news!!! Because then our tower of field extensions would not be able to recognize any information from our initial polynomials other than its degree: If α has the minimal polynomial $f(X) \in K[X]$ and β has the minimal polynomial $g(X)$ of $K[X]$, then the two problems:

Examine the solutions of the equation $f(X) = 0$

or

Examine the solutions of the equation $g(X) = 0$

have nothing to do with each other (except that f and g have the same degree). If the field extensions we end up considering f and g individually in the first floors of the tower leading to the splitting fields were isomorphic (where the roles of α and β were the same on both sides since the isomorphism maps $\alpha \mapsto \beta$), they would completely forget the equation (in other words the minimal polynomial)—then we could not use them to deduce any information about the equation (other than the degree of the polynomials)!

First, we want to deal with question (2) from the **And now?** section. The following theorem says that the issue we feared in (2) does not occur:

Theorem 10.6 (Theorem of the same minimal polynomial) Let $f(X) \in K[X]$ be irreducible, $L|K$ a field extension with a root $\alpha \in L$ of $f(X)$ and $K \subset K[\alpha] \subset L$ the corresponding intermediate field.

If $L'|K$ is another field extension and $\varphi : K[\alpha] \to L'$ is a field homomorphism with $\varphi|_K = \mathrm{id}_K$, so that we have the following commutative diagram:

$$
\begin{array}{ccc}
L & & \\
| & & \\
K[\alpha] & \xrightarrow{\ \varphi\ } & L' \\
| & & | \\
K & = & K,
\end{array}
$$

then $\varphi(\alpha) \in L'$ is also a root of the same polynomial $f(X)$.

Proof Let us write $f(X) = X^n + a_{n-1}X^{n-1} + \ldots + a_0 \in K[X]$. It follows from the homomorphism properties of φ and $\varphi(a) = a$ for $a \in K$, that

$$
\begin{aligned}
f(\varphi(\alpha)) = (\varphi(\alpha))^n + a_n(\varphi(\alpha))^{n-1} + \ldots + a_0 = \\
\varphi(\alpha^n) + \varphi(a_n)\varphi(\alpha^{n-1}) + \ldots + \varphi(a_0) = \\
\varphi(\alpha^n + a_{n-1}\alpha^{n-1} + \ldots + a_0) = \varphi(f(\alpha)) = 0.
\end{aligned}
$$

This is the claim of the theorem. □

Remark 10.7 The assumption that f is irreducible is not actually necessary. Since we usually apply the theorem to irreducible f and one can also consider the corresponding irreducible factor with root α for non-irreducible f, the theorem, as well as its variant 10.8, is formulated for irreducible f.

This solves problem (2) from the above **And now?**, which dealt with the 1st step in the construction of the tower of fields. To understand the subsequent steps, we need a variant of the theorem for which we introduce a notation first:

Let $\sigma : K \to K'$ be a field homomorphism (in the application this will be the homomorphism $\sigma : K[\alpha_1] \to K[\alpha_2]$ constructed in the first step). Then one can transport polynomials over K into polynomials over K' : the map

$$
K[X] \to K'[X]
$$
$$
f(X) = a_m X^m + \ldots + a_0 \mapsto f^\sigma(X) := \sigma(a_m)X^m + \ldots + \sigma(a_0)
$$

is a ring homomorphism.

The variant of Theorem 10.6 deals with the case where we do not consider the identity in the bottom row of the diagram as in the theorem, but we start with any given σ. As explained in Remark 10.7 we state the theorem for irreducible polynomials since this will be the important case in our applications:

Theorem 10.8 Let $f(X) \in K[X]$ be irreducible, L/K is a field extension with a root $\alpha \in L$ of $f(X)$ and $K \subset K[\alpha] \subset L$ be the corresponding intermediate field. If $\sigma : K \to K'$ is a field homomorphism, L'/K' a field extension and $\varphi : K[\alpha] \to L'$ a field homomorphism with $\varphi|_K = \sigma$, so that we have the following commutative diagram:

$$
\begin{array}{ccc}
L & & \\
| & & \\
K[\alpha] & \xrightarrow{\ \varphi\ } & L' \\
| & & | \\
K & \xrightarrow{\ \sigma\ } & K',
\end{array}
$$

then $\varphi(\alpha) \in L'$ is a root of $f^{\sigma}(X) \in K'[X]$.

> **Check it out!**
> What changes do we have to execute in the proof of Theorem 10.6 so that the proof proceeds in the same way? There are only a few changes to make!

Proof Let's write $f(X) = X^n + a_{n-1}X^{n-1} + \ldots + a_0 \in K[X]$. It follows from the homomorphism properties of φ and $\varphi(a) = \sigma(a)$ for $a \in K$, that

$$
f^{\sigma}(\varphi(\alpha)) = (\varphi(\alpha))^n + \sigma(a_n)(\varphi(\alpha))^{n-1} + \ldots + \sigma(a_0) =
$$
$$
\varphi(\alpha^n) + \varphi(a_n)\varphi(\alpha^{n-1}) + \ldots + \varphi(a_0) =
$$
$$
\varphi(\alpha^n + a_{n-1}\alpha^{n-1} + \ldots + a_0) = \varphi(f(\alpha)) = 0 \,.
$$

This completes the proof. $\square$

The changes in the last proof—see the previous **Check it out!**—only were to insert the σ in the second and third line of the proof.

> **What now?**
> The last theorems solve the issue (2) from the previous **What now?**. If we have an isomorphism
>
> $$
> \begin{array}{ccc}
> K[\alpha] & \xrightarrow[\varphi]{\cong} & K[\beta] \\
> n\,| & & n\,| \\
> K & = \!=\!= & K
> \end{array}
> $$
>
> with $\varphi(\alpha) = \beta$, then the two elements necessarily have the same minimal polynomial over K.
>   What about problem (1), i.e., the reversal: If α and β have the same minimal polynomial over K, is there always such a φ?

Theorem 10.9 (Theorem A)[1] Let $f(X) \in K[X]$ be an irreducible polynomial and let:

- $L|K$ be a field extension with a root $\alpha \in L$ of $f(X)$ and $K \subset K[\alpha] \subset L$ the corresponding intermediate field,
- $L'|K$ be another field extension that contains a root $\beta \in L'$ of $f(X)$ and $K \subset K[\beta] \subset L'$ the corresponding intermediate field.

Then there exists a unique isomorphism of fields $\varphi : K[\alpha] \to K[\beta]$ such that $\varphi(\alpha) = \beta$ and $\varphi|_K = \mathrm{id}_K$. The corresponding diagram is:

$$
\begin{array}{ccc}
L & & L' \\
| & & | \\
K[\alpha] & \xrightarrow[\cong]{\varphi} & K[\beta] \\
| & & | \\
K & =\!=\!= & K
\end{array}
\qquad
\alpha \xmapsto{\;\varphi\;} \beta
$$

Proof Existence: According to Theorem 10.2 we have the two isomorphisms:

$$
\psi_\alpha : K[X]/(f(X)) \xrightarrow{\cong} K[\alpha]
$$
$$
\psi_\beta : K[X]/(f(X)) \xrightarrow{\cong} K[\beta]
\tag{10.8}
$$

(note that the two target fields are contained in possibly different fields: $K[\alpha] \subset L$ and $K[\beta] \subset L'$. These each take on the role of Ω in the formulation of Theorem 10.2). We have $\psi_\alpha([X]) = \alpha$ and $\psi_\beta([X]) = \beta$. The composition

$$
\varphi : K[\alpha] \xrightarrow{\psi_\alpha^{-1}} K[X]/(f(X)) \xrightarrow{\psi_\beta} K[\beta]
\tag{10.9}
$$

is an isomorphism as desired.

Uniqueness: An arbitrary element $\xi \in K[\alpha]$ is of the form $\xi = b_N \alpha^N + \ldots + b_0$ with $b_j \in K$—one can choose $N = \deg(f) - 1$, then the b_j are uniquely determined by ξ, but that is not important now. It follows that any field isomorphism φ with the desired properties yields:

$$
\varphi(\xi) = b_N \varphi(\alpha)^N + \ldots + b_0 = b_N \beta^N + \ldots + b_0 \in K[\beta] ,
$$

and therefore $\varphi(\xi)$ is uniquely determined once we know its value at the element α, i.e. $\varphi(\alpha) = \beta$. $\square$

Remark 10.10 Note that in Theorem A, the irreducibility of f is important because otherwise we would not have the isomorphisms (10.8).

---

[1] the name *Theorem A* is not a generally common one, but is chosen for the purposes of this book.

And now?

When one looks at the argument for uniqueness, one might wonder why the existence is proven so complicatedly. One could also make the following attempt: *Given the element* $\xi = b_{n-1}\alpha^{n-1} + \ldots + a_0 \in K[\alpha]$ *with* $n = \deg(f)$ i.e. we have presented ξ *as its unique representation in terms of the basis, we define its image to be*

$$\varphi(\xi) := b_{n-1}\beta^{n-1} + \ldots + b_0 \in K[\beta]$$

and we are done. Unfortunately it is not so easy, because one still has to show that this definition of φ produces a field homomorphism! This is doable but not so nice to write down, especially when one wants to verify that $\varphi(\xi \cdot \eta) = \varphi(\xi) \cdot \varphi(\eta)$ holds. To this end, one has to compute the presentation of the product in the basis using the minimal polynomial of α and continue with this presentation comparing it to the ones on gets when computing the right hand side. It is essentially what happens in our proof but we elegantly avoided these computations and proved that φ is a homomorphism using (10.12).

We also prove the following variant, which is needed from the second step on in the construction of the tower of field extensions.

Theorem 10.11 (Theorem A′) Let   be an irreducible polynomial and let $\sigma : K \to K'$ be a field homomorphism. Additionally, let:

- $L|K$ be a field extension with a root $\alpha \in L$ of $f(X)$ and $K \subset K[\alpha] \subset L$ the corresponding intermediate field,
- $L'|K'$ be another field extension that contains a root $\beta \in L'$ of $f^\sigma(X)$ and $K' \subset K'[\beta] \subset L'$ the corresponding intermediate field.

Then there exists a unique field homomorphism $\varphi : K[\alpha] \to K'[\beta]$ with $\varphi(\alpha) = \beta$ and $\varphi|_K = \sigma$. Written as a diagram:

$$
\begin{array}{ccc}
L & & L' \\
| & & | \\
K[\alpha] & \xrightarrow{\ \varphi\ } & K'[\beta] \\
| & & | \\
K & \xrightarrow{\ \sigma\ } & K'
\end{array}
\qquad\qquad
\alpha \xmapsto{\ \varphi\ } \beta
$$

If σ is a field isomorphism, then so is φ.

Proof Since $\sigma : K \to K'$ is a field homomorphism, the map

$$
\begin{aligned}
K[X] &\to K'[X] \\
g(X) &\mapsto g^\sigma(X)
\end{aligned}
\tag{10.10}
$$

is itself a ring homomorphism. Now let $q(X) \in K'[X]$ be the minimal polynomial of β over K'. Then we have the field isomorphisms:

$$\psi_\alpha : K[X]/(f(X)) \overset{\cong}{\to} K[\alpha]$$

$$\psi_\beta : K'[X]/(q(X)) \overset{\cong}{\to} K'[\beta].$$

Given the assumption that $f^\sigma(\beta) = 0$, it follows that $q(X)$ is a divisor of $f^\sigma(X)$ inside $K'[X]$, let's say $f^\sigma(X) = q(X) \cdot h(X)$ for some $h(X) \in K'[X]$. Note that with σ being a field isomorphism, the ring homomorphism $K[X] \to K'[X], g \mapsto g^\sigma$ is also an isomorphism. In particular, under this assumption $f^\sigma(X)$ is already irreducible in $K'[X]$ and thus in this case: $q = f^\sigma$.

Now, in the general case, the ring homomorphism

$$K[X]/(f(X)) \to K'[X]/(q(X)) , \; [g(X)] \mapsto [g^\sigma(X)] \qquad (10.11)$$

is well-defined, because if $g_1(X) \equiv g_2(X) \bmod (f(X))$, it follows that

$$g_2(X) = g_1(X) + f(X) \cdot \ell(X) \text{ for an } \ell(X) \in K[X] \Longrightarrow$$
$$g_2^\sigma(X) = g_1^\sigma(X) + f^\sigma(X) \cdot \ell^\sigma(X) = g_1^\sigma(X) + q(X) \cdot h(X) \cdot \ell^\sigma(X) \Longrightarrow$$
$$g_1^\sigma(X) \equiv g_2^\sigma(X) \bmod (q(X)) .$$

Consequently, we now have the composition of maps:

$$\varphi : K[\alpha] \xrightarrow{\psi_\alpha^{-1}} K[X]/(f(X)) \xrightarrow{(10.11)} K'[X]/(q(X)) \xrightarrow{\psi_\beta} K'[\beta] \; (10.12)$$

and it fulfills $\varphi(\alpha) = \beta$ as well as $\varphi|_K = \sigma$.

The uniqueness follows as above since φ is already uniquely determined by these two conditions.

If σ is an isomorphism, then so is (10.11) and hence also φ. $\square$

10.4   Application for the Tower of Field Extensions

We can now successively include field isomorphisms into two versions of the tower of field extensions starting with an irreducible polynomial $f(X) \in K[X]$ with $\deg(f) = n$ using Theorem A' (10.11). Let Z be a splitting field that contains the roots $\alpha_1, \ldots, \alpha_n \in Z$. **We now assume that these are pairwise different and thus simple roots** and will examine this issue in more detail in Chap. 13.

Note that $Z = K[\alpha_1, \ldots, \alpha_n]$. Again, we construct the tower of field extensions step by step (and all field extensions will occur within Z) adjoining $\alpha_1, \ldots, \alpha_n$ in this order and investigate on the question what happens if we consider a different ordering and construct the corresponding tower of extensions simultaneously:

1. **Step:** Choose a $j = 1, \ldots, n$ and apply Theorem A to obtain a unique field isomorphism as follows (note that $f(X)$ is the minimal polynomial of all the roots α_j over the field K):

$$K[\alpha_1] \xrightarrow[\cong]{\sigma_1} K[\alpha_j] \qquad \alpha_1 \xrightarrow{\sigma_1} \alpha_j$$
$$\big| \qquad\qquad \big|$$
$$K = \!\!=\!\!=\!\!= K$$

We will simply denote this isomophism by σ_1 (it is the isomorphism σ from Theorem A), but note that for different choices of j of course we will get different σ_1. Since we assume that the roots are pairwise different, we can obtain exactly n pairwise different isomorphisms σ_1, one for each choice of $j \in \{1,\ldots,n\}$. Choosing $j=1$ for instance will give $\sigma_1 = id_{K[\alpha_1]}$.

2. **Step:** The polynomial $f(X) \in K[X]$ decomposes inside the ring $K[\alpha_1][X]$ as described in (10.4).

$$f(X) = (X - \alpha_1) \cdot h_1(X) \cdot \ell_1(X). \tag{10.13}$$

where $h_1(X)$ is the minimal polynomial of α_2 over $K[\alpha_1]$—this is exactly the 2nd step in the construction of the tower of field extensions. Note that the multiplicity ν_1 we used in the general case now is assumed to be 1.

We also have the induced isomorphism $\sigma_1 : K[\alpha_1] \to K[\alpha_j]$ and we can apply Theorem A': Consider the irreducible factor $h_1(X) \in K[\alpha_1][X]$ and its image under σ_1 (as in (10.10)):

$$h_1^{\sigma_1}(X) \in K[\alpha_j][X] .$$

Note that $f^{\sigma_1}(X) = f(X)$, because $f(X) \in K[X]$ and $\sigma_1|_K = id_K$ holds. With this, we have

$$f(X) = f^{\sigma_1}(X) = (X - \alpha_j) \cdot h_1^{\sigma_1}(X) \cdot \ell_1^{\sigma_1}(X) \in K[\alpha_j][X] . \tag{10.14}$$

According to theorem A' (where $h_1(X)$ plays the role of f in theorem A' and $K[\alpha_j]$ plays the one of K', furthermore, $L' = L = Z$) we can now choose an **arbitrary** root $\beta \in K[\alpha_j]$ **of the polynomial** $h_1^{\sigma_1}(X)$. Then (and only then) we obtain a field isomorphism σ_2 (which depends on this choice) as in the diagram:

β can be chosen
arbitrary
**but only among
those roots**
that are **roots of
the polynomial** $h_1^{\sigma_1}$

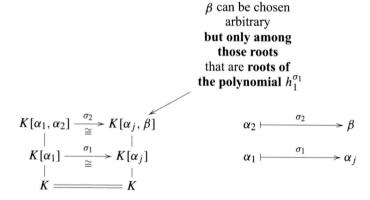

Important observation: The element $\beta \in Z$ is a root of $h_1^{\sigma_1}(X)$ in order to be able to apply Theorem A'. Because of (10.14) it is of course also a root of $f(X)$:

$$\beta \in \{\alpha_1,\ldots,\alpha_n\} ,$$

BUT: in general it is **not possible to choose every** root α_k to be this element β— one can choose β only among those roots that are roots of the polynomial $h_1^{\sigma_1}(X)$. More precisely: If, for example, in (10.13) the additional factor $\ell_1(X)$ is of degree ≥ 1, then $\ell_1^{\sigma_1}(X)$ is of the same degree and also has some roots (which are among the $\alpha_1, \ldots, \alpha_n$). These cannot be chosen to be β since these are not roots of $\ell_1^{\sigma_1}(X)$ - recall that we assumed the roots to be pairwise different.

And now?

If you think about the last step, you will (hopefully) agree that the questions: Which of the α_k belongs to which factor in (10.13) reflects exactly the question about certain relations between the roots. And this was our goal! We can now give a better explanation what we mean with relations among the roots - it is the question if they share the same minimal polynomial even after having adjoined some of the roots already, i.e. not only on the ground floor of the towers but also on higher floors.

Let us keep going on Let's say we chose $\beta = \alpha_k$ in the 2nd step. Then we can continue applying Theorem A': We examine the decomposition of $f(X) = (X - \alpha_1)(X - \alpha_2) \cdot h_2(X) \cdot \ell_2(X) \in K[\alpha_1, \alpha_2][X]$, select the irreducible factor $h_2(X)$ associated to α_3 (i.e. which has α_3 as a root) and we continue as before for each root $\beta \in Z$ of $h_2^{\sigma_2}(X)$:

$$
\begin{array}{ccc}
K[\alpha_1, \alpha_2, \alpha_3] & \xrightarrow[\cong]{\sigma_3} & K[\alpha_j, \alpha_k, \beta] \\
| & & | \\
K[\alpha_1, \alpha_2] & \xrightarrow[\cong]{\sigma_2} & K[\alpha_j, \alpha_k] \\
| & & | \\
K[\alpha_1] & \xrightarrow[\cong]{\sigma_1} & K[\alpha_j] \\
| & & | \\
K & =\!\!=\!\!=\!\!=\!\!= & K
\end{array}
\qquad
\begin{array}{ccc}
\alpha_3 & \xrightarrow{\sigma_3} & \beta \\
& & \\
\alpha_2 & \xrightarrow{\sigma_2} & \alpha_k \\
& & \\
\alpha_1 & \xrightarrow{\sigma_1} & \alpha_j
\end{array}
$$

Finally, we will arrive at a field isomorphism $\sigma : Z \to Z$ with $\sigma_K = \text{id}$.

And now?

Of course, we observe that the construction of such an isomorphism $\sigma : Z \to Z$ requires a lot of information about the relations between the roots, or in more optimistic words, hopefully stores a lot of information about these!

I would like to emphasize once again that the issues we have to solve in these proceeding, in particular: *How do the involved polynomials decompose* *and which of the $\alpha_1, \ldots, \alpha_n$ can one choose on the right hand side in the 2nd step, if one had chosen α_2 on the left hand side, what about the 3rd step, ...* are very difficult to answer in explicit applications. If one can do this explicitly, one has to understand/know the roots of $f(X)$ very well from the beginning.

In other words: If we do not have another idea of how to better understand the result of these constructions, namely all these isomorphisms $\sigma : Z \to Z$, without knowing the roots and their relationships in detail, we have not gained anything yet!

So far we have accomplished the following: We have the feeling that the set of isomorphisms $\sigma : Z \to Z$ as above stores a lot of information about the roots. But to construct these σ we need exactly this information about the roots. So we are currently still moving in circles.

10.5 The Galois Group

And now?

This *And now?* seamlessly extends the previous one.

The idea is that we want to abstractly examine the set of isomorphisms $\sigma : Z \to Z$ without having to construct all of them step by step, as explained above. In a certain sense:

- the construction of all these isomorphisms $\sigma : Z \to Z$ in the tower of field extensions as constructed above can be considered as **Bottom-Up idea**,
- now we want to proceed following a **Top-Down idea**.

We also detach ourselves from the initial situation that $Z|K$ is a splitting field of a given polynomial. So let $L|K$ be any finite field extension. This creates a problem: Previously, we were able to construct the isomorphisms $\sigma : Z \to Z$ because we knew that on the target side the desired roots, which we called β above, existed in Z. Now one has to be a little more careful—for the moment, we will assume to have a big field Ω containing all possible roots, later we will remove this assumption by a restriction (which then will be called a normal field extension).

For this, we assume that the fields $K \subset L \subset \Omega$ all lie inside a field Ω which is **algebraically closed** (i.e. in which every polynomial $f(X) \in \Omega[X]$ of degree ≥ 1 has a root).[2]

---

[2] Indeed, one can always construct such an Ω. However, we do not want to do that now. If we start with $K = \mathbb{Q}$, we can take $\Omega = \mathbb{C}$ or $\Omega = \overline{\mathbb{Q}}$. The latter is better because it is algebraic over $\mathbb{Q}$. See Sect. 12.1 and Appendix A for these issues.

Definition 10.12 Let Ω be an algebraically closed field and $K \subset L \subset \Omega$ be field extensions such that $L|K$ is algebraic. Then we call the elements of

$$\text{Hom}_K(L, \Omega) := \{\varphi : L \to \Omega \mid \varphi \text{ is a homomorphism of fields such that } \varphi|_K = \text{id}_K\}$$

the K-**homomorphisms from** L **to** Ω.
   Furthermore, we define:

Definition 10.13 If $L|K$ is an algebraic field extension, then the elements of

$$\text{Aut}_K(L) := \{\sigma : L \to L \mid \sigma \text{ is an isomorphism of fields such that } \sigma|_K = \text{id}_K\}$$

are called the K-**automorphisms of** L.
   When we are in the situation of a tower of field extensions, we obtain the following:

Theorem 10.14 If $Z|K$ is the splitting field of an irreducible polynomial $f(X) \in K[X]$ and $\Omega \supset Z$ is an algebraically closed field, then

$$\text{Hom}_K(Z, \Omega) = \text{Hom}_K(Z, Z) = \text{Aut}_K(Z).$$

Proof We have $Z = K[\alpha_1, \ldots, \alpha_n]$ with all the roots $\alpha_1, \ldots, \alpha_n \in \Omega$ of f (these do not have to be pairwise different here). If now

$$(\sigma : Z \to \Omega) \in \text{Hom}_K(Z, \Omega),$$

is any homomorphism, we know that $\sigma(\alpha_j)$ must be a root of the same minimal polynomial, namely $f(X)$—Theorem 10.6. Therefore

$$\sigma(\alpha_j) \in \{\alpha_1, \ldots, \alpha_n\} \subset Z.$$

But every element $\xi \in Z$ is of the form

$$\xi = \sum_{i_1, \ldots, i_n=0}^{N} a_{i_1, \ldots, i_n} \alpha_1^{i_1} \cdots \alpha_n^{i_n}$$

with $a_{i_1, \ldots, i_n} \in K$. Since σ is a field homomorphism and $\sigma|_K = \text{id}_K$, we obtain:

$$\sigma(\xi) = \sum_{i_1, \ldots, i_n=0}^{N} a_{i_1, \ldots, i_n} \sigma(\alpha_1)^{i_1} \cdots \sigma(\alpha_n)^{i_n} \in Z.$$

So, σ only takes up values in Z and therefore $\sigma : Z \to Z$.
   Since σ is a field homomorphism, σ is injective. Since we know that $[Z : K] < \infty$, and σ is also a K-linear map, σ is surjective due to the rank theorem of linear algebra. $\square$

Defining Lemma 10.15 If $Z|K$ is the splitting field of the irreducible polynomial $f(X) \in K[X]$, then $\text{Aut}_K(Z)$ together with the composition

$$\tau\sigma := \tau \circ \sigma : Z \xrightarrow{\sigma} Z \xrightarrow{\tau} Z$$

as a multiplication and the identity as the neutral element forms a group. It is called the **Galois group** of $Z|K$ and is written

$$\mathrm{Gal}(Z|K) := \mathrm{Aut}_K(Z).$$

Of course, we can define this notion for any field extension (it may not be very useful without further restrictions):

Definition 10.16 Let $L|K$ be an algebraic field extension, then

$$\mathrm{Gal}(L|K) := \mathrm{Aut}_K(L)$$

is called the **Galois group of** $L|K$.

And now?

So far, so good. What remains to be done?

- We have considered $\mathrm{Hom}_K(L, \Omega)$ and seen that in the case of a splitting field, the equality $\mathrm{Hom}_K(Z, \Omega) = \mathrm{Aut}_K(Z)$ holds. How can we characterize whenever $\mathrm{Hom}_K(L, \Omega) = \mathrm{Aut}_K(L)$ holds? We want to have this so that we can compose its elements and obtain a group. The answer will be given in Chap. 12.

- In observing that in the case of the splitting field the elements in $\mathrm{Aut}_K(Z)$ store the information about the roots we want to investigate, we have **assumed that the zeros are simple.** In other words, over the field Z, the polynomial decomposes as

$$f(X) = (X - \alpha_1) \cdots (X - \alpha_n)$$

  with **pairwise different** roots $\alpha_1, \ldots, \alpha_n$. The field automorphisms $\sigma : Z \to Z$ naturally only see the elements of the set $\{\alpha_1, \ldots, \alpha_n\}$ and not whether any of them occur multiple times as roots in $f(X)$. In other words: If $f(X)$ has multiple roots in Z, then $\mathrm{Aut}_K(Z)$ will loose this information. When does this problem happen, and how are we going to deal with this? The answer will be given in Chap. 13.

- How can we use our **top-down** approach to deduce information about the towers of field extensions leading to the splitting field from the knowledge of the group $\mathrm{Aut}_K(Z) = \mathrm{Gal}(Z|K)$, and how do we take advantage of the fact that $\mathrm{Gal}(Z|K)$ is a group? Answer: This will be the main theorem of Galois theory—Chap. 14.

- How do we apply the main theorem to prove interesting theorems? Answer: these will be the topics of the remaining chapters of the book.

Before we proceed, let's look at an example and determine our first Galois group:

Examples: We take a very simple example, which we have already considered in Sect. 2.2. Let $f(X) = X^4 - 2 \in \mathbb{Z}[X]$, which is irreducible by Eisenstein. We have $Z = \mathbb{Q}[\sqrt[4]{2}, i\sqrt[4]{2}]$ with the roots

$$\alpha_1 := \sqrt[4]{2}, \alpha_2 := -\sqrt[4]{2}, \alpha_3 := i\sqrt[4]{2}, \alpha_4 := -i\sqrt[4]{2}.$$

Let's go through the steps of the construction. The goal is to determine the Galois group.

1. **Step:** For each $j = 1, 2, 3, 4$ there is exactly one σ_1 with

$$\begin{array}{ccc} \mathbb{Q}[\alpha_1] & \xrightarrow[\cong]{\sigma_1} & \mathbb{Q}[\alpha_j] \\ | & & | \\ \mathbb{Q} & == & \mathbb{Q} \end{array} \qquad \alpha_1 \xmapsto{\sigma_1} \alpha_j$$

2. **Step:** Since $\alpha_2 = -\alpha_1$ applies, not much happens here. Since σ_1 is a field homomorphism, $\sigma_1(\alpha_2) = -\sigma_1(\alpha_1) = -\alpha_j$ holds.

   However, one can still see how the theory takes care of this by itself. Let's say we have chosen α_j on the right hand side in the first step. On the left hand side we have

$$f(X) = X^4 - 2 = (X - \sqrt[4]{2}) \cdot (X^3 + \sqrt[4]{2}X^2 + \sqrt{2}X + \sqrt[4]{2}^3)$$

which can be easily verified by division with remainder. The second factor is not irreducible over $\mathbb{Q}[\sqrt[4]{2}]$ (it cannot be, because it has to split off the linear factor $X + \sqrt[4]{2}$), one further obtains

$$f(X) = (X - \sqrt[4]{2}) \cdot \underbrace{(X + \sqrt[4]{2})}_{=:h_1(X)} \cdot (X^2 + \sqrt{2})$$

Indeed, $h_1(X) = X + \sqrt[4]{2} \in \mathbb{Q}[\sqrt[4]{2}][X]$ is the minimal polynomial of α_2 over $\mathbb{Q}[\sqrt[4]{2}]$.

The theory now says that on the right hand side in

$$\begin{array}{ccc} \mathbb{Q}[\alpha_1, \alpha_2] & \xrightarrow[\cong]{\sigma_2} & \mathbb{Q}[\alpha_j, \beta] \\ || & & | \\ \mathbb{Q}[\alpha_1] & \xrightarrow[\cong]{\sigma_1} & \mathbb{Q}[\alpha_j] \\ | & & | \\ \mathbb{Q} & == & \mathbb{Q} \end{array} \qquad \begin{array}{c} \alpha_2 \xmapsto{\sigma_2} \beta \\ \\ \alpha_1 \xmapsto{\sigma_1} \alpha_j \end{array}$$

we can choose for β one of the roots $\alpha_1, \ldots, \alpha_4$, **if and only if it is a root of** $h_1^{\sigma_1}(X) \in \mathbb{Q}[\alpha_j][X]$. Which of the α_i is a root of $h_1^{\sigma_1}(X)$? Now, $h_1(X) = X + \alpha_1$ and therefore

$$h_1^{\sigma_1}(X) = X + \sigma_1(\alpha_1) = X + \alpha_j \in \mathbb{Q}[\alpha_j][X].$$

This means: No matter which α_j we have chosen in the first step, the theory of the second step immediately gives that $\sigma_2(\alpha_2) = -\alpha_j$ must hold—we knew this beforehand, but it's nice that the theory confirms this automatically.

3. **Step:** Since $\mathbb{Q}[\alpha_1, \alpha_2, \alpha_3, \alpha_4] = \mathbb{Q}[\alpha_1, \alpha_2, \alpha_3]$ holds, we are done after the 3rd step anyway. So after the 2nd step, we consider the polynomial $f(X)$ in the ring $\mathbb{Q}[\alpha_1, \alpha_2][X] = \mathbb{Q}[\sqrt[4]{2}][X]$. This is nothing new, again as before:

$$f(X) = (X - \sqrt[4]{2}) \cdot (X + \sqrt[4]{2}) \cdot \underbrace{(X^2 + \sqrt{2})}_{=:h_2(X)}.$$

I already call the last factor $h_2(X)$ as in the theory, because we can assure that $h_2(X) \in \mathbb{Q}[\sqrt[4]{2}][X]$ is irreducible: Here we can apply the criterion from Lemma 9.12 due to degree reasons. The roots of $h_2(X)$ are $\alpha_3, \alpha_4 \in \mathbb{C} \setminus \mathbb{R}$ and therefore do not lie in $\mathbb{Q}[\sqrt[4]{2}] \subset \mathbb{R}$. The theory says that in

$$
\begin{array}{ccccccc}
\mathbb{Q}[\alpha_1, \alpha_2, \alpha_3] & \xrightarrow[\cong]{\sigma_3} & \mathbb{Q}[\alpha_j, -\alpha_j, \beta] & \qquad & \alpha_3 & \xrightarrow{\sigma_3} & \beta \\
\vert & & \vert & & & & \\
\mathbb{Q}[\alpha_1, \alpha_2] & \xrightarrow[\cong]{\sigma_2} & \mathbb{Q}[\alpha_j, -\alpha_j] & & \alpha_2 & \xrightarrow{\sigma_2} & -\alpha_j \\
\Vert & & \Vert & & & & \\
\mathbb{Q}[\alpha_1] & \xrightarrow[\cong]{\sigma_1} & \mathbb{Q}[\alpha_j] & & \alpha_1 & \xrightarrow{\sigma_1} & \alpha_j \\
\vert & & \vert & & & & \\
\mathbb{Q} & = & \mathbb{Q} & & & &
\end{array}
\qquad (10.15)
$$

we can choose β among the roots $\alpha_1, \alpha_2, \alpha_3, \alpha_4$ that are roots of the factor $h_2^{\sigma_2}(X)$. Here we now have to distinguish the cases, depending on which α_j we have chosen in the first step.

$j = 1$:   Then $\sigma_2 = \mathrm{id}_{\mathbb{Q}[\sqrt[4]{2}]}$. So

$$h_2^{\sigma_2}(X) = h_2(X) = X^2 + \sqrt{2} \in \mathbb{Q}[\sqrt[4]{2}][X]$$

and we can freely choose $\beta \in \{\alpha_3, \alpha_4\}$. So in this case, we ultimately obtain two possible automorphisms $\varphi_1, \varphi_2 \in \mathrm{Gal}(Z|\mathbb{Q})$, which are determined by:

$$
\begin{array}{cc}
\varphi_1 : \alpha_1 \mapsto \alpha_1 & \varphi_2 : \alpha_1 \mapsto \alpha_1 \\
\qquad \alpha_3 \mapsto \alpha_3 & \qquad \alpha_3 \mapsto -\alpha_3
\end{array}
$$

Note that $\varphi_1 = \mathrm{id}_Z$.

$j = 2$:   Then $\sigma_2(\sqrt[4]{2}) = -\sqrt[4]{2}$ and thus $\sigma_2(\sqrt{2}) = \sqrt{2}$, so again $h_2^{\sigma_2}(X) = h_2(X)$ and one can choose β again from $\{\alpha_3, \alpha_4\}$ freely. We thus again have two elements of the Galois group $\varphi_3, \varphi_4 \in \mathrm{Gal}(Z|\mathbb{Q})$, which are determined by

$$\varphi_3 : \alpha_1 \mapsto -\alpha_1 \qquad \varphi_4 : \alpha_1 \mapsto -\alpha_1$$
$$\alpha_3 \mapsto \alpha_3 \qquad\qquad \alpha_3 \mapsto -\alpha_3$$

$j = 3$: Then $\sigma_2(\sqrt[4]{2}) = i\sqrt[4]{2} = \alpha_3$ and thus

$$\sigma_2(\sqrt{2}) = \sigma_2(\sqrt[4]{2})^2 = \alpha_3^2 = -\sqrt{2}$$

Therefore, we obtain

$$h_2^{\sigma_2}(X) = X^2 + \sigma_2(\sqrt{2}) = X^2 - \sqrt{2}$$

and one can choose β freely from $\{\alpha_1, \alpha_2\}$. We thus have again two elements of the Galois group $\varphi_5, \varphi_6 \in \mathrm{Gal}(Z|\mathbb{Q})$, which are determined by

$$\varphi_5 : \alpha_1 \mapsto \alpha_3 \qquad \varphi_6 : \alpha_1 \mapsto \alpha_3$$
$$\alpha_3 \mapsto \alpha_1 \qquad\qquad \alpha_3 \mapsto -\alpha_1$$

$j = 4$: analogous to the case $j = 3$ again $h_2^{\sigma_2}(X) = X^2 - \sqrt{2}$ holds and thus $\beta \in \{\alpha_1, \alpha_2\}$ can be chosen freely. We obtain two Galois group elements $\varphi_7, \varphi_8 \in \mathrm{Gal}(Z|\mathbb{Q})$, which are determined by

$$\varphi_7 : \alpha_1 \mapsto -\alpha_3 \qquad \varphi_8 : \alpha_1 \mapsto -\alpha_3$$
$$\alpha_3 \mapsto \alpha_1 \qquad\qquad \alpha_3 \mapsto -\alpha_1$$

We observe that the Galois group $\mathrm{Gal}(Z|\mathbb{Q})$ can be embedded into the symmetric group $S(4)$

$$\eta : \mathrm{Gal}(Z|\mathbb{Q}) \to S(4), \tag{10.16}$$

as follows: given $\varphi \in \mathrm{Gal}(Z|\mathbb{Q})$ let $\eta(\varphi)$ denote the permutation such that $\varphi(\alpha_j) = \alpha_{\eta(\varphi)}(j)$ holds for all j. This is a group homomorphism (as can be easily verified) and even injective (because φ is determined by the value it associates to α_j). For example, for φ_2:

$$\varphi_2 : \alpha_1 \mapsto \alpha_1, \alpha_2 \mapsto \alpha_2, \alpha_3 \mapsto \alpha_4, \alpha_4 \mapsto \alpha_3,$$

So

$$\eta(\varphi_2) = \begin{pmatrix} 1\,2\,3\,4 \\ 1\,2\,4\,3 \end{pmatrix} = (3\,4) \in S(4).$$

If we calculate this map explicitly in our example, we get:

$$\eta(\varphi_1) = \mathrm{id} \qquad\qquad \eta(\varphi_2) = \begin{pmatrix} 1\,2\,3\,4 \\ 1\,2\,4\,3 \end{pmatrix} = (3\,4)$$

$$\eta(\varphi_3) = \begin{pmatrix} 1\,2\,3\,4 \\ 2\,1\,3\,4 \end{pmatrix} = (1\,2) \qquad \eta(\varphi_4) = \begin{pmatrix} 1\,2\,3\,4 \\ 2\,1\,4\,3 \end{pmatrix} = (1\,2)(3\,4)$$

$$\eta(\varphi_5) = \begin{pmatrix} 1\,2\,3\,4 \\ 3\,4\,1\,2 \end{pmatrix} = (1\,3)(2\,4) \quad \eta(\varphi_6) = \begin{pmatrix} 1\,2\,3\,4 \\ 3\,4\,2\,1 \end{pmatrix} = (1\,3\,2\,4)$$

$$\eta(\varphi_7) = \begin{pmatrix} 1\,2\,3\,4 \\ 4\,3\,1\,2 \end{pmatrix} = (1\,4\,2\,3) \quad \eta(\varphi_8) = \begin{pmatrix} 1\,2\,3\,4 \\ 4\,3\,2\,1 \end{pmatrix} = (1\,4)(2\,3)$$

and therefore we have an isomorphism from $\mathrm{Gal}(Z|\mathbb{Q})$ to the subgroups

$$\{1, (1\,2), (3\,4), (1\,2)(3\,4), (1\,3)(2\,4), (1\,4)(2\,3), (1\,3\,2\,4), (1\,4\,2\,3)\} \subset S(4)$$

Note that this group is not abelian, hence $\mathrm{Gal}(Z|\mathbb{Q})$ is not abelian. For example

$$(1\,2) \cdot (1\,3)(2\,4) = (1\,3\,2\,4) \neq (1\,4\,2\,3) = (1\,3)(2\,4) \cdot (1\,2).$$

Remark 10.17 By the way, no one forces us to calculate the Galois group $\mathrm{Gal}(Z|\mathbb{Q}) = \mathrm{Aut}_{\mathbb{Q}}(Z)$ in the above example exactly as we have discussed it in the construction of the tower of field extensions. One can also reach the goal in a different way: Observe that $Z = \mathbb{Q}[\sqrt[4]{2}, i]$ and that we can therefore also consider the following shorter version of a tower of field extensions arriving at the same splitting field:

$$Z = \mathbb{Q}[\sqrt[4]{2}, i]$$
$$|$$
$$\mathbb{Q}[\sqrt[4]{2}]$$
$$|$$
$$\mathbb{Q}.$$

The **1st step** remains the same: For each $j = 1, 2, 3, 4$ there is exactly one σ_1 with

$$
\begin{array}{ccc}
\mathbb{Q}[\alpha_1] & \xrightarrow{\;\sigma_1\;}_{\cong} & \mathbb{Q}[\alpha_j] \\
| & & | \\
\mathbb{Q} & =\!\!=\!\!= & K
\end{array}
\qquad
\alpha_1 \xmapsto{\;\sigma_1\;} \alpha_j
$$

For the **2nd (and thus last) step,** apply Theorem A' to the upper rectangle in the following diagram to find a corresponding $\tau \in \mathrm{Gal}(Z|\mathbb{Q})$:

$$
\begin{array}{ccc}
Z = \mathbb{Q}[\sqrt[4]{2}, i] & \xrightarrow{\;\tau\;}_{\cong} & \mathbb{Q}[\alpha_j, \beta] = Z \\
| & & | \\
\mathbb{Q}[\sqrt[4]{2}] & \xrightarrow{\;\sigma_1\;}_{\cong} & \mathbb{Q}[\alpha_j] \\
| & & | \\
\mathbb{Q} & =\!\!=\!\!= & \mathbb{Q}
\end{array}
\qquad
\begin{array}{c}
i \xmapsto{\;\tau\;} \beta \\[1ex]
\alpha_1 \xmapsto{\;\sigma_1\;} \alpha_j
\end{array}
$$

What does Theorem A' say about the possibilities for choicses of β? On the left hand side, we have to find the minimal polynomial $q(X) \in \mathbb{Q}[\sqrt[4]{2}][X]$ of $i \in \mathbb{C}$. This is easy, because it is simply $q(X) = X^2 + 1$ since it has to irreducible due to degree reasons due to the fact that $i \notin \mathbb{Q}[\sqrt[4]{2}] \subset \mathbb{R}$.

Theorem A' states that one can freely choose β from the zeros of $q^{\sigma_1}(X) \in \mathbb{Q}[\alpha_j][X]$. We have $q^{\sigma_1}(X) = X^2 + 1$ (no matter which α_j we have chosen in the first step), so one can freely choose $\beta \in \{i, -i\}$.

If you now combine all choices (in the 1st step the choice of α_j, in the 2nd step the choice of $\pm i$), you will again obtain the elements of the Galois group $\mathrm{Gal}(Z|\mathbb{Q}) = \{\varphi_1, \ldots, \varphi_8\}$ as above, but perhaps faster.

Red thread to the previous chapter
In this chapter, we have seen that

- for a given $f(X) \in K[X]$ of degree ≥ 1 one can always construct a finite field extension $L|K$ that contains a root α of $f(X)$,
- that this can be continued successively to arrive at a **splitting field** of the polynomial, in which all roots are contained,
- that the splitting field $Z|K$ (and its intermediate fields) store a lot of information about the roots of $f(X)$—for example, when examining the degrees of the field extensions in a **tower of such extensions** that leads to $Z|K$,
- that more generally for a finite extension $L|K$ the K-**homomorphisms** $\operatorname{Hom}_K(L, \Omega)$ into an algebraically closed field $\Omega \supset L \supset K$ are an abstract way to encode this information,
- that in the case of a splitting field $L = Z$, we will automatically have $\operatorname{Hom}_K(Z, \Omega) = \operatorname{Aut}_K(Z)$,
- that $\operatorname{Aut}_K(Z)$ is naturally (with composition as a group operation) a group, which we call the **Galois group** $\operatorname{Gal}(Z|K)$,
- we were able to calculate our first Galois group $\operatorname{Gal}(Z|K)$ for the splitting field of $X^4 - 2$.

Exercises:
10.1 Consider the irreducible polynomial $f(X) = X^3 - 3X + 1 \in \mathbb{Q}[X]$ (see Exercise 9.1).

(a) Show: If $\alpha \in \mathbb{C}$ is a root, then so is $\beta := \alpha^2 - 2$.
(b) Determine the degree $[Z : \mathbb{Q}]$ of the splitting field Z of f in $\mathbb{C}$.

10.2 Let p be a prime number. Consider the irreducible polynomial (according to the Eisenstein criterion) $f(X) = X^4 + pX^2 + p \in \mathbb{Z}[X]$. Let $Z \subset \mathbb{C}$ be the splitting field of f inside $\mathbb{C}$. Show:

(a) If $\pm\alpha, \pm\beta$ are the different roots of f in Z, then $\beta^2 + p + \alpha^2 = 0$ holds.
(b) We have: $[Z : \mathbb{Q}] \in \{4, 8\}$.
(c) Compute $[Z : \mathbb{Q}]$ in the case $p = 2$ (see (10.6)). Note: Keep in mind that $\mathbb{Q}[i] \subset \mathbb{Q}[\alpha]$ with $[\mathbb{Q}[\alpha] : \mathbb{Q}[i]] = 2$. Now, if $\beta \in \mathbb{Q}[\alpha]$, then one could write $\beta = z + w\alpha$ with $z, w \in \mathbb{Q}[i]$. If one squares this equation, one can deduce $\alpha \in \mathbb{Q}[i]$ and arrive at a contradiction.
(d) Compute $[Z : \mathbb{Q}]$ in the case $p = 5$. Hint: For $p = 5$ compute the element $(\alpha^3 + 3\alpha)^2 \in \mathbb{Q}[\alpha]$ for a root α (preferably, without ever calculating the root α!) in easier terms.

10.3 Consider the field $\mathbb{F}_5 := \mathbb{Z}/5\mathbb{Z}$.

(a) Consider the $\mathbb{F}_5$-vector space $V = \mathbb{F}_5^3$ and construct on it the structure of a field such that $V|\mathbb{F}_5$ is a field extension and one finds a root of the polynomial $X^3 + X + 1 \in \mathbb{F}_5[X]$ inside V.

(b) Show that inside V there also exists a root of $X^3 + X + 4$.

(c) Show that $X^2 + X + 1 \in V[X]$ is irreducible. (Hint: This is a cumbersome exercise if you calculate it by hand. See Exercise 10.4.)

10.4 Convince yourself that the following lines in SAGE realize the field $L = \mathbb{F}_5[X]/(X^3 + X + 1)$ and make the computer compute that $X^2 + X + 1$ has no root in L.

```
R.<t>=GF(5)[]
L=R.quotient(t^3+t+1,'x')
x=L.gen()
count=0
for c in L:
    w=c^2+c+1
    if (w==L.zero()): count=count+1
    print('f(',c,')=',w)
print('f has ',count,' roots in L')
```

Change the corresponding line to make SAGE calculate the number of zeros of $X^3 + X + 1$ in L. This should yield a result ≥ 1.

Intermezzo: An Explicit Example $X^5 - 777X + 7$

11

We ask ourselves the following question: Consider the polynomial $f(X) = X^5 - 777X + 7 \in \mathbb{Q}[X]$. Are there any algebraic relations between its roots (in $\mathbb{C}$), what is the Galois group $\mathrm{Gal}(Z|\mathbb{Q})$ of the splitting field $Z \subset \mathbb{C}$?

We have already developed some ideas and we will add some more in this example:

1. Since the monic (hence primitive) polynomial $f(X) \in \mathbb{Z}[X]$ fulfills the Eisenstein criterion for $p = 7$, it is irreducible in $\mathbb{Q}[X]$ according to Eisenstein+Gauss.
2. In the algebraically closed field $\mathbb{C}$ the function $f(X)$ decomposes into linear factors:

$$f(X) = \prod_{j=1}^{5}(X - \alpha_j) \in \mathbb{C}[X],$$

   where $\alpha_1, \ldots, \alpha_5 \in \mathbb{C}$ are the roots. A priori, we do not yet know whether these are pairwise different.
3. If we use a bit of standard Calculus, we can examine the function that $f(X)$ induces as a map $\mathbb{R} \to \mathbb{R}$ and find that f has exactly three real roots. So let's say $\alpha_1, \alpha_2, \alpha_3 \in \mathbb{R}$ and $\alpha_4, \alpha_5 \notin \mathbb{R}$.
   As already justified in Sect. 10.2, it follows from $f(X) \in \mathbb{Q}[X] \subset \mathbb{R}[X]$, that for any complex root $\beta \in \mathbb{C}$ of f, the complex-conjugate number $\bar{\beta} \in \mathbb{C}$ is also a root, and thus

$$\alpha_5 = \overline{\alpha_4}.$$

   Since $\alpha_4 \in \mathbb{C} \setminus \mathbb{R}$ it follows again that $\alpha_4 \neq \alpha_5$.
   Therefore, the roots $\alpha_1, \ldots, \alpha_5 \in \mathbb{C}$ are all pairwise different (Fig. 11.1).

© The Author(s), under exclusive license to Springer-Verlag GmbH, DE, part of
Springer Nature 2024
M. Hien, *Abstract Algebra*, Mathematics Study Resources 7,
https://doi.org/10.1007/978-3-662-67974-6_11

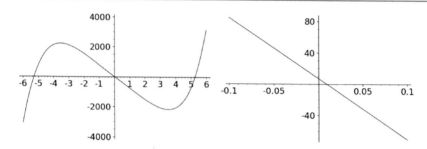

Fig. 11.1 Real graph of the polynomial $f(X)$—the right picture is there to better see the smallest positive root

4. The last observation can also be seen in the following way (anticipating Chap. 13), without having to use (real) calculus. Since $f(X)$ is irreducible, $f(X)$ is the minimal polynomial of each of its roots α_j. Now suppose that at least one of the roots is multiple, i.e., $\alpha_i = \alpha_j$ for a pair $i \neq j$. Then consider the (formal) derivative (for this we do not need analysis, we simply always replace $X^k \rightsquigarrow kX^{k-1}$):

$$f'(X) = 5X^4 - 777 \in \mathbb{Q}[X].$$

But this formal derivation fulfills the usual product formula and under our assumption we have the following decomposition in $\mathbb{C}[X]$:

$$f(X) = (X - \alpha_j)^2 \cdot g(X)$$

for a $g(X) \in \mathbb{C}[X]$. We deduce that

$$f'(X) = 2(X - \alpha_j)g(X) + (X - \alpha_j)^2 g'(X),$$

and thus the multiple root α_j is still a root of the derivative: $f'(\alpha_j) = 0$. But this is not possible, because $f(X)$ is the minimal polynomial of α_j over $\mathbb{Q}$ and $\deg(f'(X)) = 4 < \deg(f(X)) = 5$ holds.

5. We have the splitting field $Z = \mathbb{Q}[\alpha_1, \dots, \alpha_5]$ and the tower of field extenstions of Fig. 11.2.

Fig. 11.2 The tower of field extensions for the explicit example

$$
\begin{array}{c}
\mathbb{Q}[\alpha_1, \alpha_2, \alpha_3, \alpha_4, \alpha_5] =\!=\!=\!= Z \\
1 \| \\
\mathbb{Q}[\alpha_1, \alpha_2, \alpha_3, \alpha_4] \\
\leq 2 \mid \\
\mathbb{Q}[\alpha_1, \alpha_2, \alpha_3] \\
\leq 3 \mid \\
\mathbb{Q}[\alpha_1, \alpha_2] \\
\leq 4 \mid \\
\mathbb{Q}[\alpha_1] \\
5 \mid \\
\mathbb{Q}
\end{array}
$$

6. Since we know too little about the roots, it will not be possible to calculate the Galois group $\text{Gal}(Z|\mathbb{Q})$ by going up the tower of fields and installing rungs as in (10.15) for the simpler example.

7. But we do know that

$$\text{Gal}(Z|\mathbb{Q}) := \text{Aut}_{\mathbb{Q}}(Z)$$

is a **group**.

8. We even have more information about the group (as already observed in (10.16)). Each element $\varphi \in \text{Gal}(Z|\mathbb{Q})$ induces a permutation of the roots $\alpha_1, \ldots, \alpha_5$. This defines an injective group homomorphism

$$\eta : \text{Gal}(Z|\mathbb{Q}) \hookrightarrow S(5), \ \varphi \mapsto \eta(\varphi), \tag{11.2}$$

where $\eta(\varphi)$ is defined by the equation

$$\varphi(\alpha_j) = \alpha_{\eta(\varphi)(j)} \ \text{ for all } j = 1, \ldots, 5$$

Thus, $\text{Gal}(Z|\mathbb{Q})$ is isomorphic to a subgroup of $S(5)$.

9. **New idea:** Since $\mathbb{Q} \subset Z \subset \mathbb{C}$, we have an element of the Galois group given by the **complex conjugation.** Because

$$\tau : Z \to Z, \ \xi \mapsto \overline{\xi}$$

is a field automorphism (with values again in Z, because $\overline{\alpha_i} = \alpha_i$ for $i = 1, 2, 3$ and $\overline{\alpha_4} = \alpha_5$), and more precisely it is a $\mathbb{Q}$-automorphism:

$$\tau \in \text{Gal}(Z|\mathbb{Q}).$$

Note that $\tau \neq \text{id}_Z$ because $\tau(\alpha_4) = \alpha_5 \neq \alpha_4$. Since $\tau \circ \tau = \text{id}_Z$, it follows that $\text{ord}(\tau) = 2$. Considering η from (11.2), it follows that

$$\tau \overset{\eta}{\mapsto} \begin{pmatrix} 1 & 2 & 3 & 4 & 5 \\ 1 & 2 & 3 & 5 & 4 \end{pmatrix} = (4\,5).$$

10. **New Idea:** Now one delves into group theory and finds the following theorem:

Theorem 11.1 (Cauchy) If G denotes a finite group and $p \mid \#G$ is a prime number that divides the order of the group, then there is an element of order p in G.

Proof Consider the set

$$M := \{(g_1, \ldots, g_p) \in G^p \mid g_1 \cdots g_p = 1\} \subset G^p$$

of all p-tuples of group elements that yield 1 as its product. Obviously, $\#M = (\#G)^{p-1}$, because one can choose $g_1, \ldots, g_{p-1}$ arbitrarily in G and then must complete it with $g_p = (g_1 \cdots g_{p-1})^{-1}$ to obtain a tuple in M.

Now consider the equivalence relation on M given by   □

$$(g_1, \ldots, g_p) \sim (h_1, \ldots, h_p) :\Longleftrightarrow$$

there exists a k with $(g_1, \ldots, g_p) = (h_{k+1}, \ldots, h_p, h_1, \ldots h_k)$.

Claim For each class we have $\#[(g_1,\ldots,g_p)] \in \{1,p\}$.

Proof Apparently, $\#[(g_1,\ldots,g_p)] = 1 \Leftrightarrow g_1 = \ldots = g_p$.   Furthermore, $\#[(g_1,\ldots,g_p)] \leq p$ and equality holds if and only if all tuples in the class are pairwise different.

Now let $\#[(g_1,\ldots,g_p)] < p$, then not all cyclic permutations of the tuple $(g_1,\ldots,g_p)$ are pairwise different and (after an eventual cyclic permutation) there is a $k \in \{1,\ldots,p-1\}$ with

$$
\begin{array}{ccccccc}
(g_1, & g_2 & \cdots & g_{p-k}, & g_{p-k+1} & \cdots & g_p) \\
\| & \| & & \| & \| & & \| \\
(g_{k+1}, & g_{k+2} & \cdots & g_p, & g_1 & \cdots & g_k)
\end{array}
$$

It follows that $g_i = g_j$ for $i \equiv j + k \bmod p$. Hence also $g_i = g_j$ holds for $i \equiv j + qk \bmod p$ for any $q \in \mathbb{Z}$. But since $\gcd(k,p) = 1$, there exist $x,y \in \mathbb{Z}$ with $xk + yp = 1$. Therefore, for any i,j we always obtain

$$i - j = (i-j)xk + (i-j)yp \equiv (i-j)xk \bmod p$$

and thus $g_i = g_j$.

So we have shown: If $\#[(g_1,\ldots,g_p)] < p$, then already $g_1 = \ldots = g_p$ holds and thus $\#[(g_1,\ldots,g_p)] = 1$. $\square$

Now we set:

$$\varepsilon := \#\{[(g_1,\ldots,g_p)] \mid \#[(g_1,\ldots,g_p)] = 1\}$$
$$\pi := \#\{[(g_1,\ldots,g_p)] \mid \#[(g_1,\ldots,g_p)] = p\},$$

and observe that

$$(\#G)^{p-1} = \#M = \varepsilon + \pi \cdot p,$$

because the set M decomposes into the disjoint union of its classes—Corollary 4.9. Therefore we have

$$p \mid (\#G)^{p-1} = \varepsilon + \pi \cdot p \Longrightarrow p \mid \varepsilon.$$

In particular, $\varepsilon \geq 2$ and we deduce that in addition to the class $[(1,1,\ldots,1)]$ (which only has one element) there has to exist another class $[(g_1,\ldots,g_p)]$ with $\#[(g_1,\ldots,g_p)] = 1$. For this class we observe that

$$1 \neq g_1 = g_2 = \ldots = g_p =: g,$$

and finally $g^p = 1$ (because $(g,\ldots,g) \in M$) and $g \neq 1$, therefore $\mathrm{ord}(g) = p$. $\square$

11. So now we know:

- $\mathrm{Gal}(Z|\mathbb{Q})$ is isomorphic to a subgroup of $S(5)$ (so that we can identify it with this subgroup),
- has an element of order 2, thus a 2-cycle (in the interpretation as a subgroup of $S(5)$)
- and an element of order 5, thus a 5-cycle.

12. Now one has an explicit problem to solve in the group $S(5)$: What subgroups are there that fulfill the last two properties?

Lemma 11.2 *If p is a prime number and $H \subset S(p)$ is a subgroup that contains a 2-cycle and a p-cycle, then $H = S(p)$ is already the full symmetric group.*

Proof Let $\sigma = (s_1 \, s_2 \, \ldots \, s_p) \in H$ be a p-cycle and $\tau = (x \, y)$ be a 2-cycle. Since σ can start with any element (by cyclic permutation) and $\{s_1, \ldots, s_p\} = \{1, \ldots, p\}$ holds, we can assume without loss of generality that $\tau = (s_1 \, s_k)$ for a k with $2 \leq k \leq p$, so $s_1 = x$ and $s_k = y$.

Since p is a prime number, $\sigma^k \in H$ for $2 \leq k < p$ is again a p-cycle and it has the form $\sigma^k = (s_1 \, s_k \, \ldots)$. So we prepared ourselves to obtain the situation that we have

- a p-cycle of the form $\eta = (s_1 \, s_k \, x_1 \, \ldots \, x_{p-2}) \in H$
- and a 2-cycle of the form $\tau = (s_1 \, s_k) \in H$.

Conjugation with

$$\xi := \begin{pmatrix} 1 & 2 & 3 & \ldots & p \\ s_1 & s_k & x_1 & \ldots & x_{p-2} \end{pmatrix}^{-1}$$

is a group isomorphism

$$\mathrm{conj}_\xi : S(p) \to S(p), \quad g \mapsto \xi \cdot g \cdot \xi^{-1},$$

with $\mathrm{conj}_\xi(\sigma) = (1 \, 2 \, 3 \, \ldots \, p)$ and $\mathrm{conj}_\xi(\tau) = (1 \, 2)$. It is sufficient to show the assertion of the lemma after applying conj_ξ, so we can assume without loss of generality that $(1 \, 2 \, \ldots \, p), (1 \, 2) \in H$.

Now it is easy to see that

$$H \ni (1 \, 2 \, \ldots \, p)^{k-1} \cdot (1 \, 2) \cdot (1 \, 2 \, \ldots \, p)^{-(k-1)} = (k \, k + 1)$$

holds for $2 \leq k \leq p$—note that it is not assumed here that H must be a normal subgroup, the factors are all contained in H anyway. Thus, all transpositions $(k \, k + 1) \in H$ lie in H.

Left-multiplication of $(k \, k + 1)$ to any permutation has the following effect:

$$\begin{pmatrix} 1 & 2 & \ldots & k & k+1 & \ldots & p \\ * & * & \ldots & a & b & \ldots & * \end{pmatrix} \cdot (k \, k + 1) = \begin{pmatrix} 1 & 2 & \ldots & k & k+1 & \ldots & p \\ * & * & \ldots & b & a & \ldots & * \end{pmatrix}$$

and it can easily be seen that any permutation can be transformed into the identity with a finite number of such left multiplications. Therefore, the given permutation is the finite product of such special transpositions. It follows that $H = S(p)$. $\square$

In summary, we have seen:

Lemma 11.3 *If $Z \subset \mathbb{C}$ is the splitting field of*

$$f(X) = X^5 - 777X + 7 \in \mathbb{Q}[X]$$

over $\mathbb{Q}$, then

$$\mathrm{Gal}(Z|\mathbb{Q}) \cong S(5),$$

and in particular $[Z : \mathbb{Q}] = 5! = 120$.

Proof The statement about $[Z : \mathbb{Q}]$ will be proven generally later (Chap. 13), here we can justify it as follows: Since $\mathrm{Gal}(Z|\mathbb{Q}) \cong S(5)$, one can obtain **every** permutation of the roots $\alpha_1, \ldots, \alpha_5$ as an element of the Galois group. If one looks at the tower of field extensions (11.1), in each step of the construction of the tower, the maximum possible degree for the field extensions must occur . Recall that the degree of the extension in each step is the degree of the minimal polynomial of the newly adjoined element. If it had a smaller degree than the maximum degree possible, this minimal polynomial would be a proper divisor of a smaller degree than the maximum possible. This would mean that not all remaining roots are available for the choice of the new element - which we called β in the construction - at this step, but only those that are also roots of this factor.

However, this contradicts the fact that one can choose all roots on the right hand side at each step, because in the end all permutations in $S(n)$ occur. $\square$

And now?

So we have now understood that we were able to determine a Galois group through

- the fact that $\mathrm{Gal}(Z|\mathbb{Q})$ is a group,
- the observation that in the case of a splitting field Z of a polynomial of degree n with pairwise different roots there is an embedding $\mathrm{Gal}(Z|\mathbb{Q}) \hookrightarrow S(n)$ (which depends on a chosen ordering of the roots),
- and a lemma on certain subgroups in $S(n)$.

It is remarkable that we could determine this Galois group **without explicitly knowing the roots of the polynomial.**

This is a subtle but far-reaching achievement of Galois theory!

Exercises:

11.1 Consider the group

$$\left\{ \begin{pmatrix} 1 & 0 \\ x & y \end{pmatrix} \in \mathrm{GL}_2(\mathbb{F}_7) \mid x \in \mathbb{F}_7 \text{ and } y \in \mathbb{F}_7^\times \right\}.$$

(a) Show that for every $y \in \mathbb{F}_7$ the equation $y^7 = y$ holds. (Hint: given $y \neq 0$ we have $y \in \mathbb{F}_7^\times$ and the latter is a group with 6 elements.)

(b) Justify that the analogous statement holds for every prime number p: $y^p = y$ for all $y \in \mathbb{F}_p$. This statement is often called the **Fermat's Little Theorem**.

(c) Show that exactly the elements of the form $g = \begin{pmatrix} 1 & 0 \\ x & 1 \end{pmatrix}$ with $x \neq 0$ are the elements in G of order 7.

(d) Show that $\#G = 7 \cdot 6 = 2 \cdot 3 \cdot 7$ and determine all elements of order 2 and 3 in G.

11.2 Let $f(X) \in \mathbb{Q}[X]$ be irreducible and assume that f has a real root $\alpha \in \mathbb{R}$ and a complex root $\beta \in \mathbb{C} \setminus \mathbb{R}$ has. Show that $\mathrm{Gal}(f)$ is not abelian.

Normal Field Extensions

12

12.1 Algebraic Closure

And now?
In this chapter, we will need at a few places that every field L is embedded into an algebraically closed field Ω. When considering field extensions $L|\mathbb{Q}$ within $\mathbb{C}$, we know this since one can take $\Omega = \mathbb{C}$ (or better $\Omega = \overline{\mathbb{Q}} \subset \mathbb{C}$, because this field is still algebraic over $\mathbb{Q}$). In general, we can mimic this situation:

Definition 12.1 Let K be a field. Then a field Ω with $K \subset \Omega$ is called an **algebraic closure** of K, if

1. the field extension $\Omega|K$ is algebraic and
2. Ω is algebraically closed (i.e., every polynomial $f(X) \in \Omega[X]$ of degree $\deg(f) \geq 1$ has a root in Ω).

Example: As already seen in the Defining Lemma 2.23, $\overline{\mathbb{Q}} := \{z \in \mathbb{C} \mid z \text{ algebraic over } \mathbb{Q}\}$ is an algebraic closure of $\mathbb{Q}$.

Theorem 12.2 Every field has an algebraic closure.

Proof This proof is outsourced to the appendix. Theorem 10.1 plays a crucial role in it. One only needs to apply this theorem successively and in a clever way to prove the claim. However, this "clever successive application" is not so trivial (and even includes a few almost ingenious tricks) and it would interrupt the chapter too much if we wanted to carry it out now. □

© The Author(s), under exclusive license to Springer-Verlag GmbH, DE, part of
Springer Nature 2024
M. Hien, *Abstract Algebra*, Mathematics Study Resources 7,
https://doi.org/10.1007/978-3-662-67974-6_12

And now?

Of course, it is a pity that we postponed the proof, on the other hand side we are very often interested in situations over the rationals $K = \mathbb{Q}$ and then we have $\mathbb{C}$ or $\overline{\mathbb{Q}}$ available.

In Chap. 16 it will hurt us a bit. I will comment on it then.

12.2   Extension of Field Homomorphisms

And now?

We have made essential observations with Theorems A and A' on how to extend a given field homomorphism $\sigma : K \to K'$ within fields L or L' with enough roots of polynomials (see below for more details)

$$
\begin{array}{ccc}
L & & L' \\
| & & | \\
K[\alpha] & \xrightarrow{\ \varphi\ } & K'[\beta] \\
| & & | \\
K & \xrightarrow{\ \sigma\ } & K'
\end{array}
$$

to obtain an extension φ—more precisely, one has $\alpha \in L$ and one needs a root $\beta \in L'$ of $f^\sigma(X) \in K'[X]$, when $f(X) \in K[X]$ denotes the minimal polynomial of α over K. One could call this a *first level* extension.

Now one can continue this successively and finally ask whether one ultimately obtains a continuation $\psi : L \to L'$ (if L' has enough roots of polynomials). We want to briefly explain this:

Theorem 12.3 Let $L|K$ be an algebraic field extension and $\sigma : K \to K'$ a field homomorphism. Let furthermore $\Omega|K'$ be an algebraically closed field. Then there exists a field homomorphism $\psi : L \to \Omega$ with $\psi|_K = \sigma$:

$$
\begin{array}{ccc}
L & \xrightarrow{\ \psi\ } & \Omega \\
| & & | \\
K & \xrightarrow{\ \sigma\ } & K'
\end{array}
$$

Proof If $L|K$ is finite, one can find elements $\alpha_1, \ldots, \alpha_m \in L$ such that $L = K[\alpha_1, \ldots, \alpha_m] \subset L$ holds. Then one can simply apply Theorem A' 10.11 successively (i.e., climb up a tower of field extensions with finite height) and is done. In each step one has to make sure to find appropriate target elements in Ω, but this is clear since one can always find roots of the respective minimal polynomial of α_j over $K[\alpha_1, \ldots, \alpha_{j-1}]$ because Ω is algebraically closed.

If $L|K$ is not finite, one can argue with Zorn's Lemma: The set

$$\mathcal{M} := \{(M, \eta) \mid K \subset M \subset L \text{ and } \eta : M \to \Omega \text{ with } \eta|_K = \sigma\}$$

is not empty (because $(K, \sigma) \in \mathcal{M}$). If one defines

$$(M, \eta) \leq (N, \mu) :\Leftrightarrow M \subset N \text{ and } \mu|_M = \eta,$$

this is a partial order on $\mathcal{M}$. Each totally ordered subset has an upper bound in $\mathcal{M}$, because one can take the union.

So, $\mathcal{M}$ has a maximal element according to Zorn's lemma $(M, \eta) \in \mathcal{M}$. We show that $M = L$ holds. Suppose we had $M \subsetneq L$. Then there is an $\alpha \in L \setminus M$. According to Theorem A', one then has an extension μ as in the diagram:

$$
\begin{array}{ccc}
L & & \\
| & & \\
M[\alpha] & \xrightarrow{\ \mu\ } & \Omega[\beta] = \Omega \\
| & & \| \\
M & \xrightarrow{\ \eta\ } & \Omega \\
| & & | \\
K & \xrightarrow{\ \sigma\ } & K'
\end{array}
$$

when $\beta \in \Omega$ is a root of $f^\eta(X) \in \Omega[X]$ for the minimal polynomial $f(X) \in M[X]$ of α over M. Therefore, $(M[\alpha], \mu) \in \mathcal{M}$ which contradicts the maximality of (M, η). $\square$

12.3 Normal Extensions

And now?
In Chap. 10 we made the basic observation that field homomorphisms respect/recognize minimal polynomials in a certain way. For this, we started with an algebraic field extension and (cp. Definition 10.12) considered the set $\mathrm{Hom}_K(L, \Omega)$ in the situation:

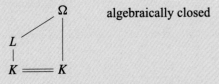

Ω algebraically closed

with an algebraically closed field Ω. The reason for considering such an Ω is that we need enough roots in the target field of such a homomorphism $\varphi \in \mathrm{Hom}_K(L, \Omega)$: If $\alpha \in L$, then $\varphi(\alpha)$ must be a root of the same minimal polynomial as α over K, so Ω must contain such roots in order to have such a $\varphi : L \to \Omega$ as desired.

On the other hand side, it is

- unsatisfactory that we have to use such an Ω because it actually plays no essential role for our interest in the equations, and
- very annoying that $\mathrm{Hom}_K(L, \Omega)$ has no multiplication operation and therefore no obvious group structure.

If $Z|K$ is a splitting field of a polynomial $f(X) \in K[X]$, we have seen that

$$\mathrm{Hom}_K(Z, \Omega) = \mathrm{Aut}_K(Z)$$

holds. This solves both problems: The right side has nothing to do with Ω any more and it is a group in a natural way.

Such a situation in which both problems are solved simultaneously, are very desirable and we give such situations a special name—see the next definition.

Definition 12.4 An algebraic field extension $L|K$ is called **normal**, if the following holds: If $f(X) \in K[X]$ is an irreducible polynomial that admits a root $\alpha \in L$, then $f(X)$ already decomposes in $L[X]$ into linear factors:

$$f(X) = c \cdot \prod_{j=1}^{n} (X - \alpha_j) \in L[X].$$

Remark 12.5 If one has an algebraically closed field Ω with $L \subset \Omega$, then a $f(X) \in K[X]$ decomposes in $\Omega[X]$ into linear factors

$$f(X) = c \cdot \prod_{j=1}^{n} (X - \omega_j) \in \Omega[X],$$

with some $c \in K$.

If $L|K$ is normal and $f(X)$ has a root in L (i.e., $\omega_j \in L$ holds for **one** $j \in \{1, \ldots, n\}$), then $\omega_i \in L$ already holds for **all** $i = \{1, \ldots, n\}$.

This solves the two problems mentioned in the *And now?* simultaneously:

Lemma 12.6 *If $L|K$ is a normal field extension and we have an embedding $L \subset \Omega$ into an algebraically closed field Ω, then*

$$\mathrm{Hom}_K(L, \Omega) = \mathrm{Aut}_K(L).$$

Proof We first need to show that for $(\varphi : L \to \Omega) \in \mathrm{Hom}_K(L, \Omega)$ automatically $\varphi(\alpha) \in L$ for all $\alpha \in L$ holds. But this is easy to see, because $\alpha \in L$ is algebraic by assumption, so it has a minimal polynomial $f(X) \in K[X]$ (irreducible). According to Theorem 10.6, $\varphi(\alpha) \in \Omega$ is also a root of $f(X)$, but according to Definition 12.4 or Remark 12.5, all roots of $f(X)$ are already in L, so $\varphi(\alpha) \in L$. Thus, we have $\mathrm{Hom}_K(L, \Omega) = \mathrm{Hom}_K(L, L)$.

It remains to show that every field homomorphism $(\varphi : L \to L) \in \mathrm{Hom}_K(L, L)$ is already an isomorphism. As a field homomorphism, φ is certainly injective. For surjectivity, consider an element $\beta \in L$. Since β is algebraic over K, it has a minimal polynomial $f(X) \in K[X]$. Let $Z \subset L$ is the splitting field of Z in L (which exists because L/K is normal and thus contains all roots of f in L since it contains β). Then $\varphi|_Z \in \mathrm{Hom}_K(Z, Z)$ holds by the same argument as above. But $[Z : K] < \infty$ and thus the injective K-linear endomorphism $\varphi|_Z$ on the finite-dimensional K-vector space is also surjective according to the rank theorem of linear algebra. In particular, $\beta \in \mathrm{Image}(\varphi|_Z) \subset \mathrm{Image}(\varphi)$. $\square$

And now?
Taking a closer look at Definition 12.4 and Lemma 12.6, one realizes that **normal** has been defined in such a way that the problems from the *And now?* at the beginning of the chapter have been solved by brute force. However, Definition 12.4 is not very handy, as one has to test all irreducible $f(X) \in K[X]$ with a root in L! These are generally infinitely many tests, which cannot be carried out explicitly. However, we have already seen that the problems for splitting fields Z/K are already solved. This can even be further specified—see the next theorem.

Theorem 12.7 A finite field extension L/K is normal if and only if it is a splitting field of a polynomial $f(X) \in K[X]$ (not necessarily irreducible).

Proof We show both directions separately:

"⇐":   Let L/K be the splitting field of a polynomial $f(X) \in K[X]$. Now let Ω be an algebraic closure of L. Then we have

$$\mathrm{Hom}_K(L, \Omega) = \mathrm{Aut}_K(L), \tag{12.1}$$

as we have already proven in Theorem 10.14 (note that the requirement in Theorem 10.14 that f should be irreducible was unnecessary. In the proof, only the remark "*namely f*" needs to be omitted). To show that L/K is normal, consider an irreducible polynomial $g(X) \in K[X]$ with a root $\alpha \in L$. Then

$$g(X) = c \cdot \prod_{j=1}^{m} (X - \omega_j) \in \Omega[X],$$

because Ω is algebraically closed. Let's say $\alpha = \omega_1 \in L$. If now $\omega_j \in \Omega$ is one of these roots, we can, with successive application of Theorem A, construct a K-homomorphism $(\varphi : L \to \Omega) \in \mathrm{Hom}_K(L, \Omega)$ which fulfills $\varphi(\alpha) = \omega_j$— Theorem 12.3. Because of (12.1) we obtain $\omega_j = \varphi(\alpha) \in L$, so all roots of $g(X)$ are already contained in L.

"$\Rightarrow$":   Since $L|K$ is finite, one finds $\beta_1, \ldots, \beta_m \in L$, so that

$$L = K[\beta_1, \ldots, \beta_m] \subset L$$

holds. Let $f_j(X) \in K[X]$ be the minimal polynomial of β_j over K. Since $L|K$ is normal, $f_j(X) \in L[X]$ splits into linear factors. Therefore, L is a splitting field of

$$f(X) = \prod_{j=1}^{m} f_j(X) \in K[X],$$

as claimed. $\square$

Finally, a trivial but important observation which will also enter the main theorem of Galois theory:

Lemma 12.8 *If $L|K$ is normal and $K \subset M \subset L$ is an intermediate field extension, then $L|M$ is normal as well.*

Check it out!

A good time to ask you to execute the proof by yourself first. It is a good exercise in precise bookkeeping about the respective minimal polynomials of a $\alpha \in L$ **over** M and **over** K, and how these two polynomials are related! The proof must begin with considering an irreducible $f(X) \in M[X]$ with root $\alpha \in L$. Then ...

Proof If $f(X) \in M[X]$ is irreducible and has a root $\alpha \in L$, then consider the minimal polynomial $q(X) \in K[X]$ of α over K. Since $L|K$ is normal, $q(X) \in L[X]$ decomposes into linear factors.

Except for a constant factor in M, $f(X) \in M[X]$ is the minimal polynomial of α over M. Since $q(\alpha) = 0$, the following divisibility relation holds:

$$f(X) \mid q(X) \in M[X] \subset L[X] \, .$$

Therefore, since $q(X)$ decomposes into linear factors in $L[X]$, so does $f(X)$. $\square$

Check it out!

Find an example $K \subset M \subset L$, such that $L|K$ is normal, but $M|K$ is not normal.

Thread to the previous chapter:

In this chapter, we have seen

- that every field has an algebraic closure (we didn't really see this, we just formulated it!),
- what a **normal** field extension is,
- that a finite field extension $L|K$ is normal if and only if it is the splitting field of
  a polynomial,
- that for a **normal** field extension $L|K$ and an algebraic closure $\Omega|L$ the equality

$$\mathrm{Hom}_K(L, \Omega) = \mathrm{Aut}_K(L)$$

holds, and that this is excellent because the right side is independent of Ω and is
also a group,
- that the following applies:

$$K \subset M \subset L \text{ and } L|K \text{ normal } \implies L|M \text{ normal.}$$

Exercises: 12.1 Let $L := \mathbb{Q}(\sqrt{3}, \sqrt{5})$.

(a) Show that $L = \mathbb{Q}(\sqrt{3} + \sqrt{5})$ holds.
(b) Determine the minimal polynomial of $\alpha := \sqrt{3} + \sqrt{5}$ over $\mathbb{Q}$ and over $\mathbb{Q}(\sqrt{3})$.
(c) Show that $L|K$ is normal.
(d) Determine the Galois group $\mathrm{Gal}(L|K)$.

12.2 Let $Z|K$ be a finite field extension. Furthermore, let L,M be intermediate
fields. We define **the composite** $LM \subset \Omega$ as

$$LM = \bigcap_{\substack{N \text{ intermediate field} \\ L,M \subset N}} N = L(M) = M(L)$$

(the latter two expressions are to be read as in the Defining Lemma 2.9). Show:

(a) If $L = K[\alpha_1, \ldots, \alpha_m]$, then $LM = M[\alpha_1, \ldots, \alpha_m]$.
(b) If $L|K$ is normal, then $LM|M$ is also normal.
(c) Let $L|K$ be normal. Then the restriction induces an injective group
     homomorphism

$$\theta : \mathrm{Gal}(LM|M) \to \mathrm{Gal}(L|L \cap M), \quad \sigma \mapsto \sigma|_L.$$

(After proving the main theorem of Galois theory, one can show that θ is even
an isomorphism).

12.3 Let $L|K$ be a finite field extension. Show that there is a field extension $N|K$
with $L \subset N$ such that $N|K$ is normal and $[N : K] < \infty$.

Such a field is called a **normal hull of** $L|K$.

12.4 Let K be a field such that $X^n - 1 \in K[X]$ decomposes into n pairwise different linear factors, and let

$$\mu_n(K) := \{\zeta \in K \mid \zeta^n = 1\} \subset K$$

be the set of its roots (according to the assumption with $\#\mu_n(K) = n$). Let $a \in K$. Let $L|K$ be a field extension such that there is an $\alpha \in L$ with $\alpha^n = a$ and such that $L = K[\alpha]$ holds. We denote the element α with $\alpha = \sqrt[n]{a} \in L$. Show that

(a) $L|K$ is normal,
(b) $\mu_n(K) \subset K^\times$ is an abelian subgroup and
(c) the map

$$\mathrm{Gal}(L|K) \to \mu_n(K), \ \sigma \mapsto \frac{\sigma(\sqrt[n]{a})}{\sqrt[n]{a}}$$

is an injective group homomorphism.

Separability

13

13.1 Motivation and Definition

And now?
Let $f(X) \in K[X]$ be an irreducible polynomial and $Z \mid K$ a splitting field, let's say in an algebraic closure Ω of Z. We want to use the Galois group

$$\mathrm{Gal}(Z|K) = \mathrm{Aut}_K(Z) \overset{\text{normal}}{=} \mathrm{Hom}_K(Z, \Omega)$$

as a tool to recognize as much information on the roots as possible. Let's say that the roots of f in Z form the set $\{\alpha_1, \dots, \alpha_m\}$ with

- **pairwise different** elements $\alpha_1, \dots, \alpha_m$,
- with multiplicities v_j, that is, in $Z[X]$ the function $f(X)$ decomposes into the factors

$$f(X) = c \cdot \prod_{j=1}^{m} (X - \alpha_j)^{v_j}.$$

For the construction of elements in $\mathrm{Gal}(Z|K)$ we have examined step by step (in the bottom-up method), see Fig. 13.1.

Where do the multiplicities occur? They do have an influence on the possible degrees of the field extensions—more precisely: In the r-th step, one has to decompose the polynomial in the polynomial ring

$$f(X) = (X - \alpha_1)^{v_1} \cdots (X - \alpha_r)^{v_r} \cdot g_r(X) \in K[\alpha_1, \dots, \alpha_r][X]$$

and in the process check the factor $g_r(X)$ for irreducible factors.

But: The field homomorphisms σ_r or finally $\sigma \in \mathrm{Gal}(Z|K)$ do not see these multiplicities **at all.** For σ_{r+1} it is only important which of the roots from

© The Author(s), under exclusive license to Springer-Verlag GmbH, DE, part of Springer Nature 2024
M. Hien, *Abstract Algebra*, Mathematics Study Resources 7,
https://doi.org/10.1007/978-3-662-67974-6_13

$\{\alpha_1, \ldots, \alpha_m\}$ are the roots of the same irreducible factor $h_r(X)|g_r(X)$ as a_{r+1} (on the left hand side)—because these are exactly the roots that $\sigma_{r+1}(\alpha_{r+1})$ can have as its value! The multiplicity of these roots is irrelevant for σ_{r+1}.

Result of these considerations: The elements of the Galois group $\sigma \in \mathrm{Gal}(Z|K)$ cannot see any possible multiplicities among the roots!

This is bad because we want to recover as much information as possible by $\mathrm{Gal}(Z|K)$. So we limit ourselves to situations where this is not a problem (and then see that this - luckily - is often the case).

Definition 13.1 Let $L|K$ be an algebraic field extension and $\Omega \supset L$ an algebraic closure. Then

1. an element $\alpha \in L$ is **separable** if its minimal polynomial $f(X) \in K[X]$ over K decomposes in $\Omega[X]$ into *pairwise different* linear factors:

$$f(X) = \prod_{j=1}^{n}(X - \alpha_j) \in \Omega[X],$$

   thus the set of roots is $\mathcal{N} := \{\alpha_1, \ldots, \alpha_n\}$ with pairwise different elements and $\#\mathcal{N} = \deg(f)$.
2. the algebraic extension $L|K$ is **separable** if every $\alpha \in L$ in it is separable. If this is not the case, the extension is called **inseparable.**

Definition 13.2 Let $f(X) \in K[X]$ be a polynomial of degree $\deg(f) \geq 1$ and Ω an algebraic closure of K. Then we call $f(X)$ **separable** if f has exactly $\deg(f)$ pairwise different roots in Ω — i.e. it only has **simple** roots.

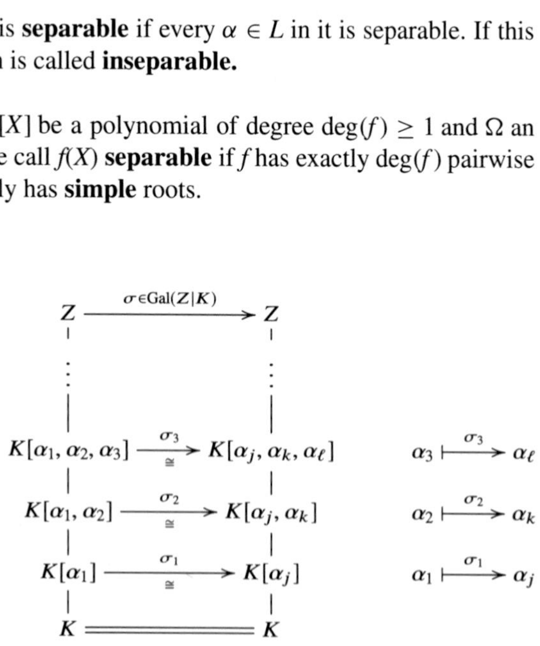

Fig. 13.1 Tower of field extensions for the bottom-up approach

13.2   Formal Derivation

Let Ω be an algebraic closure of K. Given a polynomial $f(X) \in K[X]$, how can we determine whether it has only simple or also multiple roots in Ω? And most importantly, can we decide this without having to calculate the roots!

And now?

A possible answer lies in the following observation from Calculus: A multiple root of a differentiable function $f : \mathbb{R} \to \mathbb{R}$ is a root that is even a root of the derivative $f'(X)$. When defining the derivative, one considers a limit process

$$f'(x) = \lim_{h \to 0} \frac{f(x+h) - f(x)}{h}$$

and we can't do that in an arbitrary field. But nevertheless, we only consider polynomials and we know what the derivative of polynomials should be (even without limits): It should be K-linear and satisfy

$$X^n \rightsquigarrow n \cdot X^{n-1}$$

for $n \in \mathbb{N}$. One can certainly define it this way!

Defining Lemma 13.3 Let K be a field. Then the **formal derivative** is the map

$$(.)' := D := \frac{d}{dX} : K[X] \to K[X]$$

$$f(X) = a_n X^n + \ldots + a_1 X + a_0 \mapsto f'(X) := n a_n X^{n-1} + \ldots + a_1.$$

It is K-linear and satisfies the **Leibniz rule** (=product formula)

$$(f \cdot g)' = f' \cdot g + f \cdot g'$$

Proof The K-linearity is easy to verify: $(f + g)' = f' + g'$ and $(\lambda f)' = \lambda \cdot f'$ for $\lambda \in K$ follows immediately when the polynomials are written out in detail.

For the Leibniz rule, we can already use the K-linearity. Let's show it inductively with respect to the degree of f: As a preliminary remark, let's consider the special case $f(X) = X^{n+1}$. Then, for $g(X) = b_s X^s + \ldots + b_0$, it holds that

$$\left(X^{n+1} g(X) \right)' = \left(b_s X^{s+n+1} + \ldots + b_0 X^{n+1} \right)' = (s+n+1) b_s X^{s+n} + \ldots + (n+1) b_0 X^n$$

$$(s b_s X^{n+s} + \ldots + b_1 X^{n+1}) + ((n+1) b_s X^{n+s} + \ldots + (n+1) b_1 X^{n+1} + (n+1) b_0 X^n) =$$

$$X^{n+1} g'(X) + (n+1) X^n g(X),$$

$$(13.1)$$

therefore, the Leibniz rule applies in this special case.

For $f = 0$ it applies anyway and for $\deg(f) = 0$ the Leibniz rule is clear, because it then follows from K-linearity.

Let's assume that the rule applies to all h with $\deg(h) \leq n$ (including $h = 0$) and let $f(X)$ be of degree $n + 1$, so

$$f(X) = a_{n+1}X^{n+1} + h(X) \text{ with } \deg(h) \leq n.$$

Then we have

$$(f \cdot g)' = \left((a_{n+1}X^{n+1} + h(X)) \cdot g(X)\right)' = a_{n+1}\left(X^{n+1}g(X)\right)' + \left(h(X)g(X)\right)' =$$
$$a_{n+1}\left(X^{n+1}g(X)\right)' + \left(h'(X)g(X) + h(X)g'(X)\right) \overset{(13.1)}{=}$$
$$a_{n+1}\left((n + 1)X^n g(X) + X^{n+1}g'(X)\right) + \left(h'(X)g(X) + h(X)g'(X)\right) =$$
$$f'(X)g(X) + f(X)g'(X) \,.$$

With this, the induction step is proven. $\square$

Remark 13.4 Such a mapping $D : K[X] \to K[X]$, which is K-linear and fulfills the Leibniz rule, is also called a **K-derivation on $K[X]$** .

The formal derivative is compatible with field extensions:

Lemma 13.5 *If $L|K$ is any field extension, then the following diagram is commutative:*

$$\begin{array}{ccc} L[X] & \overset{D}{\longrightarrow} & L[X] \\ \uparrow & & \uparrow \\ K[X] & \overset{D}{\longrightarrow} & K[X] \end{array} \qquad (13.2)$$

Proof This is obvious by the definition of formal derivation. $\square$

Let us use this fact and deduce the following observation. We assume that $f(X) \in K[X]$ has a **multiple** root $\alpha \in \Omega$ in an algebraic closure Ω of K, so it decomposes in the form

$$f(X) = (X - \alpha)^\nu \cdot g(X) \in \Omega[X]$$

with $\nu \geq 2$ (and when ν is chosen to be maximal with $g(\alpha) \neq 0$).

First note that

$$D((X - \alpha)^\nu) = D(\sum_{j=0}^{\nu} \binom{\nu}{j}X^j(-\alpha)^{\nu-j}) = \sum_{j=0}^{\nu} \binom{\nu}{j}jX^{j-1}(-\alpha)^{\nu-j} =$$

$$\sum_{j=0}^{\nu} \nu \cdot \binom{\nu - 1}{j - 1}X^{j-1}(-\alpha)^{\nu-1-(j-1)} = \nu(X - \alpha)^{\nu-1}$$

holds. Then we proceed invoking the product formula:

$$f'(X) \overset{(13.2)}{=} \left((X - \alpha)^\nu \cdot g(X)\right)' = \nu(X - \alpha)^{\nu-1} \cdot g(X) + (X - \alpha)^\nu \cdot g'(X), \quad (13.3)$$

and because $v - 1 \geq 1$, it follows after inserting α on both sides, that

$$f'(\alpha) = 0 .$$

holds. Therefore, we have proven:

Lemma 13.6 *If $f(X) \in K[X]$ has a multiple root in an algebraic closure $\alpha \in \Omega$, then α is also a root of the formal derivative: $f'(\alpha) = 0$.*
  Conversely, if $f(\alpha) = 0 = f'(\alpha)$, then $\alpha \in \Omega$ is a multiple root.

Proof Just a brief note on the reverse direction: This also follows immediately from (13.3), because for $v = 1$ (the maximum v), one then has

$$0 = f'(\alpha) = g(\alpha)$$

a contradiction to the maximum choice of v, because then one can again write $g(X) = (X - \alpha)h(X)$ and $f(X) = (X - \alpha)^2 h(X)$. $\square$
  We can immediately draw a conclusion from this, note the possibly unusual formulation—see the following **And now?** for more information:

Corollary 13.7 Let $L|K$ be an algebraic extension and $f(X) \in K[X]$ the minimal polynomial of $\alpha \in L$. If α is a multiple root of $f(X)$ in an algebraic closure, then

$$f'(X) = 0 \in K[X]$$

must hold.

Proof By assumption, $f(X)$ is the minimal polynomial of α over K. So by definition, there is no polynomial $g(X) \in K[X]$ with $g(X) \neq 0$, so that $\deg(g) < \deg(f)$ and $g(\alpha) = 0$. But certainly, since $\deg(f'(X)) < \deg(f(X))$ holds, we see that $f'(X) = 0$ has to hold. $\square$

And now?
In the situation of the proof, let $f(X) = X^n + \ldots + a_0$ be of degree n. When we considered $\deg(f') < \deg(f)$ in the proof, we came to this inequality by calculating

$$\deg(f'(X)) = \deg(nX^{n-1} + \ldots + a_1).$$

Apparently, we have $\deg(f') \leq n - 1$. However, one might suspect that $\deg(f') = n - 1$ is true and how does this fit to the conclusion $f'(X) = 0 \in K[X]$? The issue here is that it can happen that $n = 0 \in K$ holds in our field K.
  Example: $K = \mathbb{F}_5$ and $f(X) = X^5 + 1 \in \mathbb{F}_5[X]$ yields:

$$f'(X) = 5X^4 = 0 \in \mathbb{F}_5[X] . \tag{13.4}$$

We will briefly discuss this again in Sect. 13.3.

Corollary 13.8 If $f(X) \in K[X]$ is an irreducible polynomial such that the formal derivative $f'(X) \neq 0 \in K[X]$ does not vanish, then f is separable.

Proof We can find all roots in an algebraic closure Ω of K and f is the minimal polynomial of each of them. According to the previous corollary, one of them can only be a multiple root if $f'(X) = 0$ applies. $\square$

And now?

So far, it may not be entirely clear what we have gained by considering the formal derivative $f'(X)$. We still need the zeros $\alpha \in \Omega$ in order to insert them into the derivative $f'(X)$. If we cannot calculate these roots α it seems useless at first glance.

Or is it? The nice thing is that $f'(X) \in K[X]$ is again a polynomial in $K[X]$ (and we only briefly used the transition to $\Omega[X]$ to exploit the product formula, but due to Lemma 13.5 we don't have to stay in $\Omega[X]$). This means, we can rephrase the question:

'Does $f(X) \in K[X]$ have a multiple root in an algebraic closure Ω?'
which is equivalent to
'Do $f(X) \in K[X]$ and $f'(X) \in K[X]$ have the same roots in Ω?'
and—considering that $f(\alpha) = 0$ means that $f(X)$ is a multiple of the minimal polynomial $p(X) \in K[X]$ of α over K—is equivalent to
'Do $f(X), f'(X) \in K[X]$ have a common divisor $p(X) \in K[X]$ of degree $\deg(p) \geq 1$?'

The last reformulation is great because neither a (yet to be found) root $\alpha \in \Omega$ nor the algebraic closure Ω occur.

Lemma 13.9 *Let $f(X) \in K[X]$ be of degree $\deg(f) \geq 1$. Then the following holds:*

$$f(X) \text{ is separable} \iff \gcd(f(X), f'(X)) = 1 .$$

Proof As we have seen above, f is separable if and only if f and f' have no common zero in an algebraic closure Ω. In the factorial ring $K[X]$ (which is even Euclidean), we have the notion of a greatest common divisor. Let $d(X) := \gcd(f, f')$.

"$\Rightarrow$": If $d \neq 1$, that is $d \in K[X]$ is not a unit, then $\deg(d) \geq 1$ and $d(X)$ has a root α in the algebraic closure Ω. But then α is a common root of f and f', because $d(X)$ is a divisor of both poynomials.

"$\Leftarrow$": If f is not separable, then f and f' have a common zero $\alpha \in \Omega$. If $p(X) \in K[X]$ is the minimal polynomial of α over K, we get

$$p(X) \mid f(X) \text{ and } p(X) \mid f'(X) \Longrightarrow p(X) \mid d(X) ,$$

so $\deg(d(X)) \geq 1$ and thus $d(X)$ is not a unit in $K[X]$. $\square$

And now?

The last lemma is very helpful because the criterion $\gcd(f, f') = 1$ can easily be decided by the Euclidean algorithm. For example:

$$f(X) = X^6 + X^5 + 5X^4 + 4X^3 + 8X^2 + 4X + 4 \in \mathbb{Q}[X].$$

We have

$$f'(X) = 6X^5 + 5X^4 + 20X^3 + 12X^2 + 16X + 4$$

The following SAGE routine, which performs the Euclidean algorithm, yields as a result $\gcd = X^2 + 2$, so f is **not** separable.

```
R.<X>=PolynomialRing(QQ,'X')
f=X^6+X^5+5*X^4+4*X^3+8*X^2+4*X+4
g=f.derivative()
r=g
s=f
d=g.degree()
while (d>=0):
    h=s.quo_rem(r)[1]
    s=r
    r=h
    d=r.degree()
ggT=(s.leading_coefficient())^(-1)*s
print(ggT)
```

The SAGE routine is almost self-explanatory: After the definitions, $r := g = f'$ and $s := f$ are initially set. Then comes a loop over the degree $\deg(r)$ of the respective remainder.

With s.quo_rem(r) division with remainder is performed: $s = ar + b$ and the output is a list with the two entries s.quo_rem(r)=[a,b]. So we have s.quo_rem(r)[0] $= a$ and s.quo_rem(r)[1] $= b$. The line

$$h=s.quo\_rem(r)[1]$$

thus stores the remainder of the division as h. Then, as in the Euclidean algorithm, the new division with remainder is prepared, for which $s = r$ (the "old" remainder) and $r = h$ (the "new" remainder). The loop breaks off when Rest $= 0$ (sage assigns the degree -1 to the zero polynomial). Then the gcd is stored in the variable s which is normalized in the penultimate line (recall that the gcd is only unique up to units!). One could have also just used the SAGE command gcd(f,g) after the first two lines but it also might be enlightening to program the Euclidean algorithm by ourselves.

13.3   Characteristic of a Field and Separability

And now?

As in the example in the **And now?** around (13.4) we have encountered situations where $f(X)$ has a positive degree, but $f'(X) = 0$ applies. This phenomenon can only occur in **prime characteristic**. We want to clarify this briefly.

Defining Lemma 13.10 Let K be a field. Then there is a uniquely determined ring homomorphism

$$\varphi : \mathbb{Z} \to K$$

(Reminder: Ring homomorphisms always fulfill $1 \mapsto 1$), namely

$$n \mapsto \begin{cases} n := \overbrace{1 + \ldots + 1}^{n \text{ times}} & \text{for } n > 0 \\ 0 & \text{for } n = 0 \\ -n := \underbrace{(-1) + \ldots + (-1)}_{|n| \text{ times}} & \text{for } n < 0 \end{cases}$$

The kernel of φ is a prime ideal. We say:

- K has **characteristic zero**, written char$(K) = 0$, if ker$(\varphi) = (0)$.
- K has **characteristic p**, written char$(K) = p$, if ker$(\varphi) = (p)$.

Proof According to the homomorphism theorem, φ induces an **injective** ring homomorphism

$$\overline{\varphi} : \mathbb{Z}/\ker(\varphi) \hookrightarrow K \,,$$

and thus $\mathbb{Z}/\ker(\varphi)$ is isomorphic to a subring of K. Since K has no zero divisors, $\mathbb{Z}/\ker(\varphi)$ must also be an integral domain and therefore $\ker(\varphi) \lhd \mathbb{Z}$ is a prime ideal. $\square$

Remark 13.11 Apparently, the following holds:

- char$(K) = 0 \Leftrightarrow n \neq 0 \in K$ for all $n \in \mathbb{N}$.
- If char$(K) = p$, then $p = \min\{n \in \mathbb{N} \mid n = 0 \in K\}$.

Lemma 13.12 *If* $f(X) \in K[X]$ *is irreducible of degree* $n = \deg(f) \geq 1$ *and* char$(K) \nmid n$ *(this excludes the case* char$(K) = 0$*), then f is separable.*

Proof We have $f(X) = a_n X^n + \ldots + a_1 X + {}_0$ with $a_n \in K^\times$. Then

$$f'(X) = n a_n X^{n-1} + \ldots + a_1$$

and by assumption $n \neq 0 \in K$ (because for characteristic zero this is clear, and for characteristic p we have $n \notin (p) = \ker(\mathbb{Z} \to K)$ by assumption on n). Therefore, $na_n \neq 0$ and thus $\deg(f') = \deg(f) - 1 \geq 0$.

Since f was irreducible, f has no divisor of a degree ≥ 1, so $\gcd(f, f')$ must be a unit. $\square$

Corollary 13.13 If $\operatorname{char}(K) = 0$, every algebraic extension $L|K$ is separable.

Proof If $\alpha \in L$ with minimal polynomial $f(X) \in K[X]$, then f is separable according to the lemma above. $\square$

Definition 13.14 A field K is called **perfect** if every algebraic field extension $L|K$ is separable.

Remark 13.15 The corollary thus says that every field of characteristic zero is perfect.

And now?
The last results were the reason why I sometimes wrote that one can *often exclude the problem with multiple zeros a priori!* in previous chapters. In particular, if $K = \mathbb{Q}$ is our ground field, one does not need to worry about multiple roots for irreducible polynomials, the latter are automatically separable!

One last lemma that we will use often in the following:

Lemma 13.16 *If $K \subset M \subset L$ are finite field extensions such that $L|K$ is separable, then so are $L|M$ and $M|K$.*

Proof Let $\alpha \in L$ and let

- $f(X) \in K[X]$ be its minimal polynomial over K and
- $g(X) \in M[X]$ be its minimal polynomial over M,

then

$$g(X) \mid f(X) \text{ within } M[X] .$$

By assumption, f has only simple roots in an algebraic closure Ω hence also $g(X)$ has simple roots since it is a divisor of $f(X)$ (and it also remains a divisor in $\Omega[X]$).

Furthermore: Let $\beta \in M$ be an arbitrary element and $h(X) \in K[X]$ be the minimal polynomial of β over K. Then we can regard β as an element $\beta \in L$ and $h(X)$ is also the minimal polynomial of β (now to be understood as an element of L) over K, because the minimal polynomial depends only on $K[\beta] \subset M$). By assumption, f has only simple roots in Ω. Therefore, $M|K$ is also separable. $\square$

And now?

A small preview to Chap. 16 where we will study finite fields, i.e., fields with a finite number of elements. So far, we have seen $\mathbb{F}_p = \mathbb{Z}/p\mathbb{Z}$ for a prime number p, which has p elements. But there are also fields with p^n elements for a prime number p and a natural number $n \in \mathbb{N}$. These are the simplest fields of characteristic p. One might think that one might perhaps find inseparable field extensions among those. This is still not the case, the finite fields will turn out to be perfect—see Corollary 16.7. The simplest non-separable field extension is the subject of an exercise at the end of this chapter.

13.4   The Degree of Separability

And now?

We have encountered the problem of separability since for non-separable polynomials $f(X) \in K[X]$ the automorphisms $\text{Aut}_K(Z)$ in the tower of field extensions leading to the splitting field Z do not recognize the multiplicity of the roots—in a more general situation than the one of a splitting field, say a finite extension $L|K$ not necessarily *normal*, the automorphism group $\text{Aut}_K(L)$ does not arise in the picture in a natural way. If we think of Theorem A and A', we are lead to consider $\text{Hom}_K(L, \Omega)$ instead. The fact that this set $\text{Hom}_K(L, \Omega)$ does not capture the multiplicities of roots of potential minimal polynomial - in the sense that the possible values $\varphi(\alpha)$ for given $\alpha \in L$ and $\varphi \in \text{Hom}_K(L, \Omega)$ do not see the multiplicity of α as a root of its minimal polynomial - can be formulated more precisely as follows.

Definition 13.17 Let $L|K$ be a finite field extension and Ω be an algebraic closure of L. Then

$$[L : K]_s := \#\text{Hom}_K(L, \Omega)$$

is called the **degree of separability** of $L|K$.

Remark 13.18 Sometimes one also meets the notation $[L : K]_{\text{sep}}$.

And now?

Consider a field extension of the type $L = K[\alpha]|K$. Then, we easily understand the relation between

$$[L : K] \quad \text{and} \quad [L : K]_s. \tag{13.5}$$

The left hand side coincides with the degree of the minimal polynomial $f(X) \in K[X]$ of α over K. Let's say $\deg(f) = n$ and furthermore, assume that f has the decomposition

$$f(X) = \prod_{j=1}^{m}(X - \alpha_j)^{\nu_j} \in \Omega[X]$$

with the multiplicities ν_j (and pairwise different $\alpha_1, \ldots, \alpha_m \in \Omega$) over the algebraic closure Ω. Then we know the right hand side of (13.5) from theorem A:

$$\begin{array}{ccc} L = K[\alpha] & \xrightarrow{\;\varphi\;} & \Omega \\ | & & | \\ K & \xrightarrow{\;\text{id}\;} & K \end{array}$$

For each $\alpha_1, \ldots, \alpha_m \in \Omega$ we get exactly one $\varphi \in \mathrm{Hom}_K(L, \Omega)$ and every $\varphi \in \mathrm{Hom}_K(L, \Omega)$ arises in this way. In conclusion, we have:

$$\begin{aligned} [K(\alpha) : K] &= n = \#(\text{roots of } f \text{ counted with multiplicities}) \\ [K(\alpha) : K]_s &= m = \#(\text{roots of } f \text{ counted without multiplicities}). \end{aligned} \tag{13.6}$$

More generally, one can generalize this idea and then obtains the following two results:

Lemma 13.19 *If $K \subset M \subset L$ are finite field extensions, the following holds*

$$[L : K]_s = [L : M]_s \cdot [M : K]_s$$

Proof This is actually just a problem of counting. It's carried out most easily with the following notation: Assume we already have given the situation

that is, we have a homomorphism $\sigma \in \mathrm{Hom}_K(M, \Omega)$. We will write

$$\mathrm{Hom}_\sigma(L, \Omega) := \{\varphi \in \mathrm{Hom}_K(L, \Omega) \mid \varphi|_M = \sigma\}.$$

Now, since one can restrict each $\varphi \in \mathrm{Hom}_K(L, \Omega)$ to M and then $\varphi \in \mathrm{Hom}_\sigma(L, \Omega)$ lies in this set for the homomorphism $\sigma = \varphi|_M \in \mathrm{Hom}_K(M, \Omega)$. Obviously, this results in a pairwise disjoint decomposition:

$$\mathrm{Hom}_K(L, \Omega) = \bigsqcup_{\sigma \in \mathrm{Hom}_K(M, \Omega)} \mathrm{Hom}_\sigma(L, \Omega).$$

It follows that

$$[L : K]_s = \#\mathrm{Hom}_K(L, \Omega) = \sum_{\sigma \in \mathrm{Hom}_K(M, \Omega)} \#\mathrm{Hom}_\sigma(L, \Omega). \tag{13.7}$$

Now let $\sigma \in \mathrm{Hom}_K(M, \Omega)$ be fixed. According to Theorem 12.3 there is an extension of σ to a homomorphism $\Sigma : \Omega \to \Omega$. This is even an automorphism $\Sigma \in \mathrm{Aut}_K(\Omega)$, because the injectivity is clear, since Σ is a field homomorphism. For surjectivity, consider an $\alpha \in \Omega$ and the splitting field $Z \subset \Omega$ of the minimal polynomial of α over K. Since Z/K is normal, it follows that

$$\Sigma|_Z \in \mathrm{Hom}_K(Z, \Omega) = \mathrm{Hom}_K(Z, Z) = \mathrm{Aut}_K(Z),$$

where we have seen the last equality before - it holds due to the rank formula of linear algebra and since $[Z : K] < \infty$. Therefore, there exists a $\beta \in Z \subset \Omega$ with $\Sigma(\beta) = \alpha$.

Now consider the following map:

$$\mathrm{Hom}_M(L, \Omega) \to \mathrm{Hom}_\sigma(L, \Omega)$$
$$\eta \mapsto (\Sigma \circ \eta : L \to \Omega).$$

It is bijective with inverse mapping

$$\mathrm{Hom}_\sigma(L, \Omega) \to \mathrm{Hom}_M(L, \Omega)$$
$$\psi \mapsto (\Sigma^{-1} \circ \psi : L \to \Omega).$$

Check it out!
Convince yourself that the last assertions are true indeed. To this end, consider the diagram:

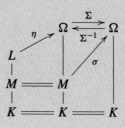

Hence, the individual sets $\mathrm{Hom}_\sigma(L, \Omega)$ all have the same cardinality, namely

$$\#\mathrm{Hom}_\sigma(L, \Omega) = \#\mathrm{Hom}_M(L, \Omega) = [L : M]_s.$$

Now, the assertion of the lemma follows easily from (13.7). $\square$

Theorem 13.20 Let $L|K$ be a finite extension. Then the following holds:

$$[L : K]_s \leq [L : K]$$

and furthermore

$$[L : K]_s = [L : K] \Longleftrightarrow L|K \text{ is separable}.$$

Proof Let Ω be an algebraic closure of L. The finite extension $L|K$ is generated by finitely many elements in the sense that there are $\alpha_1, \ldots, \alpha_m \in L$ such that

$$L = K[\alpha_1, \ldots, \alpha_m]$$

(the right hand side denoting the field constructed inside of L by adjoining these elements). For an extension of the type $M[\alpha]|M$ (let's say with $M[\alpha] \subset \Omega$) we have already considered the inequality earlier—see (13.6)—I repeat this briefly here. If $f(X) \in M[X]$ is the minimal polynomial of α over M and $f(X)$ decomposes in the algebraic closure Ω in the form

$$f(X) = \prod_{j=1}^{m}(X - \beta_j)^{\nu_j} \in \Omega[X],$$

with the pairwise different zeros $\beta_1, \ldots, \beta_m$ (then $\alpha \in \{\beta_1, \ldots, \beta_m\}$) with multiplicities $\nu_1, \ldots, \nu_m$, we have

$$[M[\alpha] : M] = \deg(f) = \sum_{j=1}^{m} \nu_j$$

$$[M[\alpha] : M]_s = m = \sum_{j=1}^{m} 1 .$$

The latter, because each of the m different roots gives rise to exactly one homomorphism $\varphi \in \mathrm{Hom}_M(M[\alpha], \Omega)$ by Theorem A and vice versa: each φ is uniquely determined by $\varphi(\alpha) = \beta_j$ and every φ arises in this way.

So, one can see that $m = [M[\alpha] : M]_s \leq [M[\alpha] : M] = \deg(f)$ and the equality holds if and only if all $\nu_j = 1$ are, that is, if and only if f is separable.

Because of the degree formula and Lemma 13.19, the inequality also follows in the general case (if we set $\alpha_0 := 1$):

$$[L : K]_s = \prod_{j=0}^{m-1}[K[\alpha_1, \ldots, \alpha_{j+1}] : K[\alpha_1, \ldots, \alpha_j]]_s \leq$$

$$\prod_{j=0}^{m-1}[K[\alpha_1, \ldots, \alpha_{j+1}] : K[\alpha_1, \ldots, \alpha_j]] = [L : K]$$

$$(13.8)$$

Now, if $L|K$ is separable, all these intermediate extensions are separable (according to Lemma 13.16) and thus equality follows in (13.8), because we have already considered this for the individual extensions of type $M[\alpha]|M$ above.

If $L|K$ is not separable, then there is an $\alpha \in L$, whose minimal polynomial has multiple roots. But then—as we have observed above:

$$[K[\alpha] : K]_s < [K[\alpha] : K]$$

and thus also

$$[L : K]_s = [L : K[\alpha]]_s \cdot [K[\alpha] : K]_s \overset{(*)}{<} [L : K[\alpha]] \cdot [K[\alpha] : K] = [L : K],$$

holds, where we have the inequality $(*)$ since we already know that $[L : K[\alpha]]_s \le [L : K[\alpha]]$. $\square$

We can now also prove the reverse statement of Lemma 13.16:

Lemma 13.21 *If $K \subset M \subset L$ are finite field extensions. Then the following holds:*

$$L|K \text{ is separable} \Longleftrightarrow \begin{cases} L|M \text{ is separable} \\ \quad\quad and \\ M|K \text{ is separable} \end{cases}$$

Proof The direction "$\Rightarrow$" is Lemma 13.16. For the reverse, we use Theorem 13.20: Since $L|M$ and $M|K$ then are separable, we have

$$[L : K]_s = [L : M]_s \cdot [M : K]_s = [L : M] \cdot [M : K] = [L : K],$$

so $L|K$ is also separable. $\square$

We will prove another lemma that we will need occasionally:

Lemma 13.22
Let K be a field and $f(X) \in K[X]$ a separable polynomial of degree $\deg(f) \ge 1$. Let $Z|K$ be a splitting field of f. Then the extension $Z|K$ is separable.

Proof Let $\alpha_1, \ldots, \alpha_n \in Z$ be the pairwise different roots of f. Then consider each extension

$$K[\alpha_1, \ldots, \alpha_{j+1}] \mid K[\alpha_1, \ldots, \alpha_j]$$

The minimal polynomial of α_{j+1} over $K[\alpha_1, \ldots, \alpha_j]$ is a divisor of $f(X)$ in the corresponding polynomial ring, so it only has simple roots. But then

$$[K[\alpha_1, \ldots, \alpha_{j+1}] : K[\alpha_1, \ldots, \alpha_j]]_s = [K[\alpha_1, \ldots, \alpha_{j+1}] : K[\alpha_1, \ldots, \alpha_j]],$$

as we have already seen above. With the product formulas for the degrees of the field extensions and the degrees of separability, it follows that also

$$[Z : K]_s = [Z : K]$$

holds and hence $Z|K$ is separable according to Theorem 13.20. $\square$

13.5 The Primitive Element Theorem

As a conclusion in this chapter, we prove a theoretically interesting and important theorem, which, however, is usually not practical for explicit computations—see the *And now?* after the theorem.

Theorem 13.23 (Primitive Element Theorem) If $L|K$ is a finite, separable extension, then there exists a $\gamma \in L$ such that $L = K[\gamma]$. Such an element is called a **primitive element** for the extension.

Proof We will assume in the following proof that K has infinitely many elements. The theorem also holds for finite fields there, which we will examine more closely in Chap. 16 and prove the theorem for it there.

Since $L|K$ is finite, we can again assume $L = K[\alpha_1, \ldots, \alpha_m]$.

1. Step: We reduce the assertion to the case $m = 2$. If we know that every finite separable extension $M[\alpha, \beta]|M$ can be written with a primitive $\gamma \in M[\alpha, \beta]$, then one can descend step by step:

$$K[\alpha_1, \ldots, \alpha_m] = K[\alpha_1, \ldots, \alpha_{m-2}][\alpha_{m-1}, \alpha_m] =$$
$$K[\alpha_1, \ldots, \alpha_{m-2}][\gamma_1] = K[\alpha_1, \ldots, \alpha_{m-3}][\alpha_{m-2}, \gamma_1] =$$
$$K[\alpha_1, \ldots, \alpha_{m-3}][\gamma_2] = \ldots = K[\alpha_1, \gamma_{m-1}] = K[\gamma]$$

2. Step: According to the 1st step we can assume without loss of generality that $L = K[\alpha, \beta]$.

And now?

The idea of the proof now is a naive approach: One looks for a $\gamma \in K[\alpha, \beta]$ as easy as possible, for example in the form

$$\gamma = \alpha + \beta \cdot x$$

for some (yet to be found) $x \in K$. One then has to show that there is such an $x \in K$ making this γ into a primitive element. How do you find such an x? This is done with a trick.

Consider an algebraic closure Ω of L and let's enumerate the elements of the set

$$\text{Hom}_K(L, \Omega) = \{\sigma_1, \ldots, \sigma_n\}$$

where

$$n := [L : K] = [L : K]_s$$

by assumption. Now, consider the polynomial

$$f(X) := \prod_{\substack{i,j=1,\dots,n \\ i \neq j}} \big((\sigma_i(\alpha) + \sigma_i(\beta) \cdot X) - (\sigma_j(\alpha) + \sigma_j(\beta) \cdot X)\big) \in \Omega[X] \,.$$

And now?
Why do we care about this polynomial $f(X) \in \Omega[X]$. Inserting an element $x \in K$ into it, one can multiply x into the various σ_j since the latter are K-homomorphisms. Let us see what we obtain:

For each $x \in K$ we have

$$\begin{aligned} f(x) &= \prod_{\substack{i,j=1,\dots,n \\ i \neq j}} \big((\sigma_i(\alpha) + \sigma_i(\beta) \cdot x) - (\sigma_j(\alpha) + \sigma_j(\beta) \cdot x)\big) = \\ &= \prod_{\substack{i,j=1,\dots,n \\ i \neq j}} \big(\sigma_i(\alpha + \beta \cdot x) - \sigma_j(\alpha + \beta \cdot x)\big). \end{aligned} \tag{13.9}$$

Claim The polynomial $f(X) \neq 0 \in \Omega[X]$ is not the zero polynomial. $\square$

Proof Apparently, $f(X) = 0$ if and only if $\sigma_i(\alpha) + \sigma_i(\beta)X = \sigma_j(\alpha) + \sigma_j(\beta)X$ for at least one pair $i \neq j$ holds. But the homomorphisms $\sigma_1,\dots,\sigma_n$ are pairwise different as elements in $\mathrm{Hom}_K(L, \Omega)$. Because of $L = K[\alpha, \beta]$ every $\eta \in \mathrm{Hom}_K(L, \Omega)$ is uniquely determined by the pair of values $(\eta(\alpha), \eta(\beta))$. In particular

$$(\sigma_i(\alpha), \sigma_i(\beta)) \neq (\sigma_j(\alpha), \sigma_j(\beta)) \text{ für } i \neq j,$$

with which the assertion is shown. $\square$

Since K has infinitely many elements, but $f(X)$ has only finitely many roots in K, there exists an $x \in K$ such that $f(x) \neq 0$. Then, due to (13.9), the n elements

$$\sigma_1(\alpha + \beta \cdot x), \sigma_2(\alpha + \beta \cdot x), \dots, \sigma_n(\alpha + \beta \cdot x) \in \Omega$$

are pairwise different.

Claim $x \in K$ We obtain $L = K[\alpha + x\beta]$. $\square$

Proof The restrictions of σ_i on $K[\alpha + x\beta]$

$$\sigma_i|_{K[\alpha+x\beta]} : K[\alpha + x\beta] \to \Omega$$

are thus pairwise different. Therefore, we deduce that

$$[K[\alpha + x\beta] : K] = [K[\alpha + x\beta] : K]_s = \#\mathrm{Hom}_K(K[\alpha + x\beta], \Omega) \geq n = [L : K]$$

and we arrive at $K[\alpha + x\beta] = L.$ $\square$

Hence, the primitive element theorem (for fields with infinitely many elements) is shown. For finite fields, the statement is proven in Lemma 16.9. $\square$

And now?
Two remarks on this:

- The proof shows how to search for a primitive element in an extension $K[\alpha, \beta]|K$ (if K has infinitely many elements): One simply takes a linear combination $\gamma = \alpha + x\beta$ for some $x \in K$ and for all but finitely many x, this will be a primitive element. One only needs to check that one has not accidentally chosen one of the not-allowed $x \in K$ - the roots of the polynomial we introduced in the proof.
- One might now ask why we have always considered towers of field extensions with possibly lots of floors before and why we needed the variant Theorem A' in addition to Theorem A, when we could have managed with **one** step for finite, separable extensions: We just have to find a primitive element γ and consider the short tower with just one extension on the right hand side in Fig. 13.2 instead of the tall one on the left hand side.
The problem is that in order to understand the right hand side one has to compute the minimal polynomial of γ over K and **this is not so easy in general and even if we manage to do so, the polynomial may become very complicated.**
Note that

$$\deg(\text{minimal polynomial of } \gamma \text{ over } K) = [L : K]$$

holds.
Example: Let $Z|\mathbb{Q}$ be the splitting field of our example from the intermezzo $f(X) = X^5 - 777X + 7$. If then $\gamma \in Z$ is a primitive element, we know that

$$\deg(\text{minimal polynomial of } \gamma \text{ over } \mathbb{Q}) = 5! = 120 .$$

If we ask SAGE to determine such a primitive element and its minimal polynomial, one obtains the following answer for such a minimal polynomial:

$X^{120}+$

$1025677296X^{118}+$

$4950330X^{117}+$

$495533876580606780X^{116}+$

$4797003138944884X^{115}+$

$150030264243635083850393243X^{114}+$

$2186627260092334984900906X^{113}+$

$319535635897845924883129385287 53107X^{112}+$

$623862068211169775662804418784544X^{111}+$

$509520390822448568155542651362631197 3478690X^{110}+$

$1250768545260983767421221991797885596 88732X^{109}+$

$6322464610612614903599432747567870575 31094039151196X^{108}+$

$\dots$

and the coefficients keep getting larger.

Thread to the previous chapter:
We have seen in this chapter

- what a separable field extension is,
- that the degree of separability $[L : K]_s$ determines whether L/K is separable or not,
- that one can tell from a polynomial whether it has multiple roots by considering $\gcd(f, f')$,
- what the characteristic of a field is and

Fig. 13.2 The theorem of the primitive element and its supposed practical application

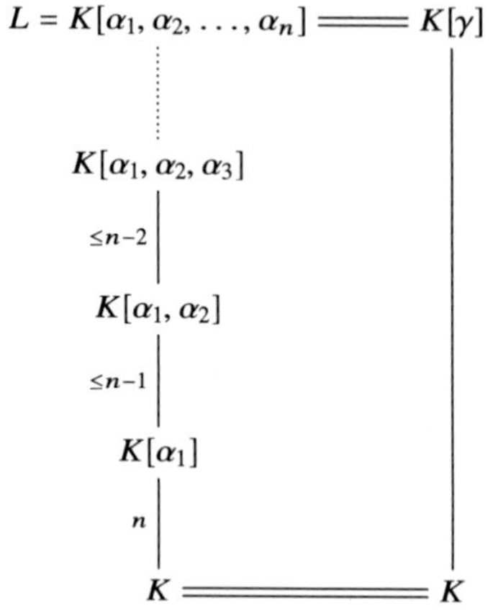

$$L = K[\alpha_1, \alpha_2, \dots, \alpha_n] \Longrightarrow K[\gamma]$$

$$K[\alpha_1, \alpha_2, \alpha_3]$$

$$\leq n-2$$

$$K[\alpha_1, \alpha_2]$$

$$\leq n-1$$

$$K[\alpha_1]$$

$$n$$

$$K \Longrightarrow K$$

- that fields K of characteristic zero are perfect, meaning that every algebraic field extension $L|K$ is separable,
- and that a finite separable extension $L|K$ an always be written in the form $L = K[\gamma]$ (Primitive Element Theorem).

Exercises: 13.1 (in preparation for Exercise 13.2):
Let $\zeta = \exp(\frac{2\pi i}{8}) = \exp(\frac{\pi i}{4}) = \frac{1}{2}(\sqrt{2} + i\sqrt{2})$. Show:

(a) ζ is a root of the irreducible polynomial $X^4 + 1 \in \mathbb{Q}[X]$ (Hint: first determine $[\mathbb{Q}(\zeta) : \mathbb{Q}]$).
(b) The map

$$(\mathbb{Z}/8\mathbb{Z})^\times \to \mathrm{Gal}(\mathbb{Q}(\zeta)|\mathbb{Q}), \ [k] \mapsto (\zeta \mapsto \zeta^k)$$

defines a group isomorphism (note: don't forget to show well-definedness).

13.2 We anticipate a definition from Chap. 14: If $L|K$ is a finite, normal, and separable extension and $H \subset \mathrm{Gal}(L|K)$ is a subgroup, then

$$L^H := \{x \in L \mid \sigma(x) = x \text{ for all } \sigma \in H\},$$

is the called the **fixed field** of H (see (14.1)), it is an intermediate field $K \subset L^H \subset L$.

(a) *A candidate for a supposedly primitive element:* Let $L|K$ be a finite Galois extension and $\alpha \in L$ a primitive element, so $L = K[\alpha]$. Consider a subgroup $H \subset \mathrm{Gal}(L|K)$. Define $\beta := \sum_{\sigma \in H} \sigma\alpha$ and show that

$$K[\beta] \subset L^H.$$

(Note that if equality holds, one has found a primitive element for L^H.)
(b) *Unfortunately, the candidate does not always meet the desired criteria:* Consider the example $L = \mathbb{Q}[\zeta]$ with $\zeta = \exp(2\pi i/8)$ from Exercise 13.1 and the subgroup $H := \{1, \varphi\}$ with

$$\varphi : \mathbb{Q}[\zeta] \to \mathbb{Q}[\zeta], \ \zeta \mapsto \zeta^5.$$

Show that the construction from (a) does not always lead to a primitive element. Also determine a primitive element of $\mathbb{Q}[\zeta]^H$.

13.3 When one applies the proof of the primitive element theorem to the extension

$$L := \mathbb{Q}(\sqrt{3}, \sqrt{5})|\mathbb{Q} =: K,$$

one indeed has $4 = [L : K]_s$ homomorphisms $\sigma_1, \ldots, \sigma_4 \in \mathrm{Hom}_K(L, \overline{\mathbb{Q}})$.

- Determine these σ_j.
- For which $x \in \mathbb{Q}$ is the candidate $\sqrt{3} + x \cdot \sqrt{5} \in L$ a primitive element?

13.4 The standard example of an inseparable extension: Let p be a prime number and $\mathbb{F}_p := \mathbb{Z}/\mathbb{Z}p$. Consider the quotient field $K := \mathbb{F}_p(t) := \mathrm{Quot}(\mathbb{F}_p[t])$, as well as another copy with a different name for the indeterminate $L := \mathbb{F}_p(w)$. Then

$$K \hookrightarrow L, \quad \frac{g(t)}{h(t)} \mapsto \frac{g(w^p)}{h(w^p)}$$

gives an injective field homomorphism and we thus consider $L|K$ as a field extension. Show:

(a) Inside the ring $L[X]$, we have $(X - w)^p = X^p - w^p$.
(b) The field L is the splitting field of the polynomial $f(X) = X^p - t \in K[X]$.
(c) We have $[L : K] = p$ and $[L : K]_s = 1$.

This shows that $L|K$ is not separable.

Galois Theory (II)—The Fundamental Theorem

14

14.1 The Fundamental Theorem of Galois Theory— Statement

Definition 14.1 An algebraic field extension $L|K$ is called **Galois** (or a **Galois extension**) if it is normal and separable.

Note that the definition also applies to infinite algebraic extensions, but we will only consider finite extensions from now on.

Theorem 14.2 (Fundamental Theorem of Galois Theory) *Let $L|K$ be a finite Galois extension with Galois group $\mathrm{Gal}(L|K)$. Consider the two sets:*

$$\mathrm{IntF}(L|K) := \{M \mid M \text{ is an intermediate field}, \ K \subset M \subset L\}$$
$$\mathrm{SubG}(L|K) := \{H \subset \mathrm{Gal}(L|K) \mid H \text{ is a subgroup}\}.$$

For each subgroup $H \in \mathrm{SubG}(L|K)$ we denote with L^H the **fixed field**

$$L^H := \{x \in L \mid \sigma(x) = x \text{ for all } \sigma \in H\}. \tag{14.1}$$

Then the two maps

$$\mathcal{IntF}(L|K) \underset{\Phi}{\overset{\Psi}{\rightleftarrows}} \mathcal{SubG}(L|K)$$

$$M \longmapsto \mathrm{Gal}(L|M)$$

$$L^H \longleftarrow\!\!\shortmid H$$

are mutually inverse bijections.

© The Author(s), under exclusive license to Springer-Verlag GmbH, DE, part of
Springer Nature 2024
M. Hien, *Abstract Algebra*, Mathematics Study Resources 7,
https://doi.org/10.1007/978-3-662-67974-6_14

Theorem 14.3 (Addendum to the fundamental theorem) In the situation of the fundamental theorem, the following holds:

1. These bijections reverse inclusion relations, that is:

$$M_1 \subset M_2 \implies \mathrm{Gal}(L|M_1) \supset \mathrm{Gal}(L|M_2)$$

   for intermediate fields M_1, M_2 and

$$L^{H_1} \supset L^{H_2} \impliedby H_1 \subset H_2$$

   for subgroups H_1, H_2.
2. The intermediate extension $M|K$ is normal if and only if the associated subgroup $\mathrm{Gal}(L|M) \subset \mathrm{Gal}(L|K)$ is a normal subgroup. If this is the case, there is a natural isomorphism

$$\mathrm{Gal}(L|K)/\mathrm{Gal}(L|M) \to \mathrm{Gal}(M|K) \, , \; [\sigma] \mapsto \sigma|_M.$$

Remark 14.4 The fact that in the above notation the set L^H is an intermediate field, of course, also needs a proof, but this is rather easy to see:

Proof Since all $\sigma \in H \subset \mathrm{Gal}(L|K)$ leave the elements of the field K fixed, we certainly have $K \subset L^H$, in particular $0, 1 \in L^H$. If $x, y \in L^H$, then also $x + y$ and xy, because $\sigma \in H$ are field homomorphisms. Similarly, $-x$ and x^{-1} (for $x \neq 0$). $\square$

14.2   Outlook on an Application—Quadratic Formula for All Degrees?

This section can also be skipped if you want to go directly to the proof of the fundamental theorem in Sect. 14.3.

And now?
Before we prove the fundamental theorem (which will turn out to be not very difficult with all our preparations), I would like to add some thoughts about the following question:

Why does the fundamental theorem deserve its name? What is so important about the intermediate fields?

We have already seen an attempt to an answer: The floors in the towers of extensions leading to a splitting field of a polynomial we have considered before are such intermediate fields. If you understand all intermediate fields, you also understand these towers.

However, this is somewhat "vague" because it is not clear what is meant by ... *understand intermediate fields, understand* ...: What does *understand* mean?

> Therefore, I'd like to give an application where the importance of the fundamental theorem also becomes formally clear—the question of a *solution formula* for higher degrees.

Consider a polynomial equation

$$f(x) = a_n x^n + \ldots + a_0 = 0$$

of degree $n \in \mathbb{N}$ over the field K (that is, with $a_j \in K$). Can we provide a formula for the solutions in terms of the coefficients and roots of algebraic expressions thereof?

For $n = 2$ and $2 \neq 0 \in K$ this is the old *quadratic formula*

$$ax^2 + bx + c = 0 \Leftrightarrow x = \frac{-b \pm \sqrt{b^2 - 4ac}}{2a}.$$

For $n = 3, 4$ there are also formulas—as already mentioned in the motivational chapter of the book. How do we want to formulate

Formula in the coefficients and roots of algebraic expressions thereof

mathematically exact? This happens as follows:

Consider an algebraic field extension $L|K$. If you have an intermediate field $K \subset M \subset L$, then you can look at a so-called **pure equation**

$$X^m = \mu \in M$$

over the field M. If it has a root in L, which we simply denote as $\sqrt[m]{\mu} \in L$, one can consider the intermediate extension

$$M[\sqrt[m]{\mu}] \mid M.$$

If, by chance, $M = K[\alpha_1, \ldots, \alpha_r]$, then $\mu \in K[\alpha_1, \ldots, \alpha_r]$ is an *algebraic expression* in the elements $\alpha_1, \ldots, \alpha_r \in L$ and $\sqrt[m]{\mu}$ is a *root of an algebraic expression in these elements*. The following definition describes exactly all elements that arise when successively adjoining such roots starting from K:

Definition 14.5 Let $L|K$ be an algebraic field extension and $\alpha \in L$. Then we say that α **is solvable by radicals** if there is a finite sequence of field extensions

$$K =: K_0 \subset K_1 \subset K_2 \subset \ldots \subset K_N$$

and respective elements $\mu_j \in K_j$ such that

$$K_{j+1} = K_j[\sqrt[m_j]{\mu_j}]$$

for an $m_j \in \mathbb{N}$ and such that finally $\alpha \in K_N$.

We say that a polynomial $f(X) \in K[X]$ is **solvable by radicals** if there exists such a sequence for a splitting field $Z|K$ of f such that finally $Z \subset K_N$ holds.

And now?

The definition does not quite describe the question of a solution **formula.** Rather, $\alpha \in L$ is solvable by radicals if and only if the element $\alpha \in L$ can be written as a polynomial expression of roots of polynomial expressions of …. of elements in K. The fact that the coefficients $a_n, \ldots, a_0 \in K$ should appear in these expressions is not required.

If f cannot be resolved by radicals, then there can be no solution **formula** with radicals. If $\alpha \in L$ is a root of a polynomial $f(X) = a_n X^n + \ldots + a_0 \in K[X]$, but cannot be resolved by radicals, then α cannot be written as roots of sums of products of roots …. let alone as such a radical expression depending on the coefficients. On the other hand side, it could be possible a priori that no solution formula can be found depending on the coefficients, but each individual root can be resolved by radicals. Section 18.3 will address this.

Example: The element

$$\alpha := \frac{\sqrt[13]{7 + \sqrt[7]{19} - \sqrt[5]{\sqrt[2]{3} + \sqrt[5]{6}}}}{\sqrt[5]{8 + \sqrt[7]{3}}} \in \overline{\mathbb{Q}}$$

is apparently solvable by radicals.

Check it out!

Is it really *apparent*? The example includes denominators, does that pose a problem? Hopefully not. Why not?

In response to the last question, note that all elements are algebraic and Theorem 2.18 applies.

We will now use the following notation:

Definition 14.6 If $f(X) \in \mathbb{Q}[X]$ is a polynomial of degree $n \geq 1$ and $Z \subset \mathbb{C}$ is the splitting field of f in $\mathbb{C}$, we will write

$$\mathrm{Gal}(f) := \mathrm{Gal}(Z|\mathbb{Q})$$

and call this group the **Galois group of** f.

How can we now use the fundamental theorem? Let $f(X) \in \mathbb{Q}[X]$ be of degree n, irreducible and $Z|\mathbb{Q}$ the splitting field inside $\mathbb{C}$. Let's assume that $f(X)$ is solvable by radicals. Then there is a finite sequence as above

$$\mathbb{Q} =: K_0 \subset K_1 \subset K_2 \subset \ldots \subset K_N$$

and elements $\mu_j \in K_j$, so that

$$K_{j+1} = K_j[\sqrt[m_j]{\mu_j}]$$

for a $m_j \in \mathbb{N}$ applies and finally $Z \subset K_N$.

And now?

Now we have many intermediate field extensions at hand, but these are not necessarily Galois! Note that $K_{j+1} = K_j[\sqrt[m_j]{\mu_j}]$ may not be normal over K_j, because the roots of $X^{m_j} - \mu_j$ are the elements

$$\sqrt[m_j]{\mu_j} \cdot \zeta_{m_j}^k$$

with $\zeta_{m_j} := \exp(\frac{2\pi i}{m_j})$ and $k = 0, \ldots, m_j - 1$. These do not have to lie inside K_{j+1}. What do we do to be able to apply the fundamental theorem? We adjoin all these unit roots a priori.

Let $m := \prod_{j=0}^N m_j \in \mathbb{N}$ and $\zeta := \exp(\frac{2\pi i}{m}) \in \mathbb{C}$. Consider the fields $L_j := K_j[\zeta]$. Still $L_{j+1} = L_j[\sqrt[m_j]{\mu_j}]$, holds but now the extensions $L_{j+1}|L_j$ are all Galois, and we have the tower of field extensions:

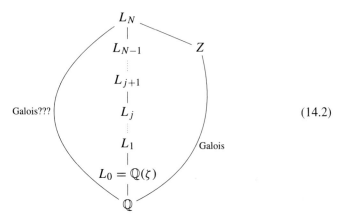

$$(14.2)$$

Again, we run into a problem because $L_N|\mathbb{Q}$ does not necessarily have to be normal again. However, one can construct the tower of estensions in such a way that $L_N|\mathbb{Q}$ is normal—one just needs to proceed with a bit more caution as before. To this end, we first introduce a notion:

Definition 14.7 Let L_0 be a field and $m \in \mathbb{N}$. Then we call a finite sequence of field extensions

$$L_0 \subset L_1 \subset \ldots \subset L_M$$

an *m*-**divisible radical extension** of L_0, if

$$L_{j+1} = L_j(\sqrt[m_j]{\mu_j})$$

for certain $\mu_j \in L_j$ and $m_j \in \mathbb{N}$ such that $m_j \mid m$ holds.

Remark 14.8 The adjective *m-divisible* is not a very common term you might not find it outside this book.

Lemma 14.9 *Let* $m \in \mathbb{N}$ *and* $\zeta = \exp(\frac{2\pi i}{m})$ *and* $L_0 = \mathbb{Q}(\zeta)$.
If then

$$L_0 \subset L_1 \subset \ldots \subset L_N$$

is an m-divisible radical extension of L_0, then there also exists an m-divisible radical extension of L_0

$$L_0 \subset L_1' \subset \ldots \subset L_M'$$

(with $M \geq N$), such that $L_N \subset L_M'$ holds and $L_M'|\mathbb{Q}$ is normal.

Proof We prove the assertion by induction on N. For $N = 1$ there is nothing to show, because then $L_1|\mathbb{Q}$ is normal, since L_1 contains all m_1-th roots of unity.

So, assume that the claim is valid by induction for all *m*-divisible radical extensions of length N. Let then

$$L_0 \subset L_1 \subset \ldots \subset L_N \subset L_{N+1}$$

be an *m*-divisible radical extension of length $N + 1$. Then, by induction assumption, we already have another *m*-divisible radical extension

$$L_0 \subset L_1' \subset \ldots \subset L_M' \tag{14.3}$$

with normal $L_M'|\mathbb{Q}$ and $L_N \subset L_M'$.

Let $\alpha := (\mu_{N+1})^{1/m_{N+1}} \in L_N$ be the element with $L_{N+1} = L_N(\alpha)$. Let further $f(X) \in \mathbb{Q}[X]$ be the minimal polynomial of α over $\mathbb{Q}$. If then $\beta_1, \ldots, \beta_d$ (with $d = \deg(f)$) are the roots of $f(X)$ in $\mathbb{C}$, then

$$L' := L_M'(\beta_1, \ldots, \beta_d)|\mathbb{Q}$$

is normal (because $L_M'|\mathbb{Q}$ is normal and all roots of an irreducible polynomial over $\mathbb{Q}$ are adjoined) and furthermore, $L_{N+1} \subset L'$, because α is one of these roots. So, we are done if we can show that L' arises from L_M' by successive adjunction of the m_{N+1}-th roots $(\mu')^{1/m_{N+1}} \in L'$ of elements $\mu' \in L_M'$. Because then we have found an *m*-divisible radical extension.

$$L_M' \subset \ldots \subset L'$$

and this sequence continues (14.3) and accomplishes the desired result. But this follows easily because β_j is of this form: According to Theorem A, for each β_j as above there is exactly one field homomorphism

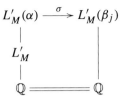

with $\sigma|\mathbb{Q} = \mathrm{id}_{\mathbb{Q}}$ and $\sigma(\alpha) = \beta_j$. Since $L'_M|\mathbb{Q}$ is normal, we have

$$\mathrm{Hom}_{\mathbb{Q}}(L'_M, \overline{\mathbb{Q}}) = \mathrm{Hom}_{\mathbb{Q}}(L'_M, L'_M)$$

and thus we have the intermediate level in the diagram:

$$
\begin{array}{ccc}
L'_M(\alpha) & \xrightarrow{\ \sigma\ } & L'_M(\beta_j) \\
\vert & \scriptstyle\sigma|_{L'_M} & \vert \\
L'_M & \xrightarrow{\hspace{1.2cm}} & L'_M \\
\vert & & \vert \\
\mathbb{Q} & = & \mathbb{Q}
\end{array}
$$

But now, since α satisfies the equation $\alpha^{m_{N+1}} = \mu_{N+1} \in L_N \subset L'_M$, it follows after applying σ, that $\beta_j^{m_{N+1}} = \sigma(\alpha^{m_{N+1}}) = \sigma(\mu_{N+1}) \in L'_M$ holds. Therefore, $\beta_j = (\sigma(\mu_{N+1}))^{1/m_{N+1}}$, as we wanted to show. $\square$

With this lemma, we can now find a radical extension for our splitting field $Z|\mathbb{Q}$, which is solvable by radicals, as in (14.2), where now indeed $L_N|\mathbb{Q}$ is normal.

So, what does the Fundamental Theorem of Galois Theory tell us about the Galois group $\mathrm{Gal}(Z|\mathbb{Q})$?

- We have, according to the addendum 14.3, that $\mathrm{Gal}(L_N|Z) \triangleleft \mathrm{Gal}(L_N|\mathbb{Q})$ is a normal subgroup and that

$$\mathrm{Gal}(Z|\mathbb{Q}) \cong \mathrm{Gal}(L_N|\mathbb{Q})/\mathrm{Gal}(L_N|Z).$$

- Similarly, according to the fundamental theorem, we have the subgroups

$$H_j := \mathrm{Gal}(L_N|L_j) \subset \mathrm{Gal}(L_N|\mathbb{Q}) =: G,$$

and thus a sequence of subgroups

$$\mathrm{Gal}(L_N|\mathbb{Q}) \supset H_0 \supset H_1 \supset \ldots \supset H_j \supset H_{j+1} \supset \ldots \supset H_{N-1} \supset H_N = \{e\}$$

(note the reversal of the inclusion relation from the addendum 14.3).
In each case, according to the addendum, $H_{j+1} \triangleleft H_j$ is a normal subgroup (because $L_{j+1}|L_j$ is normal), and likewise according to the fundamental theorem,

$$H_j/H_{j+1} \cong \mathrm{Gal}(L_{j+1}|L_j) = \mathrm{Gal}(L_j[\sqrt[m_j]{\mu_j}]|L_j).$$

holds. In the following Lemma 14.10 (see also Exercise 12.4) we see that the latter Galois group is abelian:

Lemma 14.10 *If a field L contains a primitive m-th root of unity and let $\mu \in L$ be given, then*

$$\mathrm{Gal}(L(\sqrt[m]{\mu})|L) \to \mu_m(L), \ \sigma \mapsto \frac{\sigma(\sqrt[m]{\mu})}{\sqrt[m]{\mu}}$$

is an injective group homomorphism. In particular, $\mathrm{Gal}(L(\sqrt[m]{\mu})|L)$ is abelian.

Proof Let $\alpha = \sqrt[m]{\mu}$ and $f(X) \in L[X]$ be its minimal polynomial over L. Since $\alpha^m = \mu$, it follows that $f(X) \mid (X^m - \mu)$ and the roots of $f(X)$ in $L(\alpha)$ are of the form $\alpha \cdot \zeta$ for each $\zeta \in \mu_m(L)$—this also shows that $L(\alpha)|L$ is normal.

For $\sigma \in \mathrm{Gal}(L(\alpha)|L)$ it follows that $\sigma(\alpha)/\alpha \in \mu_m(L)$ and the above map is well-defined. It is a group homomorphism, as one can easily see. Furthermore, it is injective, because σ is uniquely determined by the image $\sigma(\alpha)$. $\square$

One step is still missing, namely the first: According to the addendum we have

$$\mathrm{Gal}(L_N|\mathbb{Q})/H_0 \cong \mathrm{Gal}(L_0|\mathbb{Q}) = \mathrm{Gal}(\mathbb{Q}(\zeta)|\mathbb{Q})$$

and we will see in Chap. 15 that this group is also **abelian**—Theorem 15.20.

So we have obtained the following observation:

Lemma 14.11 *If $f(X) \in \mathbb{Q}[X]$ is irreducible of degree $n \geq 1$ and f is solvable by radicals, then there exists a group G,*

- *which admits a finite sequence of subgroups*

$$G \supset H_0 \supset H_1 \supset \ldots \supset H_{N-1} \supset H_N = \{e\}$$

such that $H_0 \lhd G$ and each $H_{j+1} \lhd H_j$ is normal inside its precursor such that the factor groups

$$G/H_0 \text{ and } H_j/H_{j+1}$$

are all abelian,
- *and such that there is a surjective group homomorphism*

$$G \twoheadrightarrow \mathrm{Gal}(f)$$

(hence $\mathrm{Gal}(f) \cong G/N$ for a normal subgroup $N \lhd G$).

Proof see above—where we have anticipated the result of Theorem 15.20 from Chap. 15. $\square$

The conditions on $\mathrm{Gal}(f)$ from this lemma are very restrictive. Not every group fulfills this. We will later introduce a term (that of a **solvable group**) and examine this in more detail.

For example, the group $S(5)$ cannot fulfill these conditions.

So, if we have an f with $\mathrm{Gal}(f) \cong S(5)$ (like our intermezzo example), its roots cannot be solved by radicals!

14.3   Proof of the Fundamental Theorem

Proof (of the fundamental theorem) Let's recall the situation from the fundamental theorem: We need to show that the maps

$$\mathcal{I}nt\mathcal{F}(L|K) \underset{\Phi}{\overset{\Psi}{\rightleftarrows}} \mathcal{S}ub\mathcal{G}(L|K)$$

$$M \longmapsto \mathrm{Gal}(L|M)$$

$$L^H \longleftarrow\!\!\shortmid H$$

are mutually inverse bijections.

$\Phi \circ \Psi = \mathrm{id}$:   Let $M \in \mathcal{ZK}(L|K)$ be an intermediate field. It needs to be shown that $M = \Phi \circ \Psi(M) = L^{\mathrm{Gal}(L|M)}$ holds. Note that $L|M$ is a finite Galois extension. We need to show the following claim:

Claim Given a finite Galois extension $L|M$, the following holds true

$$L^{\mathrm{Gal}(L|M)} = M \qquad\qquad (14.4)$$

Proof (In other words: The base field K no longer plays a role, we can consider M as the base field).

We show the two inclusions $\subset$ and $\supset$ separately.
"$\supset$"(the easy direction):

> **Check it out!**
> First, think about it yourself. This follows almost trivially from the definitions of $\mathrm{Gal}(L|M)$ and L^H for a subgroup $H \subset \mathrm{Gal}(L|M)$.

This is trivial, because by definition

$$L^{\mathrm{Gal}(L|M)} = \{x \in L \mid \sigma(x) = x \text{ for all } \sigma \in \mathrm{Gal}(L|M) = \mathrm{Aut}_M(L)\}$$

and since $\sigma \in \mathrm{Gal}(L|M) = \mathrm{Aut}_M(L)$ and hence the condition $\sigma|_M = \mathrm{id}_M$ is fulfilled, it follows for all $x \in M$, that $\sigma(x) = x$ holds. Therefore, $M \subset L^{\mathrm{Gal}(L|M)}$.

"$\subset$": We have to show that every $x \in L$, such that $\sigma(x) = x$ for all $\sigma \in \mathrm{Gal}(L|M)$, must already lie in the intermediate field $x \in M$.

We do this by contraposition. If $x \in L \setminus M$, we must find a $\sigma \in \mathrm{Gal}(L|M)$ with $\sigma(x) \neq x$.

Since $x \in L \setminus M$, the minimal polynomial $f(X) \in M[X]$ of x over M has
$\deg(f) \geq 2$. Furthermore, f decomposes into linear factors in L (because $L|M$ is
normal) and L thus contains all $\deg(f)$ many roots, which are pairwise different
(because $L|M$ is separable). In particular, there is at least one other root $y \in L$ of
$f(X)$ with $x \neq y$. According to Theorem A, there is a unique field homomorphism
$\tau : M[x] \to M[y]$ with $\tau|_M = \mathrm{id}_M$ and $\tau(x) = y$, so

$$
\begin{array}{ccc}
L & & L \\
| & & | \\
M[x] & \xrightarrow{\;\tau\;} & M[y] \\
| & & | \\
M & =\!=\!= & M
\end{array}
$$

According to Theorem 12.3 we know that τ admits an extension $\sigma : L \to L$ (in
Theorem 12.3 an algebraic closure Ω of L was necessary, but we don't need it here
because $L|M$ is normal and thus the proof of Theorem 12.3 works with L instead of
Ω). This is an element $\sigma \in \mathrm{Gal}(L|M)$ with $\sigma(x) = \tau(x) = y \neq x$. This is what we
had to prove. $\square$

Hence, the assertion (14.4) and therefore $\Phi \circ \Psi = \mathrm{id}$ are proven. $\square$

$\Psi \circ \Phi = \mathrm{id}$:   We need to prove the following

Claim Given a subgroup $H \subset \mathrm{Gal}(L|K)$, the following holds:

$$H = \Psi \circ \Phi(H) = \mathrm{Gal}(L|L^H) . \tag{14.5}$$

Proof Again, we show both inclusions separately.

"$\subset$":   We need to show $H \subset \mathrm{Gal}(L|L^H)$.

> **Check it out!**
> This is again the easy direction, almost trivial from the definitions. Try it!
> Write out the definition of $\mathrm{Gal}(L|L^H)$ combined with the one of L^H?

By definition,

$$\mathrm{Gal}(L|L^H) = \mathrm{Aut}_{L^H}(L) = \{\sigma : L \to L \mid \sigma|_{L^H} = \mathrm{id}_{L^H}\}$$
$$L^H = \{x \in L \mid \sigma(x) = x \text{ for all } \sigma \in H\}.$$

So if $\sigma \in H$, then by definition of L^H, it follows that $\sigma(x) = x$ for all $x \in L^H$. Therefore, $\sigma|_{L^H} = \mathrm{id}_{L^H}$ and thus $\sigma \in \mathrm{Gal}(L|L^H)$. So we have shown $H \subset \mathrm{Gal}(L|L^H)$.

"$\supset$":   This is the place in the proof that is the most delicate but it can be solved with the following idea. We have already shown that $H \subset \mathrm{Gal}(L|L^H)$ holds and both sides of this inclusion are finite groups, so it suffices to show that

$$\#H = \#\mathrm{Gal}(L|L^H) \overset{(*)}{=} [L : L^H]$$

holds, where the equation $(*)$ follows because $L|L^H$ is separable.

Now $H \subset \mathrm{Gal}(L|L^H)$, hence we already know that $\#H \le \#\mathrm{Gal}(L|L^H) = [L : L^H]$. In order to show $\ge$ we make our lives a little easier and apply the Primitive Element Theorem (Theorem 13.23). The extension $L|L^H$ is finite and separable, so there is a primitive element $\alpha \in L$ for it:

$$L = L^H[\alpha] .$$

Let $f(X) \in L^H[X]$ be its minimal polynomial over L^H.
    Now consider the set

$$A := \{\sigma(\alpha) \in L \mid \sigma \in H\} \subset L.$$

Let us define the polynomial

$$g(X) := \prod_{\beta \in A} (X - \beta) \in L[X],$$

which has exactly the elements from A as its roots (each as a simple root). It is a polynomial $g(X) \in L[X]$ with coefficients in L (a priori by definition).

Claim We have $g(X) \in L^H[X]$.

Proof For an arbitrary $\tau \in H$ (using the notation g^τ as introduced before Theorem 10.8):

$$g^\tau(X) = \prod_{\beta \in A}(X - \tau(\beta)) = g(X),$$

because $\tau(\beta) = (\tau \circ \sigma)(\alpha)$ and thus the index set is only permuted by applying τ:

$$A = \{\sigma(\alpha) \mid \sigma \in H\} = \{(\tau \circ \sigma)(\alpha) \mid \tau \circ \sigma \in H\}$$

(whoever still has some doubts, remember that $H \to H$, $\sigma \mapsto \tau\sigma$ is bijective). Therefore, if we denote by $a_j \in L$ the coefficients of $g(X)$, we deduce that

$$\tau(a_j) = a_j \text{ for all } \tau \in H,$$

that is $a_j \in L^H$. $\square$

What have we achieved up to now? We have found a polynomial $g(X) \in L^H[X]$ that has exactly the elements of A as simple roots in L. In particular, $\alpha = \operatorname{id}(\alpha) \in A$ (because $\operatorname{id} \in H$) and thus $g(\alpha) = 0$. Since $f(X)$ is the minimal polynomial of α over L^H, we have the divisibility relation

$$f(X) \mid g(X) \text{ in the ring } L^H[X].$$

It follows that $\deg(f) \leq \deg(g) \leq \#H$ (the latter because $\deg(g) = \#A \leq \#H$). But since α is a primitive element of the separable extension $L = L^H[\alpha] \mid L^H$, we know that

$$[L : L^H] = \deg(f) \leq \deg(g) \leq \#H,$$

which was to be shown. $\square$

Therefore, the Fundamental Theorem of Galois Theory is proven. $\square$

14.4   Proof of the Addendum

We want to prove the addendum 14.3 as well.

Proof to (1):   We can now already use the main theorem. It is then easy to see for two subgroups $H, H' \subset \operatorname{Gal}(L|K)$, that

$$H \subset H' \overset{(*)}{\Rightarrow} L^H \supset L^{H'} \overset{(*)}{\Rightarrow} \underbrace{\operatorname{Gal}(L|L^H)}_{=H} \subset \underbrace{\operatorname{Gal}(L|L^{H'})}_{=H'}$$

so the conditions are equivalent.

> **Check it out!**
> Consider why the two conclusions marked with $(*)$ are obvious!

to (2):   It needs to be shown that $M = L^H|K$ is normal if and only if $H \lhd \mathrm{Gal}(L|K)$ is a normal subgroup.

Assume that $L^H|K$ is normal and let $\sigma \in \mathrm{Gal}(L|K)$ be any element. We need to show that $\sigma H \sigma^{-1} \subset H$ holds. Let's assume that $h \in H$. Then it must be shown that

$$\sigma h \sigma^{-1} \in H \overset{(*)}{=} \mathrm{Gal}(L|L^H),$$

where we used the fundamental theorem at the place $(*)$. So we need to show that

$$\sigma h \sigma^{-1}|_{L^H} = \mathrm{id}_{L^H}$$

holds.

To this end: For any $x \in L^H$ we know that $\sigma^{-1}x \in L$ is a root of the same minimal polynomial $f(X) \in K[X]$ of x over K *(cp. Theorem 10.6)*, and since $L^H|K$ is normal, it already lies in L^H. Therefore, $\sigma^{-1}(x) \in L^H$ and thus we deduce that $h\sigma^{-1}(x) = \sigma^{-1}(x)$ (because $h \in H$ acts as an identity on L^H). Thus,

$$\sigma h \sigma^{-1}(x) = \sigma \sigma^{-1}(x) = x \text{ for all } x \in L^H,$$

which was to be shown.

If, conversely, $H \lhd \mathrm{Gal}(L|K)$ is a normal subgroup and $f(X) \in K[X]$ is irreducible with a root $\alpha \in L^H$, then we must show that all other roots (which lie in the normal extension $L|K$) are already in L^H.

Let $\beta \in L$ be a root of $f(X)$. Then, according to Theorem A and Theorem 12.3 (just like in the proof of the fundamental theorem), there is a $\sigma \in \mathrm{Gal}(L|K)$ with $\sigma(\alpha) = \beta$:

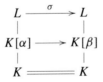

For all $h \in H$ we know that $\sigma^{-1}h\sigma =: h' \in H$ (because $H \lhd \mathrm{Gal}(L|K)$) and thus

$$h(\beta) = h\sigma(\alpha) = \sigma h'(\alpha) \overset{(*)}{=} \sigma(\alpha) = \beta,$$

(at the place $(*)$, note that $\alpha \in L^H$) so $\beta \in L^H$. This was to be demonstrated.

It remains to show the following assertion:

Claim If $M|K$ is a normal intermediate extension (and anyway separable, hence Galois), then we have the canonical isomorphism

$$\mathrm{Gal}(L|K)/\mathrm{Gal}(L|M) \overset{\cong}{\to} \mathrm{Gal}(M|K)$$

$$[\sigma] \mapsto \sigma|_M.$$

Proof Consider the group homomorphism

$$\psi : \mathrm{Gal}(L|K) \to \mathrm{Gal}(M|K), \quad \sigma \mapsto \sigma|_M,$$

which is well-defined (which here means that $\sigma|_M \in \mathrm{Gal}(M|K)$ lies in the right hand side), because according to the assumption $M|K$ is normal and thus

$$\sigma|_M \in \mathrm{Hom}_K(M,L) = \mathrm{Hom}_K(M,M) = \mathrm{Aut}_K(M) = \mathrm{Gal}(M|K)$$

(Lemma 12.6—again in a variant, because one can use the normal $L|K$ instead of Ω). Its kernel is

$$\ker(\psi) = \{\sigma \in \mathrm{Gal}(L|K) \mid \sigma|_M = \mathrm{id}_M\} = \mathrm{Gal}(L|M) .$$

Furthermore, ψ is surjective, because an arbitrary $\tau \in \mathrm{Gal}(M|K)$ can be extended to a $\sigma \in \mathrm{Gal}(L|K)$ according to Theorem 12.3. The assertion of the Claim follows from the homomorphism theorem for groups. □
 With that, the addendum is proven. □

Red thread to the previous chapter:
In this chapter, we have

- formulated the Fundamental Theorem of Galois Theory,
- seen how, with its help, the question of *solution formulas* for polynomial equations of higher degrees can be reformulated into a question for groups, and
- proven the fundamental theorem together with its addendum.

Exercises: 14.1 Prove the remaining statement from Exercise 12.2: Let $Z|K$ be finite and Galois and let $K \subset M \subset Z$ and $K \subset L \subset Z$ be field extensions, such that $L|K$ is Galois. Then $LM|M$ is Galois (Exercise 12.2) and the homomorphism

$$\theta : \mathrm{Gal}(LM|M) \to \mathrm{Gal}(L|L \cap M) , \quad \sigma \mapsto \sigma|_L.$$

is an isomorphism.
 This statement is also known as the **Translation Theorem of Galois Theory.**
 14.2 Let L be the splitting field of $X^{12} - 729 \in \mathbb{Q}[X]$ in $\mathbb{C}$ and ζ the primitive 12th root of unity $\zeta = \exp(\frac{2\pi i}{12}) = \frac{\sqrt{3}+i}{2} \in \mathbb{C}$.

(a) Show: $\sqrt{3} \in \mathbb{Q}(\zeta)$ and $L = \mathbb{Q}(\zeta)$.
(b) Show that $\mathrm{Gal}(L|\mathbb{Q})$ is an abelian group of order four that contains exactly three elements of order two.
(c) Describe all proper intermediate fields of the extension $L|\mathbb{Q}$, by specifying a primitive element for each proper intermediate field.

14.3 Let $Z \subset \mathbb{C}$ be the splitting field of the polynomial $f(X) = X^4 + 2X^2 + 2 \in \mathbb{Q}[X]$ and $\alpha \in Z$ a root. Determine the Galois group $\mathrm{Gal}(f) = \mathrm{Gal}(Z|\mathbb{Q})$ and the subgroup belonging to the intermediate field $\mathbb{Q}(\alpha)$. (Hint: Consider Exercise 10.2.)

14.4 Let $L|K$ be a finite Galois extension. Furthermore, let M and M' be intermediate fields of $L|K$ and $\sigma \in \mathrm{Gal}(L|K)$. Show:

$$\sigma(M) = M' \iff \sigma \mathrm{Gal}(L|M)\sigma^{-1} = \mathrm{Gal}(L|M').$$

Cyclotomic Fields

15

15.1 Roots of Unity

Definition 15.1 Let Ω be a field and $n \in \mathbb{N}$. Then an element $\zeta \in \Omega$ is called an *n-th root of unity*, if $\zeta^n = 1$ holds.

If the order of $\zeta \in \Omega^\times$ in the group $\Omega^\times$ is equal to n, that is $\operatorname{ord}(\zeta) = n$, then ζ is called a **primitive n-th root of unity**.

We will write

$$\mu_n(\Omega) := \{\zeta \in \Omega \mid \zeta^n = 1\} \subset \Omega^\times$$

for the subgroup (verify!) of the n-th roots of unity in Ω.

Remark 15.2 Since these elements $\zeta \in \mu_n(\Omega)$ are the roots of $X^n - 1$, there are at most n of them, so

$$\#\mu_n(\Omega) \leq n .$$

Example:
- If $\zeta \in \mu_n(\Omega)$ is an nth root of unity, then also $\zeta \in \mu_{nk}(\Omega)$ is true for all multiples of n, because $\zeta^{nk} = (\zeta^n)^k = 1$.
- In $\Omega = \mathbb{Q}(i)$ one indeed has all four possible 4th roots of unity:

$$\mu_4(\mathbb{Q}(i)) = \{1, -1, i, -i\} ,$$

but for example, no further 8th roots of unity

$$\mu_8(\mathbb{Q}(i)) = \{1, -1, i, -i\} = \mu_4(\mathbb{Q}(i)).$$

- In $\Omega = \mathbb{C}$ the complex number

$$\zeta = \exp\left(\frac{2\pi i}{n}\right)$$

© The Author(s), under exclusive license to Springer-Verlag GmbH, DE, part of
Springer Nature 2024
M. Hien, *Abstract Algebra*, Mathematics Study Resources 7,
https://doi.org/10.1007/978-3-662-67974-6_15

is a primitive n-th root of unity and $\mu_n(\mathbb{C})$ is the cyclic group

$$\mu_n(\mathbb{C}) = \langle \zeta \rangle = \{1, \zeta, \zeta^2, \dots, \zeta^{n-1}\}$$

with n elements. The generator ζ is not uniquely determined, more precisely we have the next lemma.

Lemma 15.3 *Let $n \in \mathbb{N}$. Then, $\xi = \exp\left(\frac{2\pi ik}{n}\right) \in \mathbb{C}$ for a given $k \in \mathbb{N}$ is a primitive n-th root of unity if and only if $\gcd(k, n) = 1$ holds.*

Proof Observe the following:

$$1 = \xi^\ell = \exp\left(\frac{2\pi ik\ell}{n}\right) \iff k\ell \in n\mathbb{Z} \iff k\ell = an \text{ for some } a \in \mathbb{Z}$$

If k, n are coprime, it follows that $n | \ell$ must hold, hence ord $(\xi) = n$ and ξ is a primitive nth root of unity.

On the other hand, if $d = \gcd(k, n) \neq 1$, WLOG $d > 1$, let $k = xd$, and $n = yd$. Then $0 < y < n$ and

$$\xi^y = \exp\left(\frac{2\pi iky}{n}\right) = \exp\left(\frac{2\pi ixdy}{n}\right) = \exp\left(\frac{2\pi ixn}{n}\right) = 1 \,,$$

and thus ord $(\xi) < n$, so ξ is not primitive. $\square$
Also note the following lemma.

Lemma 15.4 *The group of units in the ring $\mathbb{Z}/n\mathbb{Z}$ consists exactly of the residue classes.*

$$(\mathbb{Z}/n\mathbb{Z})^\times = \{[a] \in \mathbb{Z}/n\mathbb{Z} \mid \gcd(a, n) = 1\}$$

Proof First note that the expression $\gcd(a, n)$ is well-defined for each class $[a]$ (as always up to sign, $\mathbb{Z}^\times = \{\pm 1\}$), because for $[a] = [b]$ there is a $k \in \mathbb{Z}$ with $b = a + kn$ and then $\gcd(b, n) = \gcd(a, n)$.

If now $[a] \in (\mathbb{Z}/n\mathbb{Z})^\times$, then there exists $[b] \in \mathbb{Z}/n\mathbb{Z}$ with $1 = [a][b] = [ab]$, thus

$$ab = 1 + kn \text{ for some } k \in \mathbb{Z}.$$

But then every prime divisor of a is certainly not a divisor of n, so a and n are coprime.

Conversely, if a, n are coprime, then we have a representation $1 = \gcd(a, n) = xa + yn$ and thus $[x] \in \mathbb{Z}/n\mathbb{Z}$ is an inverse to $[a]$. $\square$

Definition 15.5 The map

$$\varphi : \mathbb{N} \to \mathbb{N}, \ n \mapsto \varphi(n) := \#(\mathbb{Z}/n\mathbb{Z})^\times$$

is called the **Euler φ-function**.

Remark 15.6
- According to Lemma 15.4, we also have
$$\varphi(n) = \#\{k \in \{1, \dots, n-1\} \mid \gcd(k, n) = 1\} \,. \tag{15.1}$$

- Calculating the value $\varphi(n)$ is complicated and usually involves first determining the prime factorization of n (then you can apply the following lemma), but: Determining the prime factorization is very complicated for huge numbers. Some crypto systems rely on the fact that it takes too long to determine the prime factorization of large numbers—those who are more interested in this will find a lot of literature on the subject under the keyword RSA cryptography.

As just mentioned, it is complicated (for large numbers) to explicitly calculate the Euler φ function. However, the following two properties of $\varphi(n)$ easily allow its computation if you know the prime factorization of n.

Lemma 15.7 *The following holds:*
1. *The Euler φ-function is a so-called arithmetic function, which means it is multiplicative for coprime products: If*

$$n, m \text{ are coprime}, \quad \varphi(n \cdot m) = \varphi(n) \cdot \varphi(m) .$$

2. *If p is a prime number and $\ell \in \mathbb{N}$, then the following holds:*

$$\varphi(p^\ell) = p^{\ell-1} \cdot (p - 1) .$$

Proof to (1):

> **Check it out!**
> Do you see an argument? Do you know a sentence that would be applicable to $\mathbb{Z}/nm\mathbb{Z}$ for coprime n,m?

This follows because $\mathbb{Z}/nm\mathbb{Z} \cong \mathbb{Z}/n\mathbb{Z} \times \mathbb{Z}/m\mathbb{Z}$ (Chinese Remainder Theorem) holds and this is an isomorphism of rings. This induces an isomorphism of the unit groups and hence

$$(\mathbb{Z}/n\mathbb{Z} \times \mathbb{Z}/m\mathbb{Z})^\times = (\mathbb{Z}/n\mathbb{Z})^\times \times (\mathbb{Z}/m\mathbb{Z})^\times.$$

to (2): One can use (15.1) and count the number of elements of the complement of this set—note that $\gcd(p^\ell, k) = 1 \Leftrightarrow p \nmid k$, and since p is a prime number:

$$\#\big(\{1,\ldots,p^\ell - 1\} \smallsetminus \{k \in \{1,\ldots,p^\ell - 1\} \mid \mathrm{ggT}(k, p^\ell) = 1\}\big) =$$
$$\#\{k \in \{1,\ldots,p^\ell - 1\} \mid p \mid k\} =$$
$$\#\{p, 2p, 3p, \ldots, (p^{\ell-1} - 1)p\} = p^{\ell-1} - 1$$

(the next multiple of p, namely $p^{\ell-1}p = p^\ell$ is no longer in the chosen set of representatives). Therefore, it follows that

$$\varphi(p^\ell) = p^\ell - 1 - (p^{\ell-1} - 1) = p^\ell - p^{\ell-1} = p^{\ell-1}(p - 1) ,$$

as claimed. $\square$

Example: For small n, for which one knows the prime factor decomposition, one can calculate $\varphi(n)$ using these rules:

$$\varphi(1125) = \varphi(9 \cdot 125) = \varphi(3^2)\varphi(5^3) = (3 \cdot 2)(25 \cdot 4) = 600.$$

Because it will play a role now, let's determine the automorphism group of $\mu_n(\mathbb{C})$, so

$$\text{Aut}(\mu_n(\mathbb{C})) := \{\psi : \mu_n(\mathbb{C})) \to \mu_n(\mathbb{C}) \mid \psi \text{ is a group isomorphism}\}$$

In $\mathbb{C}$ we know the roots of unity very well, we have seen above that $\mu_n(\mathbb{C})$ is cyclic, the generators are exactly the *primitive n-th* roots of unity. In particular, there exists a generator, for example $\zeta = \exp\left(\frac{2\pi i}{n}\right) \in \mathbb{C}$, but any other generator is also suitable for the following. Therefore, let's say that ζ is an arbitrary generator of $\mu_n(\mathbb{C})$.

Now, consider an element $\psi \in \text{Aut}(\mu_n(\mathbb{C}))$, then ψ is uniquely determined by its value for the generator ζ, let's say $\psi(\zeta) = \zeta^\ell$ for a $\ell \in \mathbb{Z}/n\mathbb{Z}$ (note that $\zeta^n = 1$, so the value ζ^k for a given $k \in \mathbb{Z}$ depends only on the class $\ell = [k] \in \mathbb{Z}/n\mathbb{Z}!$). Conversely, every $\ell \in \mathbb{Z}/n\mathbb{Z}$ induces a group **homomorphism**

$$\eta : \mu_n(\mathbb{C}) \to \mu_n(\mathbb{C}) , \ \zeta \mapsto \zeta^\ell$$

(also $\eta(\zeta^k) = \zeta^{\ell k}$). The question is whether η is an automorphism. Apparently *(check!)* this is the case if and only if ζ^ℓ is again a generator, so according to the considerations above if and only if $\ell \in (\mathbb{Z}/n\mathbb{Z})^\times$. Let's summarize:

Lemma 15.8 *Let $n \in \mathbb{N}$. After choosing a generator $\zeta \in \mu_n(\mathbb{C})$ one obtains a group isomorphism*

$$(\mathbb{Z}/n\mathbb{Z})^\times \xrightarrow{\cong} \text{Aut}(\mu_n(\mathbb{C})), \ \ell \mapsto (\zeta \mapsto \zeta^\ell).$$

Proof This is a group homomorphism, because $\psi_2 \circ \psi_1(\zeta) = \psi_2(\zeta^{\ell_1}) = \zeta^{\ell_1 \ell_2}$ if we used the notation $\psi_j(\zeta) = \zeta^{\ell_j}$. We have considered bijectivity above. $\square$

And now?

In the last lemma, note that the fact that $\mu_n(\mathbb{C})$ is cyclic, plays an essential role. This allowed every automorphism $\psi : \mu_n(\mathbb{C}) \to \mu_n(\mathbb{C})$ to be characterized by its value for the generator. We now want to pass from $\mathbb{C}$ to any arbitrary field Ω. By the way, I write Ω, because the letter K was often used for the base field of our situation and we will continue to do so, but then we want to study $\mu_n(\Omega)$ or $\mu_n(Z)$ for an algebraic closure Ω or a suitable splitting field Z/K.

Let Ω be an arbitrary field and $\mu_n(\Omega)$ the group of nth roots of unity. Is it cyclic again? The answer is yes and follows from the following theorem in which we prove an even more general statement:

Theorem 15.9 Let Ω be a field and $G \subset \Omega^\times$ a **finite** subgroup of the group of units. Then G is a cyclic group.

Proof Firstly, G is abelian (since $\Omega^\times$ is abelian). Let $g \in G$ be an element with maximum order ord $(g) = n$ among all elements (which is possible because there are only finitely many elements in G).

Claim For all $x \in G$ we have ord $(x) \mid$ ord (g). $\square$

Proof One can easily convince oneself that given two elements $g, x \in G$ in an abelian group G **with the condition** gcd$($ ord $(x),$ ord $(g)) = 1$, we have

$$\text{ord}\,(gx) = \text{lcm}(\,\text{ord}\,(g),\,\text{ord}\,(x)) = \text{ord}\,(x) \cdot \text{ord}\,(g).$$

Apparently, the left hand side divides the right one. On the other hand side, $(gx)^r = g^r x^r = 1$ and hence $x^r = g^{-r} \in \langle g \rangle$. Therefore, $x^r \in \langle x \rangle \cap \langle g \rangle$. The order of this group must divide ord (x) and ord (g) and is therefore 1. Thus, we have $x^r = 1$ and then also $g^r = 1$. It follows that $r \in \mathbb{Z} \cdot$ ord (x) and $r \in \mathbb{Z} \cdot$ ord (g). Due to the coprimality, $r \in \mathbb{Z} \cdot$ lcm$($ ord $(x),$ ord $(g))$.
   If we assume that there is a $x \in G$ with ord $(x) \nmid$ ord (g), we want to deduce a contradiction. If ord (x) and ord (g) are coprime, we have

$$\text{ord}\,(gx) = \text{lcm}(\,\text{ord}\,(g),\,\text{ord}\,(x))\,\text{ord}\,(g) = n\,,$$

in contradiction to the maximality of ord (g).
   If ord (x) and ord (g) are not coprime, we proceed as follows. Then $d := \text{gcd}($ ord $(x),$ ord $(g)) \neq 1$. Let's write

$$\text{ord}\,(x) = k \cdot d$$

with $k \neq 1$ (because ord $(x) \nmid$ ord (g)). It follows that k and ord (g) are coprime. Then $y := x^d$ has order ord $(y) = k$. Therefore, y and g are two elements with coprime order and thus according to the above

$$\text{ord}\,(yg) = \text{ord}\,(y) \cdot \text{ord}\,(g) = k\,\text{ord}\,(g) > \text{ord}\,(g) = n$$

a contradiction. $\square$
   Therefore, $x^n = 1$ holds for all $x \in G$, that is, the group elements $x \in G$ are roots of the polynomial $f(X) = X^n - 1 \in \Omega[X]$. Since Ω is a field, $f(X)$ has at most n roots in Ω, so

$$\#G \leq \#(\text{roots of } f(X) \text{ in } \Omega) \leq n\,.$$

But $g \in G$ has order n and thus

$$n = \#\langle g \rangle \leq \#G \leq n,$$

and equality holds, so $\langle g \rangle = G$. $\square$

Corollary 15.10 If Ω is a field, the group $\mu_n(\Omega)$ is cyclic. If $\mu_n(\Omega) = n$, then a primitive nth root of unity $\zeta \in \Omega$ exists and after choosing a primitive nth root of unity ζ one has the group isomorphism

$$(\mathbb{Z}/n\mathbb{Z})^{\times} \xrightarrow{\cong} \mathrm{Aut}(\mu_n(\Omega)) \,, \ \ell \mapsto (\zeta \mapsto \zeta^{\ell}) \,.$$

Proof Firstly, $\mu_n(\Omega)$ is cyclic according to the lemma, so $\mu_n(\Omega) = \langle \zeta \rangle$ for a $\zeta \in \mu_n(\Omega)$. Since, by assumption, $\#\mu_n(\Omega) = n$ (meaning that Ω has the maximum possible number of n th roots of unity), we have ord $(\zeta) = n$ and thus ζ is a primitive n-th root of unity. We can now proceed literally as in the case of $\Omega = \mathbb{C}$, because there we only used that we have a primitive n-th root of unity. $\square$

15.2   Cyclotomic Fields and Cyclotomic Polynomials

We now start with an arbitrary base field K and want to adjoin roots of unity (if necessary):

Definition 15.11 Let K be a field and $n \in \mathbb{N}$. The[1] splitting field $Z_n | K$ of the polynomial $f(X) = X^n - 1$ is then called the n-**th cyclotomic field** over K.

What now?

What do we already know about $\mu_n(Z_n)$? We know that it is a cyclic group $\mu_n(Z_n) = \langle \zeta \rangle$ with an element $\zeta \in \mu_n(Z_n)$. Is it true that ζ automatically a primitive nth root of unity, so ord $(\zeta) = n$? This does not have to be the case, it could happen that there are fewer nth roots of unity. This is a very similar phenomenon to non-separability (and indeed it is connected with it).

  (Extreme) Example: $K = \mathbb{F}_5 = \mathbb{Z}/5\mathbb{Z}$. Then, for $n = 5$ the polynomial

$$(X - 1)^5 = X^5 + \sum_{j=1}^{4} \binom{5}{j}(-1)^j X^{5-j} - 1 = X^5 - 1,$$

(the coefficients in the sum are all divisible by 5) thus $X^5 - 1$ already decomposes in $\mathbb{F}_5[X]$ in linear factors and $Z = \mathbb{F}_5$ is the 5th cyclotomic field. However, $\mu_5(\mathbb{F}_5) = \{1\}$

More generally, the following applies:

---

[1] So far, I have always avoided the question of the uniqueness of a splitting field. It is indeed unique, but only up to isomorphism. I would like to always assume that we have fixed an algebraic closure Ω and then we take the splitting field within Ω. Then it is the unique splitting field inside Ω.

Lemma 15.12 *If K has a prime characteristic, $p = \mathrm{char}(K)$, and if $p \mid n$, let's say $n = pm$, then*
$$Z_{mp} = Z_m \quad \text{and} \quad \mu_{mp}(Z_m) = \mu_m(Z_m) \, .$$

Proof As in the example in *And now?* one has
$$X^p - 1 = (X - 1)^p \in K[X] \, .$$

From this the assertion easily follow. $\square$

So, successively, when considering $n = mp^k$ with $p \nmid m$, we know that $Z_n = Z_m$ and $\mu_n(Z_n) = \mu_m(Z_m)$, so we can neglect p-powers when working over a base field with $\mathrm{char}(K) = p$.

When we consider the n-th cyclotomic field over K, we will assume in the following that $\mathrm{char}(K) \nmid n$ holds (this includes the case $\mathrm{char}(K) = 0$).

Then the situation is the following:

Theorem 15.13 Let K be a field. If $n \in \mathbb{N}$ and $\mathrm{char}(K) \nmid n$, then we have the following statement for the n-th cyclotomic field $Z_n | K$:

1. The extension $Z_n | K$ is Galois.
2. The group $\mu_n(Z_n)$ is cyclic of order n. In particular, there are primitive n-th roots of unity.
3. If $\zeta \in \mu_n(Z_n)$ is a primitive root of unity, then $Z_n = K[\zeta]$.

Proof Since $Z_n | K$ is a splitting field, it is a normal extension. Furthermore, $X^n - 1$ is separable for $\mathrm{char}(K) \nmid n$, so $Z_n | K$ is separable. Thus, $X^n - 1$ also has n pairwise different roots in Z_n and thus $\#\mu_n(Z_n) = n$ and a generator of this cyclic group has order n. Furthermore, the roots of $X^n - 1$ are the elements
$$1, \zeta, \zeta^2, \ldots, \zeta^{n-1}$$
and all of these already lie in $K[\zeta]$. $\square$

And now?
Now we want to calculate the Galois group $\mathrm{Gal}(K[\zeta] | K)$. This is easy according to (3) in the above theorem, because we only need to apply theorem A: If $f(X) \in K[X]$ is the minimal polynomial of the (chosen) primitive nth root of unity ζ, then for each root β of f in Z_n we will find exactly one element $\sigma \in \mathrm{Gal}(K[\zeta] | K)$ with $\sigma(\zeta) = \beta$.

But: What is the minimal polynomial $f(X) \in K[X]$ of ζ over K? It must be different to $X^n - 1$ because this polynomial is not irreducible (it has the root 1 in K). However, it is certainly a divisor of $X^n - 1$. The exact form of $f(X)$ will of course depend on K, because there may already be n-th roots of unity in K, which then also fall out as a factor from $X^n - 1$.

Let's fix a primitive n-th root of unity $\zeta \in Z_n$. Let's think about a candidate for the minimal polynomial $f(X)$ **backwards**, by considering what this means for the Galois group and what we therefore expect. If $f(X) \in K[X]$ is this minimal polynomial, we know that $f(X)|(X^n - 1)$. The roots of $f(X)$ in Z_n are certain nth roots of unity, but which ones? Let's say $\beta \in \mu_n(Z_n)$ is such a root of $f(X)$ in Z_n. Then, Theorem A provides a Galois group element:

$$
\begin{array}{ccc}
Z_n = K(\zeta) & \xrightarrow{\ \sigma\ } & Z_n = K(\beta) = K(\zeta) \\
| & & | \\
K & =\!=\!=\!=\!=\!=\!= & K
\end{array}
$$

with $\sigma(\zeta) = \beta$. Since $\sigma \in \mathrm{Aut}_K(Z_n)$ is an automorphism, the element $\zeta \in Z_n^\times$ of order n must be mapped to an element of order n again (because σ is a field isomorphism and thus induces a group isomorphism of the unit group), hence β must again be a **primitive** nth root of unity! In other words: The only candidates for a root β of $f(X)$ among the roots of unity are the **primitive** nth roots of unity (we don't know if all primitive candidates will be good candidates but any root of the minimal polynomial must be a primitive root of unity).

Therefore, one simply defines a polynomial that has exactly these primitive n-th roots of unity as a zero. This is a candidate for the minimal polynomial (and we will see that this will indeed be the case for $K = \mathbb{Q}$).

Definition 15.14 Let $n \in \mathbb{N}$. Let $\zeta_1, \ldots, \zeta_{\varphi(n)} \in Z_n$ denote the primitive n-th roots of unity. Then the n-**th cyclotomic polynomial over** K is defined as the polynomial

$$
\Phi_n(X) := \prod_{j=1}^{\varphi(n)} (X - \zeta_j) \in Z_n[X] .
$$

It is clear that $\deg(\Phi_n) = \varphi(n)$.

And now?
One can immediately see a first problem. By definition, $\Phi_n(X) \in Z_n[X]$. However, if it should serve as a candidate for the minimal polynomial of this ζ_j over K, it must be a polynomial in $K[X]$. This is indeed the case, as can be seen with the Fundamental Theorem of Galois Theory.

Check it out!
Do you see how to apply the fundamental theorem in order to show $\Phi_n(X) \in K[X]$? Remember that $K = Z_n^{\mathrm{Gal}(Z_n|K)}$ and that you can consider the transported polynomials Φ_n^σ for any $\sigma \in \mathrm{Gal}(Z_n|K)$.

Lemma 15.15 *The n-th cyclotomic polynomial* $\Phi_n(X)$ *has coefficients in K, i.e.*

$$\Phi_n(X) \in K[X].$$

Proof Let $\mathrm{Gal}(Z_n|K)$ be the Galois group of $Z_n|K$ (the latter is a Galois extension). Then for every $\sigma \in \mathrm{Gal}(Z_n|K)$, we know that

$$\Phi_n^\sigma(X) = \prod_{j=1}^{\varphi(n)}(X - \sigma(\zeta_j)) = \Phi_n(X),$$

because the homomorphisms σ are isomorphisms and thus $\mathrm{ord}(\varphi(\zeta_j)) = n$ as an element in $Z_n^\times$ for all ζ_j, hence σ maps the set $\{\zeta_1, \ldots, \zeta_{\varphi(n)}\}$ bijectively onto itself.

Therefore, $\Phi_n(X) \in Z_n^{\mathrm{Gal}(Z_n|K)}[X] = K[X]$. $\square$

One can calculate the $\Phi_n(X) \in K[X]$ inductively as follows:

Lemma 15.16 *For* $n \in \mathbb{N}$ *the following holds:*

$$X^n - 1 = \prod_{d|n, d>0} \Phi_d(X),$$

where the product is taken over all positive divisors $d \in \mathbb{N}$ *of n (including* $d = 1$ *and* $d = n$).

Proof This is easy to see, because the roots of $X^n - 1$ in Z_n are all the n-th roots of unity (we have always assumed $\mathrm{char}(K) \nmid n!$) and every n-th root of unity ζ is a primitive d-th root of unity for $d = \mathrm{ord}(\zeta) \mid n$ (the order in the group $\mu_n(Z_n)$). Therefore, in the group $Z_n[X]$, we have

$$X^n - 1 = \prod_{\zeta \in \mu_n(Z_n)}(X - \zeta) = \prod_{d|n, d>0} \underbrace{\prod_{\substack{\zeta \text{ with} \\ \mathrm{ord}(\zeta)=d}}(X - \zeta)}_{=\Phi_d(X)}.$$

We have already seen special cases for $K = \mathbb{Q}$:

Lemma 15.17 *Let p be a prime number. Then all* $p - 1$ *elements of* $\mu_p(\mathbb{C}) \setminus \{1\}$ *are primitive p-th roots of unity and*

$$\Phi_p(X) = X^{p-1} + X^{p-2} + \ldots + X + 1 \in \mathbb{Q}[X]$$

is the minimal polynomial of each of them.

Proof Because $\mu_p(\mathbb{C})$ is a cyclic group of order p, every element $\neq 1$ is a generator. We know that $X^{p-1} + \ldots + 1$ has exactly these as roots, thus coinciding with $\Phi_p(X)$. The irreducibility of the polynomial was seen in Lemma 9.17. $\square$

Note: If one has already calculated $\Phi_d(X)$ for all divisors $0 < d < n$, it follows from the formula that:

$$\Phi_n(X) = \frac{X^n - 1}{\prod_{d|n, 0<d<n} \Phi_d(X)} \in K[X] . \tag{15.2}$$

Examples:

- For $K = \mathbb{Q}$ and $n = 15$, the set of divisors is $\{1, 3, 5, 15\}$. We know that

$$\Phi_1(X) = X - 1 \qquad\qquad \Phi_3(X) = X^2 + X + 1$$
$$\Phi_5(X) = X^4 + X^3 + X^2 + X + 1$$

So,

$$\Phi_{15}(X) = \frac{X^{15} - 1}{(X - 1)(X^2 + X + 1)(X^4 + X^3 + X^2 + X + 1)} =$$
$$\frac{X^{15} - 1}{X^7 + X^6 + X^5 - X^2 - X - 1} =$$
$$X^8 - X^7 + X^5 - X^4 + X^3 - X + 1$$

Easy to calculate by hand, or via:

```
S.<x>=PolynomialRing(QQ)
h=(x-1)*(x^2+x+1)*(x^4+x^3+x^2+x+1)
Phi=(x^15-1)/h
print('denominator=',h,' cyclotomic polynomial=',Phi)
```

Or simply because SAGE knows these polynomials:

```
Ph=cyclotomic_polynomial(15)
print ('cyclotomic polynomial=',Phi)
```

- For $K = \mathbb{Q}$ and $n = 8$ one has the divisors $\{1, 2, 4, 8\}$ and one knows a priori that

$$\Phi_1(X) = X - 1 \text{ and } \Phi_2(X) = X + 1 .$$

Then

$$\Phi_4(X) = \frac{X^4 - 1}{(X - 1)(X + 1)} = \frac{X^4 - 1}{X^2 - 1} = X^2 + 1 ,$$

and furthermore

$$\Phi_8(X) = \frac{X^8 - 1}{(X - 1)(X + 1)(X^2 + 1)} = \frac{X^8 - 1}{(X^2 - 1)(X^2 + 1)} = \frac{X^8 - 1}{X^4 - 1} = X^4 + 1 . \tag{15.3}$$

- A small list: The SAGE routine

```
with open ('cyclo.txt','a') as fil:
    for k in range (1,17):
        Phi=cyclotomic_polynomial(k)
        fil.write(r'& \Phi_{')
        fil.write(latex(k))
        fil.write(r'}=')
        fil.write(latex(Phi))
        fil.write(r'\\ ')
```

has generated the LaTeX lines, which then deliver the following lines in the book (embedded in an *align* ∗ environment):

$$\Phi_1 = x - 1$$
$$\Phi_2 = x + 1$$
$$\Phi_3 = x^2 + x + 1$$
$$\Phi_4 = x^2 + 1$$
$$\Phi_5 = x^4 + x^3 + x^2 + x + 1$$
$$\Phi_6 = x^2 - x + 1$$
$$\Phi_7 = x^6 + x^5 + x^4 + x^3 + x^2 + x + 1$$
$$\Phi_8 = x^4 + 1$$
$$\Phi_9 = x^6 + x^3 + 1$$
$$\Phi_{10} = x^4 - x^3 + x^2 - x + 1$$
$$\Phi_{11} = x^{10} + x^9 + x^8 + x^7 + x^6 + x^5 + x^4 + x^3 + x^2 + x + 1$$
$$\Phi_{12} = x^4 - x^2 + 1$$
$$\Phi_{13} = x^{12} + x^{11} + x^{10} + x^9 + x^8 + x^7 + x^6 + x^5 + x^4 + x^3 + x^2 + x + 1$$
$$\Phi_{14} = x^6 - x^5 + x^4 - x^3 + x^2 - x + 1$$
$$\Phi_{15} = x^8 - x^7 + x^5 - x^4 + x^3 - x + 1$$
$$\Phi_{16} = x^8 + 1$$

One might conjecture that the coefficients are always $0, \pm 1$ when looking at the list and possibly calculating more examples. But this is not the case: $\Phi_{105}(X)$ is the first example that has a coefficient -2:

$$\Phi_{105}(x) = x^{48} + x^{47} + x^{46} - x^{43} - x^{42} - \mathbf{2} \cdot x^{41} - x^{40} - x^{39} + x^{36} + x^{35} +$$
$$x^{34} + x^{33} + x^{32} + x^{31} - x^{28} - x^{26} - x^{24} - x^{22} - x^{20} + x^{17} + x^{16}$$
$$+ x^{15} + x^{14} + x^{13} + x^{12} - x^9 - x^8 - 2x^7 - x^6 - x^5 + x^2 + x + 1$$

From now on, we consider the situation with base field $K = \mathbb{Q}$. Then Φ_n is indeed the minimal polynomial we are looking for:

Theorem 15.18 The n-th cyclotomic polynomial $\Phi_n(X) \in \mathbb{Q}[X]$ has coefficients in $\mathbb{Z}$ and is irreducible.

Proof We fix a primitive n-th root of unity $\zeta \in \mathbb{C}$ (it will be important later that we allow any primitive one in the argument).

Let $q(X) \in \mathbb{Q}[X]$ be the minimal polynomial of ζ over $\mathbb{Q}$. If $\xi \in Z_n$ is a root of $q(X)$, there exists (according to Theorem A) a $\sigma \in \mathrm{Gal}(Z_n|\mathbb{Q})$ with $\sigma(\zeta) = \xi$. Since σ is a field automorphism, $\xi = \sigma(\zeta) \in Z_n^\times$ has the same order as ζ, hence it is again a primitive nth root of unity. Therefore, all roots of $q(X)$ are (certain?) primitive nth roots of unity and since $\Phi_n(X)$ was the polynomial with all these as roots/zeroes, we know that

$$q(X) \mid \Phi_n(X) \in \mathbb{Q}[X]$$

holds (we have already considered this at the beginning of the section, which is why we chose Φ_n as a candidate in the first place).

We now prove the following claim:

Claim Every primitive nth root of unity $\xi \in Z_n$ appears as a root of $q(X)$. □

Proof We multiply $q(X)$ by a (possible) common denominator and obtain $p(X) = c \cdot q(X) \in \mathbb{Z}[X]$ as a primitive polynomial with the same roots as $q(X)$ (we see a posteriori that $p(X) = q(X) = \Phi_n(X) \in \mathbb{Z}[X]$ is monic, but that's what we want to show). We further write

$$X^n - 1 = p(X)g(X) \text{ with a } g(X) \in \mathbb{Q}[X]. \tag{15.4}$$

We know that $g(X) \in \mathbb{Z}[X]$ (Theorem 9.8). We now prove the assertion with $p(X) \in \mathbb{Z}[X]$ instead of $q(X) \in \mathbb{Q}[X]$.

An arbitrary primitive nth root of unity ξ has the form $\xi = \zeta^m$ with $\gcd(m, n) = 1$ (according to Lemma 15.4). We prove the assertion by induction on the number of prime factors of m.

Beginning of Induction: So let $m = p^\ell$ be a power of a prime number p. Suppose ξ is not a root of $p(X)$. Then it must be a root of $g(X)$ (because it is a root of $X^n - 1$). Therefore,

$$0 = g(\xi) = g(\zeta^m)$$

and thus ζ is a root of $g(X^m) \in \mathbb{Z}[X]$.

It follows that $p(X) \mid g(X^m) \in \mathbb{Q}[X]$ (because $p(X)$ is the minimal polynomial of ζ up to multiplication with $c \in \mathbb{Z}$). Let's say

$$g(X^m) = p(X) \cdot h(X)$$

with $h(X) \in \mathbb{Z}[X]$ (again using the Gauss Theorem).

Now we use that $m = p^\ell$ and reduce modulo p:

$$\overline{p(X) \cdot h(X)} = \overline{g(X^{p^\ell})} = \overline{g(X)}^{p^\ell} \in \mathbb{Z}/p\mathbb{Z}[X] = \mathbb{F}_p[X] \,,$$

where in the last step we apply lemma 15.19 which we will prove after this proof
Also note that the degrees of $p(X)$ and $g(X)$ must remain the same due to (15.4)
after reduction, because $X^n - 1$ also has degree n after reduction. It follows that
the polynomials $\overline{p(X)}$ and $\overline{g(X)}$ have a common divisor in $\mathbb{F}_p[X]$ which is of
degree ≥ 1. But

$$X^n - 1 = \overline{X^n - 1} = \overline{p(X)} \cdot \overline{g(X)}$$

is separable, because we had assumed $p \nmid n$ (we had $\gcd(m, n) = 1$, since ζ^m is a
primitive n-th root of unity). This is a contradiction.

Induction step: Consider the case that m has more than one prime factor. Let p
be one of them and we write $m = p^\ell \cdot m'$ with $p \nmid m'$. Since m' has one less prime
divisor than m, the induction assumption applies to m' and we know that the primitive nth root of unity $\zeta^{m'}$ appears as a root of $p(X)$. Therefore, $p(X)$ is not only the
minimal polynomial of ζ but also of $\zeta' = \zeta^{m'}$.

We can now perform the same argument of the induction start, because if one
starts the whole proof with ζ' instead of ζ (that's why it was important that at the
beginning ζ was any primitive n-th root of unity), one considers the same $p(X)$ and
the induction start shows that the primitive n-th root of unity $(\zeta')^{p^\ell}$ occurs as a root
of $p(X)$, but

$$(\zeta')^{p^\ell} = \zeta^{p^\ell m'} = \zeta^m.$$

With this, the claim is proven. $\square$

So now we know that the monic polynomial $\Phi_n(X) \in \mathbb{Q}[X]$ is the minimal polynomial of every primitive nth root of unity $\zeta \in \mathbb{C}$.

With (15.2), one can see inductively (induction according to the num-
ber of divisors of n) that $\Phi_n(X) \in \mathbb{Z}$ has integer coefficients. Induction
start: n has exactly two positive divisors. Then $n = p$ is a prime number and
$\Phi_p(X) = X^{p-1} + \ldots + X + 1 \in \mathbb{Z}[X]$.

Induction step: If n has more than two divisors, then

$$X^n - 1 = \Phi_n(X) \cdot \prod_{d \mid n, 0 < d < n} \Phi_d(X) \in \mathbb{Z}[X]$$

and by the induction hypothesis, the second factor is in $\mathbb{Z}[X]$ and monic, so by
Gauss's theorem (Theorem 9.8) we obtain $\Phi_n(X) \in \mathbb{Z}[X]$. $\square$

We still need to fill in the gap we postponed to the following lemma:

Lemma 15.19 *Let p be a prime number. Then exponentiation with p (and thus the
same holds for with p^ℓ for any $\ell \in \mathbb{N}$)*

$$\mathbb{F}_p[X] \to \mathbb{F}_p[X] \,, \quad a(X) \mapsto (a(X))^p$$

is a ring homomorphism and we have

$$a(X)^p = a(X^p) \in \mathbb{F}_p[X] .$$

Proof The ring $R = \mathbb{F}_p[X]$ satisfies $p = 0$ and therefore

$$(a+b)^p = a^p + \sum_{j=1}^{p-1} \binom{p}{j} a^j b^{p-j} + b^p = a^p + b^p$$

for all $a, b \in R$, because the binomial coefficients are all divisible by p. Furthermore, $(ab)^p = a^p b^p$, so it is a ring homomorphism.

For $a(X) = a_n X^n + \ldots + a_1 X + a_0 \in \mathbb{F}_p[X]$ we obtain

$$a(X)^p = (a_n X^n + \ldots + a_1 X + a_0)^p =$$
$$a_n^p (X^p)^n + \ldots + a_1^p X^p + a_0^p = a_n (X^p)^n + \ldots + a_1 X^p + a_0 = a(X^p) ,$$

where we remark that given any $a_j \in \mathbb{F}_p$ we have $a_j^p = a_j$ – the latter, because it is certainly true for $a_j = 0$ and if $a_j \in \mathbb{F}_p^\times$, then Lagrange's theorem yields

$$1 = a_j^{\#\mathbb{F}_p^\times} = a_j^{p-1} \implies a_j = a_j^p .$$

Hence, the lemma is proven. $\square$

Let us summarize the Galois theory of cyclotomic fields over $\mathbb{Q}$:

Theorem 15.20 Let $n \in \mathbb{N}$ and $\zeta \in \mathbb{C}$ be a primitive n-th root of unity. Then $\mathbb{Q}[\zeta]|\mathbb{Q}$ is a Galois extension of degree $[\mathbb{Q}(\zeta) : \mathbb{Q}] = \varphi(n)$ and

$$(\mathbb{Z}/n\mathbb{Z})^\times \to \mathrm{Gal}(\mathbb{Q}[\zeta]|\mathbb{Q})$$
$$[k] \mapsto (\sigma_k : \zeta \mapsto \zeta^k)$$

is a group isomorphism.

Proof There is nothing more to show, the assertion about the degree of the field and the Galois group follows immediately from Theorem 15.18 and Theorem A . $\square$

Remark 15.21 Note that the isomorphism in the theorem depends on the choice of the primitive nth root of unity.

And now?

The extension $\mathbb{Q}[\zeta]|\mathbb{Q}$ is one of the explicit field extensions that one should know by heart. It is the topic of many exercises and explicit results in Galois theory—especially in connection with the Fundamental Theorem of Galois Theory. For example, take a natural number n, then

$$\mathrm{Gal}(\mathbb{Q}(\zeta)|\mathbb{Q}) \cong (\mathbb{Z}/n\mathbb{Z})^\times \tag{15.5}$$

after choosing a primitive nth root of unity (see the previous theorem). Now, for small n, one can determine the group $(\mathbb{Z}/n\mathbb{Z})^{\times}$ and then find subgroups in it—according to the fundamental theorem, this determines uniquely an intermediate field, ...—one can see that this can be used to construct explicit examples very nicely.

And now?

Sometimes one is confronted with the task to construct/find a Galois extension with a given Galois group: this is known as the *inverse Galois problem*, that is, given a finite group G, is there a Galois extension $L|\mathbb{Q}$ such that

$$\mathrm{Gal}(L|\mathbb{Q}) \cong G$$

holds? This is a very difficult problem (and still open in general). In easy examples one can use (15.5).

Example: Let $G = \mathbb{Z}/n\mathbb{Z}$ be the cyclic group of order n (up to isomorphism there is only one of these). How do you represent G as a Galois group $\mathrm{Gal}(L|\mathbb{Q})$? You look for a prime number p with $p \equiv 1 \bmod n$ (which one can always find due to a theorem by Dirichlet, but in the explicit case you usually find one explicitly, see the example in the example). Then we have that the Galois group

$$\mathrm{Gal}(\mathbb{Q}(\zeta)|\mathbb{Q}) \cong (\mathbb{Z}/p\mathbb{Z})^{\times} = \mathbb{F}_p^{\times} \cong \mathbb{Z}/(p-1)\mathbb{Z}$$

(for $\zeta = \exp(2\pi i/n)$) is the cyclic group of order $p - 1$ (cyclic due to Theorem 15.9). Every subgroup and every factor group of a cyclic group is again cyclic and moreover, for every divisor of the group order $\#G' = p - 1$ there exists a unique subgroup. $H' \subset G'$ with order d. So here we take $d := (p - 1)/n$ and find a subgroup

$$H' \subset (\mathbb{F}_p)^{\times} \cong \mathrm{Gal}(\mathbb{Q}(\zeta)|\mathbb{Q})$$

of order $(p - 1)/n$. Let $H \subset \mathrm{Gal}(\mathbb{Q}(\zeta)|\mathbb{Q})$ be the isomorphic subgroup in $\mathrm{Gal}(\mathbb{Q}(\zeta)|\mathbb{Q})$, then the fixed field $L := \mathbb{Q}(\zeta)^H$ is Galois over $\mathbb{Q}$ (because H is a normal subgroup) and it follows from the addendum to the Fundamental Theorem of Galois Theory that

$$\mathrm{Gal}(L|\mathbb{Q}) \cong \mathrm{Gal}(\mathbb{Q}(\zeta)|\mathbb{Q})/\mathrm{Gal}(\mathbb{Q}(\zeta)|L) \cong \mathrm{Gal}(\mathbb{Q}(\zeta)|\mathbb{Q})/H \cong (\mathbb{Z}/(p-1)\mathbb{Z})/H'$$

and the latter is a cyclic group of order $\frac{p-1}{(p-1)/n} = n$, as desired.

Example within the example: Let $n = 14$. Then choose $p = 29$ and we have with the choice of $\zeta = \exp(2\pi i/29)$, that

$$\mathrm{Gal}(\mathbb{Q}(\zeta)|\mathbb{Q}) \underset{\psi}{\overset{\cong}{\longleftarrow}} \mathbb{F}_{29}^{\times} \underset{\eta}{\overset{\cong}{\longleftarrow}} (\mathbb{Z}/28\mathbb{Z}).$$

In this group we find the subgroup $\mathbb{Z}/2\mathbb{Z}$ in the form

$$\mathbb{Z}/2\mathbb{Z} \hookrightarrow \mathbb{Z}/28\mathbb{Z}, \ 1 \mapsto 14.$$

So if we define $H := \psi \circ \eta(\mathbb{Z}/2\mathbb{Z})$, we obtain the solution $L := \mathbb{Q}(\zeta)^H$. If one wants to determine L more explicitly, one must specify the group isomorphisms η and ψ more precisely. Here, this is not really necessary, because $14 \in \mathbb{Z}/28\mathbb{Z}$ is the only element of order 2 and it is mapped to the only element of order 2 by $\psi \circ \eta$. But that element is obviously the following:

$$\sigma : \mathbb{Q}(\zeta) \to \mathbb{Q}(\zeta), \ \zeta \mapsto \zeta^{-1} = \zeta^{28}$$

since it is certainly of order 2. Thus, $H = \{1, \sigma\}$ and

$$L = \mathbb{Q}(\zeta)^H = \mathbb{Q}(\zeta)^{\{1,\sigma\}}$$

Note that $\zeta + \zeta^{-1} \in \mathbb{Q}(\zeta)^H$ (because $\sigma(\zeta + \zeta^{-1}) = \zeta^{-1} + \zeta$) and ζ is a root of the polynomial

$$X^2 - (\zeta + \zeta^{-1})X + 1 \in \mathbb{Q}(\zeta + \zeta^{-1})[X]$$

and we have the following field extensions (and we know some of its degrees):

$$
\begin{array}{c}
\mathbb{Q}(\zeta) \\
2 \Big| \\
\leq 2 \Big(\quad \mathbb{Q}(\zeta)^H \\
\Big| \\
\mathbb{Q}(\zeta + \zeta^{-1}) \\
\Big| \\
\mathbb{Q}
\end{array}
$$

and thus it follows that $\mathbb{Q}(\zeta + \zeta^{-1}) = \mathbb{Q}(\zeta)^H = L$.

Thread to the previous chapter:
In this chapter, we have

- introduced the cyclotomic fields,
- seen that every finite subgroup $G \subset K^\times$ in the unit group of a field is cyclic,
- that the n-th cyclotomic polynomial (let's say over $\mathbb{Q}$)

$$\Phi_n(X) = \prod_{\zeta \ \text{primitive} \ n \ \text{th root of unity}} (X - \zeta) \in \mathbb{Z}[X]$$

already has coefficients in $\mathbb{Q}$, even in $\mathbb{Z}$, and
- that $\Phi_n(X) \in \mathbb{Q}[X]$ can be calculated inductively through $X^n - 1 = \prod_{d \mid n} \Phi_d(X)$,

- that it has the degree

$$\deg(\Phi_n) = \varphi(n),$$

because the Euler φ function precisely indicates the number $\varphi(n) = \#(\mathbb{Z}/n\mathbb{Z})^{\times}$ of units in $\mathbb{Z}/n\mathbb{Z}$ and
- for a fixed primitive nth root of unity ζ the primitive nth roots of unity are exactly the elements ζ^m with $m \in (\mathbb{Z}/n\mathbb{Z})^{\times}$,
- that $\Phi_n(X) \in \mathbb{Q}[X]$ is irreducible and
- that after choosing a primitive nth root of unity, one has the isomorphism

$$(\mathbb{Z}/n\mathbb{Z})^{\times} \xrightarrow{\cong} \mathrm{Gal}(\mathbb{Q}(\zeta)|\mathbb{Q}) , \ [k] \mapsto (\zeta \mapsto \zeta^k).$$

Exercises: 15.1 Show that there is a Galois extension $L|\mathbb{Q}$ whose Galois group is isomorphic to $\mathbb{Z}/23\mathbb{Z}$.

(Hint: Read the last *And now?* of this chapter.)

15.2 Consider the splitting field Z of $f(X) = X^3 - 7 \in \mathbb{Q}[X]$ over $\mathbb{Q}$ in $\mathbb{C}$. Let $\alpha := \sqrt[3]{7} \in \mathbb{R}$. Show the following statements:

(a) Let $\zeta := \exp(2\pi i/3) \in \mathbb{C}$. Then $Z = \mathbb{Q}(\alpha, \zeta)$ and $[Z : \mathbb{Q}] = 6$.
(b) If $\sigma \in \mathrm{Gal}(Z|K)$, then there exists $a(\sigma) \in \mathbb{F}_3$ and $b(\sigma) \in \mathbb{F}_3^*$, such that

$$\sigma(\alpha) = \alpha \cdot \zeta^{a(\sigma)} \text{ and } \sigma(\zeta) = \zeta^{b(\sigma)}.$$

(c) The map

$$\mathrm{Gal}(Z|K) \to \mathrm{GL}_2(\mathbb{F}_3) , \ \sigma \mapsto \begin{pmatrix} 1 & 0 \\ a(\sigma) & b(\sigma) \end{pmatrix}$$

is an injective group homomorphism, whose image is given by

$$\left\{ \begin{pmatrix} 1 & 0 \\ x & y \end{pmatrix} \mid x \in \mathbb{F}_3, y \in \mathbb{F}_3^{\times} \right\}.$$

15.3 Let $Z \subset \mathbb{C}$ be the splitting field of $f(X) = X^5 - 2$ over $\mathbb{Q}$.

(a) Show (analogous to Exercise 15.2), that

$$\mathrm{Gal}(Z|\mathbb{Q}) \xrightarrow{\cong} \left\{ A(x, y) := \begin{pmatrix} 1 & 0 \\ x & y \end{pmatrix} \in \mathrm{GL}_2(\mathbb{F}_5) \mid x \in \mathbb{F}_5, y \in \mathbb{F}_5^{\times} \right\} =: G$$

(b) Show that $N := \{A(x, 1) \mid x \in \mathbb{F}_5\}$ is a normal subgroup in G. What is its fixed field?
(c) Show that $U := \{A(0, y) \mid y \in \mathbb{F}_5^{\times}\}$ is not a normal subgroup in G. What is its fixed field?
(d) Determine a subgroup $V \subset G$ of order 2 and determine its fixed field.

15.4 Let $p \geq 3$ be a prime number and $\zeta \in \mathbb{C}$ be a primitive p-th root of unity.

(a) Let $a \in \mathbb{N}$. Show that the polynomial $X^{a+1} - 1$ is a divisor of the polynomial $X^{2a} - X^{a+1} - X^{a-1} + 1$ in $\mathbb{Q}[X]$ and determine the quotient.

(b) Show that the field extension $\mathbb{Q}(\zeta + \zeta^{-1}) \mid \mathbb{Q}$ is Galois with Galois group

$$\mathrm{Gal}(\mathbb{Q}(\zeta + \zeta^{-1}) \mid \mathbb{Q}) \cong \mathbb{Z}/\left(\frac{p-1}{2}\right)\mathbb{Z}.$$

Finite Fields

16

16.1 Prime Fields, Finite Fields, and Frobenius

We want to understand which **finite fields** K, that is with $\#K < \infty$, exist.

Observation For every field K, we have a ring homomorphism

$$\mathbb{Z} \to K, \; n \mapsto n := n \cdot 1.$$

If K is finite, this cannot be injective, so K has a prime number characteristic $\mathrm{char}(K) = p$. Thus, the above ring homomorphism induces an injective field homomorphism

$$\mathbb{Z}/p\mathbb{Z} = \mathbb{F}_p \hookrightarrow K.$$

So we have a field extension $K|\mathbb{F}_p$. Since K is finite, $K|\mathbb{F}_p$ must also be finite. In summary:

Lemma 16.1 *Any finite field K contains the prime field $\mathbb{F}_p$ for the prime number $p = \mathrm{char}(K)$. Furthermore, $K|\mathbb{F}_p$ is a finite field extension and thus for the number of elements we have*

$$\#K = p^{\ell}$$

for some $\ell \in \mathbb{N}$.

Proof The only remaining issue is the one on $\#K$. As a $\mathbb{F}_p$-vector space, K is isomorphic to $(\mathbb{F}_p)^{[K:\mathbb{F}_p]}$ and thus has $p^{[K:\mathbb{F}_p]}$ elements. $\square$

We now denote by $\overline{\mathbb{F}}_p$ a fixed chosen algebraic closure of $\mathbb{F}_p$.

© The Author(s), under exclusive license to Springer-Verlag GmbH, DE, part of Springer Nature 2024
M. Hien, *Abstract Algebra*, Mathematics Study Resources 7,
https://doi.org/10.1007/978-3-662-67974-6_16

And now?

Let us make a small comment: For the rest of the chapter it is important that one fixes $\overline{\mathbb{F}}_p$. To do this, one must use the theorem that every field has an algebraic closure. One can avoid this general theorem and show that every field $\mathbb{F}_p$ has such an algebraic closure. This is still non-trivial, but not as elaborate as Theorem 12.2. Nevertheless, we will use Theorem 12.2.

In characteristic p exponentiation with p is a map compatible with addition (as we have used several times already). It even has a name:

Defining Lemma 16.2 Let K be a field of prime characteristic p. Then the **Frobenius homomorphism**

$$\text{Frob} : K \to K, \; x \mapsto x^p$$

is a field homomorphism.

Considering the algebraic closure $\overline{\mathbb{F}}_p$, we see:

Lemma 16.3 *The Frobenius homomorphism is an automorphism*

$$\text{Frob} : \overline{\mathbb{F}}_p \overset{\cong}{\to} \overline{\mathbb{F}}_p$$

with $\text{Frob}|_{\mathbb{F}_p} = \text{id}_{\mathbb{F}_p}$.

Proof As a field homomorphism, Frob is injective. Now, if $y \in \overline{\mathbb{F}}_p$ is arbitrary, then $X^p - y \in \overline{\mathbb{F}}_p$ has a root (because the field is algebraically closed), so y has a pre-image.

For $x \in \mathbb{F}_p = \mathbb{Z}/p\mathbb{Z}$, $x^p = x$ holds, because for $x = 0$ this is clear and for $x \in \mathbb{F}_p^\times$ it follows because then $x^{p-1} = 1$ in the group $\mathbb{F}_p^\times$ with $p - 1$ elements. $\square$

And now?

If we define the Galois group for infinite (but still algebraic) field extensions

$$\text{Gal}(L|K) := \text{Aut}_K(L) \tag{16.1}$$

as usual, then $\text{Frob} \in \text{Gal}(\overline{\mathbb{F}}_p|\mathbb{F}_p)$ is an element of this group:

$$
\begin{array}{ccc}
\overline{\mathbb{F}}_p & \xrightarrow{\ \text{Frob}\ } & \overline{\mathbb{F}}_p \\
| & & | \\
\mathbb{F}_p & =\!\!= & \mathbb{F}_p.
\end{array}
$$

This is indeed the correct definition, but if one wants to formulate and prove the Fundamental Theorem of Galois theory for algebraic, but not necessarily finite Galois extensions, one must consider additional structures on $\mathrm{Gal}(L|K)$ —namely a topology—and in the generalized version of the fundamental theorem, this topology will have to taken into account. But as a group (without considering the topology on it), (16.1) is unproblematic.

This is interesting insofar as: If we have a (finite) intermediate field $\mathbb{F}_p \subset M \subset \overline{\mathbb{F}}_p$ such that $M|\mathbb{F}_p$ is also normal, then we also have the Frobenius

$$
\begin{array}{ccc}
\overline{\mathbb{F}}_p & \xrightarrow{\ \text{Frob}\ } & \overline{\mathbb{F}}_p \\
\big| & & \big| \\
M & \xrightarrow{\ \text{Frob}\ } & M \\
\big| & & \big| \\
\mathbb{F}_p & =\!=\!= & \mathbb{F}_p .
\end{array}
$$

and thus immediately an element from $\mathrm{Gal}(M|\mathbb{F}_p)$. We will now examine this in more detail.

16.2   Finite Fields

And now?

We know from Lemma 16.1 that a finite field K carries a prime power p^ℓ of elements and it is a field extension $K|\mathbb{F}_p$. If we fix an algebraic closure of K, which then is also an algebraic closure of $\mathbb{F}_p$ hence we denote it by $\overline{\mathbb{F}}_p$, we find K as an intermediate field

$$
\mathbb{F}_p \subset K \subset \overline{\mathbb{F}}_p. \tag{16.2}
$$

The question one can ask is the following: If one fixes an algebraic closure $\overline{\mathbb{F}}_p$ of $\mathbb{F}_p$ and gives an arbitrary $\ell \in \mathbb{N}$, can we find such a field K with p^ℓ elements? And how many such fields K with (16.2) exist? The answer is the result of this chapter.

For the rest of the chapter, p denotes a fixed prime number.

And now?

Let's briefly consider how we can employ an idea from Galois theory to construct such fields. Suppose we already have a field $\mathbb{F}_p \subset K \subset \overline{\mathbb{F}}_p$ with $\#K = p^\ell =: q$, thus $[K : \mathbb{F}_p] = \ell$. **Let's hope** that $K|\mathbb{F}_p$ is normal. Then we have the Galois group

$$\mathrm{Gal}(K|\mathbb{F}_p)$$

at our disposal. In it, we already know an element, namely the Frobenius:

$$\left(\mathrm{Frob} : K \to K, x \mapsto x^p\right) \in \mathrm{Gal}(K|\mathbb{F}_p).$$

We do not yet know whether $1 \neq \mathrm{Frob}$ holds, but we **hope** so.

We hope furthermore that $K|\mathbb{F}_p$ is also separable, since then we know that $\#\mathrm{Gal}(K|\mathbb{F}_p) = [K : \mathbb{F}_p] = \ell$ holds. Then, according to Lagrange's theorem, the order of Frob as a group element is a divisor of ℓ, in particular

$$\mathrm{Frob}^\ell = \mathrm{id}_K.$$

That means that every $x \in K$ satisfies the equation

$$x = \mathrm{id}_K(x) = \mathrm{Frob}^\ell(x) = x^{p^\ell} = x^q.$$

Result of these considerations: This means that **all elements of K** are roots of $X^q - X$.

This is already an idea or an approach: Given ℓ and hence $q = p^\ell$ and the chosen algebraic closure $\overline{\mathbb{F}}_p$, we can take a look at the splitting field Z of $X^q - X$ in $\overline{\mathbb{F}}_p$. If all the three above **hopes** are fulfilled, this is a good candidate for such a field K. Why should they be fulfilled?

- $Z|\mathbb{F}_p$ is normal because it is a splitting field.
- $Z|\mathbb{F}_p$ is separable because $f(X) = X^q - X \in \mathbb{F}_p[X]$ is separable (because $f'(X) = qX^{q-1} - 1 = -1$ has no roots),
- the hope that $\mathrm{Frob} \neq 1$ holds is not immediately apparent—one could consider an example for this.

Remark: actually, the idea of the proof which we will present below is a little bit different: The result of the considerations says that (whenever $\overline{\mathbb{F}}_p$ is an algebraic closure of K)

$$K \subset \{x \in \overline{\mathbb{F}}_p \mid x^q = x\}$$

holds. The perhaps more naive idea is thus to define $K =$ (right hand side). Is this even a field? Yes, see the proof of the next theorem.

Example: For instance, we can construct a field K with $5^2 = 25$ elements, because $f(X) = X^2 + X + 1 \in \mathbb{F}_5[X]$ is irreducible (for degree reasons, $f(X)$ would have to have a root in $\mathbb{F}_5$ if it were reducible). Therefore,

$$K := \mathbb{F}_5[X]/(f(X))$$

is a field extension over $\mathbb{F}_5$ of degree 2, thus with 5^2 elements.

The Frobenius on K is

$$\text{Frob} : K = \mathbb{F}_5[X]/(f(X)) \to K = \mathbb{F}_5[X]/(f(X))$$

$$[g(X)] = [aX + b] \mapsto [g(X)]^5 = [aX^5 + b] \overset{(*)}{=} [-aX + b - a]$$

(again because $a^5 = a$ and $b^5 = b$ for $a, b \in \mathbb{F}_5$), where one must note at
(∗) that $[X^5] = [-X - 1] \in K$ holds (which can easily be seen by division with remainder). Apparently here $1 \neq \text{Frob}$, because for example
$\text{Frob}([X]) = [X^5] = [-X - 1] \neq [X]$ applies.
    Note, howerver, that $\text{Frob}^2 = 1$ holds because

$$\text{Frob}^2([aX + b]) = \text{Frob}([-aX + b - a]) = [aX + (b - a) - (-a)] = [aX + b].$$

Using this idea, one can prove the following theorem which states that indeed for
every ℓ there is exactly one such K in $\overline{\mathbb{F}}_p$:

Theorem 16.4 Let $\overline{\mathbb{F}}_p$ be a fixed algebraic closure of $\mathbb{F}_p$ and $\ell \in \mathbb{N}$. Then there is
exactly one intermediate field

$$\mathbb{F}_p \subset K \subset \overline{\mathbb{F}}_p$$

with $\#K = p^\ell$. Additionally, the following holds:

1. $K|\mathbb{F}_p$ is a Galois extension,
2. the Galois group is cyclic and has the Frobenius as a generator—in other
   words:

$$\mathbb{Z}/\ell\mathbb{Z} \overset{\cong}{\to} \text{Gal}(K|\mathbb{F}_p), \ [k] \mapsto \text{Frob}^k$$

is a well-defined group isomorphism,

Proof Let $q = p^\ell$. Let $Z \subset \overline{\mathbb{F}}_p$ be the splitting field of the polynomial $\quad f(X) := X^q - X \in \mathbb{F}_p[X]$. It is separable because
$f'(X) = qX^{q-1} - 1 = -1 \in \mathbb{F}_p[X]$ has no roots in the algebraic closure anyway.
    Now consider the set

$$K := \{x \in \overline{\mathbb{F}}_p \mid x^q = x\} \subset \overline{\mathbb{F}}_p$$

of all roots of $f(X) = X^q - X$ in the algebraic closure $\overline{\mathbb{F}}_p$. Since f is separable, we
know that

$$\#K = \deg(f) = q = p^\ell$$

holds. Furthermore, $K \subset Z$, because all roots of f lie in the splitting field. We now show that K itself is already a field. This holds because

$$K = \{x \in \overline{\mathbb{F}}_p \mid x^q = x\} = \{x \in \overline{\mathbb{F}}_p \mid \mathrm{Frob}^\ell(x) = x\} =$$

$$\{x \in Z \mid \mathrm{Frob}^\ell(x) = x\} = Z^{\langle \mathrm{Frob}^\ell \rangle}$$

is the fixed field in the Galois extension $Z|\mathbb{F}_p$ associated to the subgroup $\langle \mathrm{Frob}^\ell \rangle \subset \mathrm{Gal}(Z|\mathbb{F}_p)$, thus a field. This proves the existence of an intermediate field K with #K=q.

For uniqueness: Now if $\mathbb{F}_p \subset M \subset \overline{\mathbb{F}}_p$ is any intermediate field with $\#M = q$ elements, the unit group $M^\times$ has the cardinality $\#M^\times = q - 1$. According to Lagrange's theorem, it follows that $1 = x^{q-1}$ for any $x \in M^\times$. Hence $x^q = x$ and the latter equation also holds for $x = 0$. We therefore obtain

$$M \subset \{x \in \overline{\mathbb{F}}_p \mid x^q = x\} =: K$$

and, since $\#M = \#K = q$, equality follows.

Since $K = Z$ is the splitting field of a separable polynomial, $Z|\mathbb{F}_p$ is both normal and separable—Lemma 13.22.

Furthermore, we have the element $\mathrm{Frob} \in \mathrm{Gal}(K|\mathbb{F}_p)$ and $\mathrm{ord}(\mathrm{Frob}) = \ell$: Apparently, $\mathrm{ord}(\mathrm{Frob}) \mid \ell$ since $\#\mathrm{Gal}(K|\mathbb{F}_p) = \ell$. Assuming that $\mathrm{ord}(\mathrm{Frob}) < \ell$, there would be a $k < \ell$, such that $\mathrm{Frob}^k = 1$, thus

$$x^{p^k} = x \text{ för all } x \in K.$$

Since the polynomial $X^{p^k} - X$ admits only $p^k < p^\ell$ roots in $\overline{\mathbb{F}}_p$, this provides a contradiction.

Therefore, $\mathrm{Gal}(K|\mathbb{F}_p)$ is cyclic with the generator Frob. $\square$

Remark 16.5 We must make one important observation: At one place in the proof of the Fundamental Theorem of Galois Theory, we used the Primitive Element Theorem, which we have so far only proved for infinite fields. However, the only place it was used was to prove $H \supset \mathrm{Gal}(L|L^H)$ in the notation of the fundamental theorem. We did not use this conclusion in the proof of Theorem 16.4.

Corollary 16.6 (from the proof of Theorem 16.4) For $\ell \in \mathbb{N}$ and $q = p^\ell$ the set

$$K := \{x \in \overline{\mathbb{F}}_p \mid x^q = x\} \subset \overline{\mathbb{F}}_p$$

is the uniquely determined intermediate field with p^ℓ elements.

Additionally, we have;

Corollary 16.7 (to Theorem 16.4) Every finite field is perfect.

Proof Let $L|K$ be an algebraic extension. We must show that it is separable, that is, every $\alpha \in L$ is separable over K. Now, $K[\alpha]$ is a finite field and according to the above, both $K[\alpha]|\mathbb{F}_p$, as well as $K|\mathbb{F}_p$ are separable, so also $K[\alpha]|K$ according to Lemma 13.21. $\square$

Notation 16.8 When we consider the finite fields $K \subset \overline{\mathbb{F}}_p$ in a fixed chosen algebraic closure, we write

- $K = \mathbb{F}_{p^\ell}$ for the uniquely determined intermediate field with p^ℓ elements and
- sometimes emphasize in the notation of the Frobenius homomorphism on which field we consider it to act:

$$\text{Frob}_\ell : \mathbb{F}_{p^\ell} \to \mathbb{F}_{p^\ell}, \ x \mapsto x^p.$$

Now we can also show the Primitive Element Theorem for finite fields:

Lemma 16.9 (Primitive Element Theorem for Finite Fields) *Let K be a finite field and $L|K$ a finite field extension (which is then separable). Then there exists an $\alpha \in L$ with $L = K[\alpha]$.*

Proof Let $p = \text{char}(K)$ and $\overline{\mathbb{F}}_p$ be an algebraic closure of L. Then we have

$$\mathbb{F}_p \subset K \subset L \subset \overline{\mathbb{F}}_p.$$

In the notation from above, $K = \mathbb{F}_{p^\ell}$ and $L = \mathbb{F}_{p^m}$ for a pair $\ell \leq m$ (we will examine this in more detail in Lemma 16.13). WLOG, let $L \neq K$.

Let $q = p^m$. Then the unit group $\mathbb{F}_{p^m}^\times$ has order $q - 1$ and is cyclic according to Theorem 15.9. Therefore, there is an element $\alpha \in \mathbb{F}_q^\times$ with $\text{ord}(\alpha) = q - 1$. This element cannot be in any proper subfield, because every $0 \neq x \in \mathbb{F}_{p^k}$ with $k < m$ already fulfills

$$x = \text{Frob}_k(x) = x^{p^k} \implies x^{p^k - 1} = 1.$$

So in particular, $\mathbb{F}_{p^\ell}[\alpha] \subset \mathbb{F}_{p^m}$ is not a proper intermediate field and equality follows. $\square$

Corollary 16.10 Let p be a prime number and $n \in \mathbb{N}$ be arbitrary. Then there exists an irreducible polynomial $f(X) \in \mathbb{F}_p[X]$ of degree n.

Proof Consider the field extension $\mathbb{F}_{p^n}|\mathbb{F}_p$ in $\overline{\mathbb{F}}_p$ of degree n. According to Lemma 16.9 there is a $\alpha \in \mathbb{F}_{p^n}$ with $\mathbb{F}_{p^n} = \mathbb{F}_p[\alpha]$. Then the minimal polynomial $f(X)$ of α over $\mathbb{F}_p$ is as desired. $\square$

Remark 16.11 Note that for finite fields $\mathbb{F}_q \subset \overline{\mathbb{F}}_p$ often a **primitive element** in $\mathbb{F}_q$ is defined as a generator of the cyclic group $\mathbb{F}_q^\times$. Every such is also an element α as in the preceding lemma (we have used the corresponding argument in the proof), but not every α as above is a primitive element in this new sense. This sometimes leads to misunderstandings.

Remark 16.12 Returning again to Remark 16.5: So far in Chapter 16, we have not used the statement from the Fundamental Theorem of Galois Theory, which has not yet been proven for finite fields. From now on, we have seen that the Primitive Element Theorem also holds for finite fields, and now have no problem applying the Fundamental Theorem of Galois Theory in its full statement also for finite fields.

The question of how the finite fields with characteristic p relate to each other is answered by the next lemma:

Lemma 16.13 *Let $\overline{\mathbb{F}}_p$ be a fixed algebraic closure of $\mathbb{F}_p$. Then for the respective uniquely determined intermediate fields, we have*

$$\mathbb{F}_{p^\ell} \subset \mathbb{F}_{p^m} \Longleftrightarrow \ell \mid m.$$

Proof Obviously, for $\mathbb{F}_{p^\ell} \subset \mathbb{F}_{p^m}$ necessarily $\ell \le m$. We know that

$$\mathrm{Gal}(\mathbb{F}_{p^m} | \mathbb{F}_p) = \langle \mathrm{Frob}_m \rangle$$

is a cyclic group of order m. All intermediate fields are, according to the Fundamental Theorem of Galois Theory, fixed fields of subgroups $H \subset G := \mathrm{Gal}(\mathbb{F}_{p^m} | \mathbb{F}_p)$—here we use that we can now apply the fundamental theorem unrestrictedly also for finite fields. For finite cyclic groups, however, it is easy to see which subgroups exist: For each divisor $d \mid \#G = m$ there is exactly one subgroup H_d of order d. Therefore, all intermediate fields of $\mathbb{F}_{p^m} | \mathbb{F}_p$ are exactly the fixed fields

$$(\mathbb{F}_{p^m})^{H_d} \text{ for } d|m.$$

Note that

$$[\mathbb{F}_{p^m} : (\mathbb{F}_{p^m})^{H_d}] = \#\mathrm{Gal}(\mathbb{F}_{p^m} | (\mathbb{F}_{p^m})^{H_d}) = \#H_d = d,$$

hence

$$[(\mathbb{F}_{p^m})^{H_d} : \mathbb{F}_p] = \frac{m}{d}.$$

and thus, with the above notation,

$$(\mathbb{F}_{p^m})^{H_d} = \mathbb{F}_{p^\ell} \text{ for } \ell = \frac{m}{d}.$$

With this, the lemma is proven. $\square$

And now?

The Galois theory of finite fields is thus very well understood. Everything takes place in a fixed chosen algebraic closure. Then for $\ell \mid m$ there are the field extensions and the respective Frobenius maps:

$$
\begin{array}{ccc}
\mathbb{F}_{p^m} & \xrightarrow{\mathrm{Frob}_m} & \mathbb{F}_{p^m} \\
| & & | \\
\mathbb{F}_{p^\ell} & \xrightarrow{\mathrm{Frob}_\ell} & \mathbb{F}_{p^\ell} \\
| & & | \\
\mathbb{F}_p & = & \mathbb{F}_p
\end{array}
$$

We have $\text{ord}(\text{Frob}_m) = m$ (as a group element of $\text{Gal}(\mathbb{F}_{p^m}|\mathbb{F}_p)$) and furthermore, the following holds:

$$\text{Frob}_\ell = \text{Frob}_m|_{\mathbb{F}_{p^\ell}}$$

(after all, it is the same mapping rule: $x \mapsto x^p$) with $\text{ord}(\text{Frob}_\ell) = \ell$ (as a group element in $\text{Gal}(\mathbb{F}_{p^\ell}|\mathbb{F}_p)$) and we get

$$\mathbb{F}_{p^\ell} = \{x \in \mathbb{F}_{p^m} \mid \text{Frob}^\ell(x) = x\} = \{x \in \mathbb{F}_{p^m} \mid x^{p^\ell} = x\}$$

the fixed field of the subgroup generated by Frob_m^ℓ, namely $\langle \text{Frob}_m^\ell \rangle \subset \text{Gal}(\mathbb{F}_{p^m}|\mathbb{F}_p)$.

And now?

Note that the divisibility relation in $\mathbb{N}$ is a **partial** order on the natural numbers. Therefore, the set of intermediate fields M with

$$\mathbb{F}_{p^\ell} \subset M \subset \mathbb{F}_{p^m}$$

is also partially ordered due to Lemma 16.13. Note that there are intermediate fields that do not lie within each other:

with $\ell|r|m$ and $\ell|s|m$ but neither $r|s$ nor $s|r$, e.g. $\ell = 1, m = 15$ and $r = 3, s = 5$. See also Fig. 16.1.

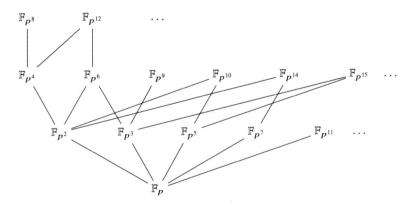

Fig. 16.1 The forest of all finite fields of characteristic p

The uniqueness of K with a given cardinality **within** a fixed algebraic closure results in the following more general uniqueness statement up to isomorphism:

Lemma 16.14 *Let $\ell \in \mathbb{N}$ and let K_1 and K_2 be fields with*

$$\#K_1 = p^\ell = \#K_2.$$

Then there exists a field isomorphism $K_1 \xrightarrow{\cong} K_2$ which is the identity on $\mathbb{F}_p$.

Proof We can take an algebraic closure Ω_1 of K_1 or Ω_2 of K_2 for each. If we then apply Lemma 16.9 to K_1, we obtain an element $\alpha \in K_1$ with

$$K_1 = \mathbb{F}_p[\alpha] \cong \mathbb{F}_p[X]/(f(X)),$$

when $f(X)$ is the minimal polynomial of α over $\mathbb{F}_p$. Let $n := \deg(f)$.

Now, $f(X)$ has a root $\beta \in \Omega_2$ and we thus consider the field extension

$$\Omega_2$$
$$|$$
$$\mathbb{F}_p[\beta]$$
$$|$$
$$\mathbb{F}_p.$$

We know that $\mathbb{F}_p[\beta] \cong \mathbb{F}_p[X]/(f(X)) \cong K_1$. Now, $[\mathbb{F}_p[\beta] : \mathbb{F}_p] = n$, so according to the uniqueness of finite fields in the fixed algebraic closure Ω_2 we obtain that $\mathbb{F}_p[\beta] = K_2$.$\square$

Notation 16.15 (to be used with caution) If $\ell \in \mathbb{N}$, then one often writes $\mathbb{F}_{p^\ell}$ for the (due to the above lemma, unique up to isomorphism) finite field with p^ℓ elements.

Attention If you explicitly want to work with $\mathbb{F}_{p^\ell}$, it is better to choose an algebraic closure $\overline{\mathbb{F}}_p$ and then consider $\mathbb{F}_{p^\ell} \subset \overline{\mathbb{F}}_p$—then $\mathbb{F}_{p^\ell}$ is properly uniquely defined.

And now?

For example, a statement

$$\mathbb{F}_{p^\ell} \subset \mathbb{F}_{p^m} \text{ for } \ell \mid m$$

is not really meaningful if $\mathbb{F}_{p^\ell}$ is only well-defined up to isomorphism! If one wants to write or use something like this, one should really pin down $\mathbb{F}_{p^\ell}$ by stating that one considers $\mathbb{F}_{p^\ell}$ and $\mathbb{F}_{p^m}$ in the same algebraic closure $\overline{\mathbb{F}}_p$.

Anyone who has already seen the fundamental group $\pi_1(X, x_0)$ of a topological space with base point $x_0 \in X$ (for example, in function theory for $X \subset \mathbb{C}$), has already encountered a similar phenomenon. If X is path-connected, then $\pi_1(X, x_0)$ is not essentially dependent on the choice of the base point x_0 in the sense that there is always an isomorphism $\pi_1(X, x_0) \cong \pi_1(X, x_1)$ for different choices—but the isomorphism is also not unique! So one could write: Let $\pi_1(X)$ be the fundamental group of X (uniquely determined up to isomorphism). However, when working explicitly with it, one must specify the base point.

The analogy to the fundamental group is not far-fetched: In fact, the choice of $\overline{\mathbb{F}}_p$ from a certain standpoint of algebraic geometry is the choice of a "base point".

And now?
Finite fields often find application, for example, in cryptography, but also in integer optimization. The field with $q = p^n$ elements (unique up to isomorphism) is also called the **Galois field with q elements** and one often writes GF(q) for it. In SAGE, for example, the command GF(q) initializes such a field, see the following example.

The SAGE lines:

```
K.<a>=GF(125)
f=a.minpoly()
print(K,'With minimal polynomial',f)
```

define K as a field with 125 elements and also return a primitive element α as in Lemma 16.9, whose minimal polynomial SAGE gives as follows:

```
Finite field in a of size 5^3 With minimal polynomial x^3+3*x+3
```

One can continue to calculate with it. Here, R is defined as the polynomial ring in one variable (named x) over $K = \mathbb{F}_{125}$ and the polynomial $h(x) = x^3 + x + 1$ is factorized.

```
R.<x>=PolynomialRing(K,'x')
h=x^3+x+1
h.factor()
```

delivers as a result $(x+a^2+3*a+2)*(x+a^2+4*a+2)*(x+3*a^2+3*a+1)$
One can also run a usual-loop over all elements in $K = \mathbb{F}_{125}$:

```
Nst=[]
n=0
for y in K:
    if (h(y)==0):
        n=n+1
        Nst.append(y)
print('h has ',n,'roots,namely')
print(Nst)
```

These lines generate an empty list *Nst*, then the loop runs through all elements in $y \in K$ and attaches the element y to the list *Nst* if it is a root of $h(x)$. The elements y are given by SAGE in the base $1, a, a^2$ of K over $\mathbb{F}_5$:

```
h has 3 roots,namely')
[2*a^2+2*a+4,4*a^2+a+3,4*a^2+2*a+3]
```

Check it out!
See how the two outputs from SAGE correspond to each other: the one giving the factorization of the polynomial and the one giving the roots?

Red thread to the previous chapter
We have seen in this chapter,

- that given a field K of characteristic p, we always have the Frobenius homomorphism $\mathrm{Frob} : K \to K, x \mapsto x^p$,
- that Frob is the identity on the prime field $\mathbb{F}_p$,
- that for a fixed algebraic closure $\overline{\mathbb{F}}_p$ of $\mathbb{F}_p$ there is exactly one intermediate field K for every $\ell \in \mathbb{N}$ with $[K : \mathbb{F}_p] = \ell$ and thus $\#K = p^\ell$,
- that therefore for every prime power $q = p^\ell$ a field $\mathbb{F}_{p^\ell}$ with p^ℓ elements exists and it is unique up to isomorphism,
- that one should better only use the notation $\mathbb{F}_{p^\ell}$ when one has fixed an algebraic closure $\overline{\mathbb{F}}_p$ and then considers $\mathbb{F}_{p^\ell} \subset \overline{\mathbb{F}}_p$,
- that then

$$\mathbb{F}_{p^\ell} \subset \mathbb{F}_{p^m} \iff \ell \mid m$$

holds,
- that for $\ell \in \mathbb{N}$ the extension $\mathbb{F}_{p^\ell} | \mathbb{F}_p$ is Galois and the Galois group is cyclically generated by the Frobenius:

$$\mathbb{Z}/\ell\mathbb{Z} \xrightarrow{\cong} \mathrm{Gal}(\mathbb{F}_{p^\ell} | \mathbb{F}_p), \ [k] \mapsto \mathrm{Frob}^k.$$

Exercises: 16.1 (Quadratic formula in finite fields): Let p be a prime number, $q = p^n$ for a $n \in \mathbb{N}$ and let $\mathbb{F}_q$ be a field with q elements. Consider the equation $X^2 + bX + c = 0$ with $b, c \in \mathbb{F}_q$.

(a) Let $p \neq 2$. Show that the above equation has a solution in $\mathbb{F}_q$ if and only if $b^2 - 4c$ is a square in $\mathbb{F}_q$.
(b) Let $p = 2$ and $c \neq 0$. Show that the above equation has a solution in $\mathbb{F}_q$ if and only if b/c^2 is of the form $y^2 + y$ for some $y \in \mathbb{F}_q$.

16.2 Let $p \neq 2$ be a prime number and $q = p^n$ for $n \in \mathbb{N}$. Let further $\mathbb{F}_q$ be a finite field with q elements. Show that $\mathbb{F}_q$ has exactly $\frac{q+1}{2}$ squares, that is, that $\{x^2 \mid x \in \mathbb{F}_q\} = \frac{q+1}{2}$ holds.

16.3 Let $\mathbb{F}_7$ be a field with 7 elements.

(a) Show that the residue class rings $\mathbb{F}_7[X]/(X^2 + 1)$ and $\mathbb{F}_7[X]/(X^2 + X + 3)$ each define a field with 49 elements.

(b) Determine an explicit isomorphism

$$F_7[X]/(X^2 + 1) \to \mathbb{F}_7[X]/(X^2 + X + 3)$$

by specifying the image of $[X]$ in $\mathbb{F}_7[X]/(X^2 + X + 3)$.

16.4 Let $\mathbb{F}_{11}$ be a field with 11 elements.

(a) Show that $f(X) = X^3 + 2X + 2 \in \mathbb{F}_{11}[X]$ is irreducible[1] , and that

$$F_{1331} = \mathbb{F}_{11}[X]/(f(X))$$

is a field with 1331 elements.

(b) Present the Frobenius homomorphism Frob $: \mathbb{F}_{1331} \to \mathbb{F}_{1331}$ with respect to the representation of $\mathbb{F}_{1331}$ from (a) as a 3×3 matrix over $\mathbb{F}_{11}$ in the basis $1, [X], [X^2]$.

16.5 Consider the polynomial $f(X) = X^2 + X + 1 \in \mathbb{F}_5[X]$.

(a) Show that $K := \mathbb{F}_5[X]/(f(X))$ is a field with 25 elements.

(b) Determine an element $w \in K$, with $w^2 = 2$.

(c) Show that the matrix

$$A := \begin{pmatrix} 1 & 2 \\ 3 & 4 \end{pmatrix} \in M(2 \times 2, \mathbb{F}_5)$$

is diagonalizable over K. (For this, you need knowledge about diagonalizability from Linear Algebra. The exercise combines Linear Algebra with statements about finite fields.)

16.6 Let p be a prime number.

(a) How many generators does the cyclic group $(\mathbb{Z}/p\mathbb{Z})^\times$ have?

(b) Determine a generator of $(\mathbb{Z}/17\mathbb{Z})^\times$.

---

[1] Please do not make the mistake of using the name Eisenstein in an (incorrect) attempt to solve this question! You may have to do some calculations here.

Fig. 16.2 For Task 16.6

```
def ordng(x):
    k=1
    y=x
    while y!=1:
        y=y*x%p
        k=k+1
    return(k)
##end of function ordng(x)
p=997
erz=[]
for x in range(1,p):
    if ordng(x)==p-1:
        erz.append(x)
print(erz)
print("Number =",len(erz))
```

(c) Try to reproduce the program from Fig. 16.2 in SAGE. What does it do? Use it to determine a generator of $(\mathbb{Z}/997\mathbb{Z})^\times$ and modify the program so that it stops after the first hit.

More Group Theory—Group Actions and Sylow's Theorems

17

17.1 Group Actions

And now?
It's a bit unfortunate that this term only comes up so late in this book because actually many groups only exist because they act on (for example, geometric) objects.

Definition 17.1 Let G be a group. A **group action of G on a set M** is a map

$$\Phi : G \times M \to M, \ (g,m) \mapsto \Phi(g,m) =: gm,$$

(where gm is just a shorthand notation) which has the following properties:

1. $1m = m$ for all $m \in M$,
2. $g(hm) = (gh)m$ for all $g, h \in G$ and $m \in M$,

We then also say, G **acts on M,** and for given $g \in G$ and $m \in M$, we say that g acts on the element m **and** gives gm as a result.

Remark I mostly use the notation gm. If one wants to formulate the conditions of the definition with the help of the map Φ, one gets

$$\Phi(1,m) = m \text{ and } \Phi(g, \Phi(h,m)) = \Phi(gh,m).$$

Remark 17.2 The above definition is more precisely the definition of a **left group action.** Similarly, a **right group action** is a map

© The Author(s), under exclusive license to Springer-Verlag GmbH, DE, part of
Springer Nature 2024
M. Hien, *Abstract Algebra*, Mathematics Study Resources 7,
https://doi.org/10.1007/978-3-662-67974-6_17

$$M \times G \to M, \ (m, g) \mapsto mg$$

with the conditions:

1. $m1 = m$ for all m and
2. $(mg)h = m(gh)$ for all $g, h \in G$ and $m \in M$.

Note that the two definitions really make a difference in axiom no. 2. : If $m \in M$ is fixed and we first let g and then let h act on it, this results in

$$\text{for a left action: } m \mapsto gm \mapsto h(gm) = (hg)m$$
$$\text{for a right action: } m \mapsto mg \mapsto (mg)h = m(gh),$$

meaning that in a left action, the result will be the action of the group element $hg \in G$ on m and in a right action the result will be the action of $gh \in G$ on m .

Notation 17.3 Given a set $M \neq \emptyset$, we write

$$S(M) := \{\sigma : M \to M \mid \sigma \text{ is bijective}\}$$

for the group of **permutations of M (with composition as group multiplication).**

Lemma 17.4 *Let G be a group and $M \neq \emptyset$ a non-empty set. Then a group action Φ induces a group homomorphism*

$$\varphi : G \to S(M), \ g \mapsto (g._- : M \to M, m \mapsto gm).$$

Conversely, if $\varphi : G \to S(M)$ is a group homomorphism, then

$$\Phi : G \times M \to M, \ (g, m) \mapsto (\varphi(g))(m)$$

defines a group action on M.

Proof The properties of a homomorphism or an action respectively are easy to verify. $\square$

A somewhat "generic" formulation, which becomes clearer by the examples that will follow:

Notation 17.5 If M is not just a set but carries an (algebraic) structure, we will write

$$\text{Aut}(M) := \{\sigma \in S(M) \mid \sigma, \sigma^{-1} \text{are compatible with this structure}\} \subset S(M)$$

for the so-called **automorphisms of M.**

Examples
1. If M is a group, $\sigma \in Aut(M)$ are the **group isomorphisms**.
2. If M is a K-vector space for a field K, $Aut(M)$ are the **linear isomorphisms**.
3. If M is a commutative ring with one, $Aut(M)$ are the **ring isomorphisms**.

Definition 17.6 Consider a group action of the group G on a set M and let M have an additional algebraic structure as in the examples. We say that the group action **preserves the algebraic structure of** M, if the group homomorphism $\varphi : G \to S(M)$ associated to it according to Lemma 17.4 is a group homomorphism to $Aut(M)$, i.e. if we have the diagram

$$G \xrightarrow{\varphi} S(M)$$
$$\varphi \searrow \quad \uparrow$$
$$Aut(M)$$

Remark 17.7 Depending on the algebraic structure on M, there are also special terms:

- for a group M, we say that G **acts by group homomorphisms,**
- for a vector space M, we say that G **acts linearly,**...

Example of Group Actions
1. $G = S(n)$ acts on $M = \{1, \ldots, n\}$ in an obvious way.
2. $G = GL_n(\mathbb{R})$ acts linearly on $\mathbb{R}^n$.
3. The group

$$G = O(n) := \{A \in GL_n(\mathbb{R}) \mid A \cdot {}^t A = 1\}$$

of orthogonal matrices acts linearly on $\mathbb{R}^n$. The same group also acts on $S^{n-1} = \{x \in \mathbb{R}^n \mid ||x|| = 1\}$.

And now?
In the examples above, it is already apparent that many groups are defined precisely by the fact that they act on a certain set. The next example is perhaps even clearer:

Example of a group that is defined by its (geometric) operation—the **Dihedral groups**

Definition 17.8 For $n \geq 3$ let $P_n \subset \mathbb{R}^2 = \mathbb{C}$ be the regular n-gon, that is, the polygon in the complex plane with the points $\exp(\frac{2\pi i k}{n})$ for $k = 0, \ldots, n-1$ as vertices. The **dihedral group** D_{2n} is the group

$$D_{2n} := \{T \in O(2) \mid T(P_n) = P_n\}$$

of all orthogonal 2×2-matrices (these are rotations and reflections in $\mathbb{R}^2$), which map the polygon P_{2n} onto itself.

> **And now?**
> There is no general agreement on the notation. Some authors write D_n instead of D_{2n} in order to reflect that it is the group of symmetries of the regular n-gon. The index D_{2n} also fits, because D_{2n} has exactly $2n$ elements—see the next lemma.

Lemma 17.9 *The dihedral group D_{2n} has $2n$ elements and is generated by the two elements*

$$\rho := \begin{pmatrix} \cos(\frac{2\pi}{n}) & -\sin(\frac{2\pi}{n}) \\ \sin(\frac{2\pi}{n}) & \cos(\frac{2\pi}{n}) \end{pmatrix} \quad and \quad \sigma := \begin{pmatrix} 1 & 0 \\ 0 & -1 \end{pmatrix}$$

For these generators, the relations hold:

$$\rho^n = 1, \ \sigma^2 = 1, \ \rho\sigma\rho = \sigma.$$

Proof These are elementary geometric considerations: D_{2n} contains exactly

- the n rotations by the angles $0, \frac{2\pi}{n}, \frac{4\pi}{n}, \dots, \frac{2(n-1)\pi}{n}$ (where the identity is the rotation by the angle 0),
- the n reflections each around an axis through $(0, 0)$, which encloses an angle $0, \frac{\pi}{n}, \frac{2\pi}{n}, \dots, \frac{(n-1)\pi}{n}$ with the x-axis.

Each rotation is a power of the first non-trivial rotation by $\frac{2\pi}{n}$.

Now consider the reflection τ around the axis with angle $\frac{k\pi}{n}$. If $k = 2\ell$ is even, then $\tau = \rho^\ell \sigma \rho^{-\ell}$. If $k = 2\ell + 1$ is odd, then we distinguish again:

- If $n \equiv 1 \bmod 2$, then writing $\nu = \frac{n+k}{2} \in \mathbb{Z}$, it follows that $\tau = \rho^\ell \sigma \rho^{-\ell}$.
- If $n \equiv 0 \bmod 2$, then $\tau = \rho^{\ell+1} \sigma \rho^{-\ell}$.

The relations can also be easily derived by calculation with the given matrices. $\square$

A Theoretically Important Example Let G be a group. Then G acts by conjugation on itself and this is an action by group homomorphisms:

$$G \times G \to G, \ (g, x) \mapsto gxg^{-1}.$$

The group homomorphism associated to this action (as in Lemma 17.4) is

$$G \to Aut(G), \ g \mapsto \mathrm{conj}_g,$$

which we have already encountered in Sect. 3.2.

A variant that will become important later: If $H \lhd G$ is a normal subgroup in the group G and $U \subset G$ is any subgroup, then U also acts as a group on H by conjugation (which strictly speaking takes place in G even if we don't see the group G in the homomorphism:

$$U \to Aut(H), \ u \mapsto conj_u.$$

Now let's continue with general group actions:

Definition 17.10 Consider a group action of the group G on the set M and an element $x \in M$

1. The set

$$Gx := \{gx \in M \mid g \in G\} \subset M$$

   is called the **orbit** or the **trajectory** of x,
2. The subgroup

$$G_x := St(x) := \{g \in G \mid gx = x\} \subset G$$

   is called the **stabilizer group** or **isotropy group** of x.

Check it out!
Show that $G_x \subset G$ is really a subgroup.

Defining Lemma 17.11 Consider a group action of the group G on the set M. Then

$$x \sim y :\Longleftrightarrow \text{ there exists a } g \in G \text{ with } y = gx$$

defines an equivalence relation on M . The class of an element $x \in M$ is the orbit $[x] = Gx$. We will write

$$_G\backslash^M$$

for the quotient after this relation (we put G on the left hand side in order to reflect that G acts from the left).

Check it out!
Show that this is indeed an equivalence relation.

Corollary 17.12 Given a group action of the group G on M, the set M decomposes into the disjoint union of the orbits:

$$M = \bigsqcup_{[x] \in_G \backslash M} Gx.$$

Example Consider the conjugation operation of a group G on itself (as above). Given an element $x \in G$, the orbit Gx is the so-called **conjugacy class of** x:

$$Gx = [x] = \{gxg^{-1} \in G \mid g \in G\}.$$

Definition 17.13 A group action of G on $M \neq \emptyset$ is called **transitive,** if for one (then all) $x \in M$ the orbit already is the entire set $M = Gx$.

For arbitrary group actions, we have the following

Lemma 17.14 *Consider a group action of the group G on $M \neq \emptyset$ and let $x \in M$ be an element. Then the map*

$$G/G_x \rightarrow Gx, \quad [g] \mapsto gx$$

is a well-defined bijection of sets.

Proof In order to prove well-definedness note tha $[g] = [h]$ holds if and only if there is a $\tau \in G_x$ with $h = g\tau$. Then, however,

$$hx = (g\tau)x = g(\tau x) = gx.$$

Injectivity follows, because if $gx = hx$, then we deduce (using the properties of a group action):

$$h^{-1}gx = h^{-1}hx = x \Longrightarrow h^{-1}g \in G_x \Longrightarrow [g] = [h].$$

Surjectivity follows directly from the definition of the orbit Gx. $\square$

And now?

The orbits $Gx \subset M$ generally do not carry any further structure. Therefore, the lemma just proven is particularly useful when M is a finite set. Because then $\#Gx$ is also finite and the lemma makes a statement about this cardinality.

But it can also be useful in other areas. For example, when a group acts transitively on a geometric object (e.g., a manifold) M, because then one has the bijection

$$G/G_x \rightarrow M$$

Such groups, which operate in this way, usually also carry the structure of a manifold (so-called Lie groups) by themselves and then one has certain geometric/topological structures on both sides of the bijection which one can compare to each other or which one can use if one already knows that the same structures arise to get two point of views on the same structure.

Reminder Given a group G and a subgroup $H \subset G$, we call

$$(G : H) := \#(G/H)$$

the **index of H in G.**

From Corollary 17.12 and Lemma 17.14 we immediately obtain the following Theorem:

Theorem 17.15 (Orbit-Stabilizer Theorem) Consider a group action of the finite group G on the finite set M and let $x \in M$ be an element of M. Then:

$$\#(Gx) = (G : G_x).$$

Let furthermore $R \subset M$ be a representative system of the equivalence relation from the Defining Lemma 17.11, let $R_{\geq 2} := \{x \in R \mid \#(Gx) \geq 2\}$ be the subset of representatives each of whose orbit has more than one point, and let

$$M^G := \{x \in M \mid gx = x \text{ for all } g \in G\} \subset M$$

be the set of **fixed points** of the group action. Then the following formula holds:

$$\#M = \sum_{x \in R}(G : G_x) = \#M^G + \sum_{x \in R_{\geq 2}}(G : G_x). \tag{17.1}$$

This statement is called the **Orbit-Stabilizer Theorem.**

What now?
The reason why one considers the fixed points separately in (17.1) is that one usually wants to know the number of these. The Orbit-Stabilizer Theorem thus provides the formula

$$\#M^G = \#M - \sum_{x \in R_{\geq 2}}(G : G_x).$$

In some applications, one can say something about the right hand side. An example of this is Corollary 17.19 below.

Definition 17.16 Let G be a group. Then the subgroup

$$Z(G) := \{g \in G \mid gh = hg \text{ for all } h \in G\} \subset G$$

is called the **center of** G.

If $g \in G$, then the subgroup

$$Z(g) := \{h \in G \mid gh = hg\} \subset G$$

is called the **centralizer of** g.

The two terms just defined can also be obtained from the group action of G on G by conjugation:

Lemma 17.17 *Consider the conjugation operation of a group G on itself. Then we have:*

- *$Z(G)$ coincides with the set of fixed points of the action,*
- *For each $g \in G$ the subgroup $Z(g)$ coincides with the stabilizer group of the element $g \in G$ (where this last mentioned G takes on the role of the set M).*

Proof The fixed points of the conjugation action are

$$\{x \in G \mid gxg^{-1} = x \text{ for all } g \in G\} = Z(G).$$

Given an element $x = g \in G$ (in the set) G, its stabilizer group is

$$St(g) = \{h \in G \mid hgh^{-1} = g\} = Z(g),$$

as claimed. □

Thus, we deduce from the orbit-stabilizer equation:

Theorem 17.18 (Class Equation) Let G be a finite group and $R \subset G$ be a representative system of all conjugacy classes, and let $R_{\geq 2} := \{g \in R \mid \#[g] \geq 2\}$ be the subset of representatives in R whose conjugacy class has more than one element. Then the following **Class Equation** holds:

$$\#G = \#Z(G) + \sum_{g \in R_{\geq 2}} (G : Z(g)).$$

Proof There is nothing more to show. □

Corollary 17.19 Let G be a finite group and $\#G = p^r$ for a prime number p. Then G has a non-trivial center:

$$Z(G) \neq \{1\}.$$

Proof Let $R_{\geq 2}$ be as given in the class equation. Then for every $g \in R_{\geq 2}$, we have

$$\#[g] = (G : Z(g)) \geq 2. \tag{17.2}$$

On the other hand, according to Lagrange's theorem, we know that

$$\#Z(g) \mid \#G = p^r \implies \#Z(g) = p^s \text{ for some } s \leq r.$$

Due to (17.2), if $g \in R_{\geq 2}$, the inequality $s < r$ must hold. Thus,

$$(G : Z(g)) = \#(G/Z(g)) = p^{r-s}$$

with $r - s > 0$ and therefore $(G : Z(g)) \equiv 0 \bmod p$.

With the class equation we get

$$\#Z(G) = \underbrace{\#G}_{\equiv 0 \bmod p} - \sum_{g \in R_{\geq 2}} \underbrace{(G : Z(g))}_{\equiv 0 \bmod p} \equiv 0 \bmod p$$

and since $1 \in Z(G)$, the number of elements $\#Z(G)$ in the center cannot be 0, hence $\#Z(G) \geq p$. □

17.2   The Sylow Theorems

And now?
The aim of this chapter is as follows: Ideally, one would like to classify all finite groups of a fixed group order n. The idea is to find subgroups $H \subset G$ (preferably normal subgroups) in G and then to examine H and G/H (if H is a normal subgroup), and hopefully successively reduce the group order until one reaches such small orders that one knows the groups. The whole idea has several drawbacks:

1. There are **simple** groups of large order where this idea does not work from the very start because one won't find any normal subgroups except the trivial ones.
2. Even if G has a proper normal subgroup $1 \neq H \lhd G$, one obtains the so-called **exact sequence**

$$1 \to H \overset{\alpha}{\to} G \overset{\beta}{\to} G/H \to 1,$$

which simply means that α is injective (it is the inclusion of the subgroup) and β is surjective (the projection), and $\ker(\beta) = \mathrm{im}(\alpha)$. The idea from above only works perfectly if one could determine the group G up to isomorphism by the knowledge of the two groups H and G/H. But this is generally **not the case!** A group G is not already determined by knowing that G fits into an exact sequence

$$1 \to H \to G \to H' \to 1$$

with given H and H'. On the other hand side, one can further investigate this problem and therefore this idea is not completely hopeless.

How can this idea be pursued? For example, by decomposing the group order $n = p_1^{n_1} \cdots p_r^{n_r}$ into its prime factors and then looking for normal subgroups that have an order $p_j^{n_j}$, because then

$$\#(G/H) = p_1^{n_1} \cdots p\!\!\!/_j^{n\!\!\!/_j} \cdots p_r^{n_r}$$

and thus the quotient would have one less prime divisor—this could lead to a successive reduction? As we have already mentioned before, this will not lead to a general algorithm for arbitrary groups but nevertheless will often turn out to be useful. Therefore, we continue with this train of thoughts .

Definition 17.20 Let p be a prime number. A p-**group** is by definition a finite group G such that $\#G = p^n$ is a power of p.
Considering subgroups of a given group, one defines the following:

Definition 17.21 Let G be a finite group and $\#G = p^n \cdot m$ with $p \nmid m$. A subgroup $H \subset G$

- is called a p-**subgroup in** G if $\#H = p^s$ for some $s \in \mathbb{N}$,
- is called a p-**Sylow group in** G if $\#H = p^n$ holds.

Note A priori, it is not clear whether such p-Sylow groups always exist.

There are the so-called three **Sylow theorems** (after Peter Ludwig Sylow 1832–1918) for these concepts. I will first introduce their statements and then we will discuss the proofs.

Theorem 17.22 (1st Sylow Theorem) Let G be a finite group and p a prime number with $p \mid \#G$. Then G admits a p-Sylow group.

The various p-Sylow groups for a given prime divisor p of the group order have a relation to each other:

Theorem 17.23 (2nd Sylow Theorem) Let G be a finite group and p a prime number with $p \mid \#G$. Then the following holds:

1. If $H \subset G$ is a p-subgroup of G, there exists a p-Sylow group $P \subset G$ with $H \subset P$.
2. Any two p-Sylow groups P and P' are conjugate to each other, that is, there exists a $g \in G$, such that

$$P' = gPg^{-1}$$

holds.

The last of the three theorems is a statement about the possible number of p-Sylow groups in a given group:

Theorem 17.24 (3rd Sylow Theorem) Let G be a finite group and $\#G = p^n \cdot m$ with $p \nmid m$. Let $s_p \in \mathbb{N}$ denote the the number of p-Sylow groups in G. Then the following holds

$$s_p \equiv 1 \bmod p \quad \text{and} \quad s_p \mid m.$$

As an important consequence of Sylow's Theorems, we obtain the following:

Corollary 17.25 Let G be a finite group, $p \mid \#G$ a prime number and $H \subset G$ a p-Sylow group in G. Then H is a normal subgroup in G if and only if H is the only p-Sylow group in G.

Proof If there is another p-Sylow group $H' \subset G$ with $H \neq H'$, then first note that this already implies $H' \not\subset H$ because the two subgroups have the same cardinality. According to the 2nd Sylow theorem, there is a $g \in G$ with

$$gHg^{-1} = H' \not\subset H,$$

so H is not a normal subgroup.

Conversely, if H is the only p-Sylow group with $\#H = p^n$ and $g \in G$ is an arbitrary group element, the subgroup $gHg^{-1} \subset G$ has exactly p^n elements, so it is also a p-Sylow group. Since H is the only one, it follows that $gHg^{-1} = H$. $\square$

And now?
Now, given a finite group G, one first decomposes the group order $n = \#G$ into its prime factors. Pick a prime divisor p and let $\#G = p^n \cdot m$ with $p \nmid m$. We then know that the number of p-Sylow groups s_p fulfills the statements of the 3rd Sylow theorem:

$$s_p \equiv 1 \bmod p \quad \text{and} \quad s_p \mid m.$$

If we are lucky, $s_p = 1$ remains as the only possibility and the only p -Sylow subgroup is a normal subgroup of order p^n in G. If this is not the case, there are still finitely many possibilities for s_p and one can try to determine the exact number using other methods.

Example Let G be a group of order $\#G = 70 = 2 \cdot 5 \cdot 7$. Then we know for the number s_7 of the 7-Sylow groups, that

$$s_7 \equiv 1 \bmod 7 \implies s_7 \in \{1, 8, 15, 22, \ldots\}$$
$$s_7 \mid 10 \implies s_7 \in \{1, 2, 5, 10\},$$

so $s_7 = 1$ must hold. It follows that G has a normal subgroup of order 7.

17.3   Applications of Sylow's Theorems and Some Common Tricks

We give a small compilation of usual and common arguments and tricks one can employ trying to apply Sylow's Theorems in Appendix B.

17.4   Proof of Sylow's Theorems

And now?
Before we conduct these proofs, let's start with a small nremark: The proofs of the theorems are not really trivial and at one point or another also very technical. If you decide not to follow the proofs in full detail, I would nevertheless like to make some brief comments on the basic ideas and how they are used in many exercises.

Important fundamental idea: one considers the group action of the given group G (or a suitable subgroup) by conjugation, but this time **on the set of all p-Sylow groups.**

More precisely: If $\mathfrak{S} := \{S \subset G \mid S$ p -Sylow$\}$ is the set of all p-Sylow groups in G, then we have (if we accept Sylow's 1st Theorem):

- $\mathfrak{S} \neq \emptyset$ due to the 1st Sylow theorem,
- for each $S \in \mathfrak{S}$ and each $g \in G$ the subgroup gSg^{-1} is again a p-Sylow group (since $\#(gSg^{-1}) = \#S$), hence $gSg^{-1} \in \mathfrak{S}$.

Note that in this point of view, each individual p-Sylow group $S \subset G$ is interpreted as an **element** of the set $\mathfrak{S}$. So, we obtain the action **of the group G on the set** $\mathfrak{S} \neq \emptyset$:

$$G \times \mathfrak{S} \to \mathfrak{S}, \ (g, S) \mapsto gSg^{-1}.$$

The 2nd Sylow theorem (2) states that this operation is transitive, thus having exactly one orbit.

We can now examine all notions we introduced in the section on *Group Actions*: isotropy groups, orbits, orbit-stabilizer formula, ...– you can imagine that many exercises can be constructed this way.

Example: Let G be a group with $\#G = 300 = 2^2 \cdot 3 \cdot 5^2$. Then, the number of 5-Sylow groups satisfies:

$$s_5 \equiv 1 \bmod 5 \text{ and } s_5 \mid 12 \Rightarrow s_5 \in \{1, 6\}.$$

Now, assuming that G is supposed to be **simple**, it follows that $s_5 = 6$ and $\mathfrak{S}_5 := \{S \subset G \mid S$ is a 5 -Sylow group$\}$ has exactly 6 elements. We enumerate these as follows

$$\mathfrak{S}_5 = \{S_1, \ldots, S_6\}.$$

The action of G on $\mathfrak{S}_5$ induces a group homomorphism

$$\varphi : G \to S(6), \ g \mapsto \sigma \text{ such that } gS_jg^{-1} = S_{\sigma(j)} \text{ for all } j = 1, \ldots, 6.$$

Since G is **simple**, $\ker(\varphi) \lhd G$ being a normal subgroup has to be either $\{1\}$ or G. The latter cannot occur because the action is transitive according to the 2nd Sylow theorem. Therefore, φ must be injective but this is impossible since

$$300 = \#G \nmid 6! = 720$$

and φ being injective would produce a subgroup of order 300 in $S(6)$ which contradicts Lagrange's Theorem. Therefore, we have solved the exercise: "Show that there is no simple group of order 300.".

The 1st Sylow Theorem

Proof (of the 1st Sylow theorem—Theorem 17.22)

Let p be the prime number under consideration with $p \mid \#G$.

We prove the theorem by induction on $\#G$. Let $\#G = p^r \cdot m$ with $p \nmid m$. If $m = 1$, we are done.

Let $m \geq 2$. If we find a subgroup $H \subsetneq G$ such that $p \nmid (G : H)$ holds, then $H = p^r \cdot m'$. But then every p-Sylow group in H is also one in G (***think about it briefly!***). So we are done by induction assumption in this case.

We therefore only need to consider the case that every subgroup $H \subsetneq G$ has an index with $p \mid (G : H)$. In particular, $p \mid (G : Z(g))$ for every element $g \in G$ with $Z(g) \neq G$. From the class equation (Theorem 17.18) - with the notation used there - it follows that

$$\underbrace{\#G}_{\equiv 0 \bmod p} = \#Z(G) + \sum_{g \in R_{\geq 2}} \underbrace{(G : Z(g))}_{\equiv 0 \bmod p},$$

so we deduce that $p \mid \#Z(G)$. Thus, $Z(G)$ is a non-trivial subgroup with $p \mid \#Z(G)$ and according to Cauchy's theorem (Theorem 11.1) there exists an element $x \in Z(G)$ of order p. Now consider the subgroup $H := \langle x \rangle \subset Z(G)$. Then $H \triangleleft G$, because $H \subset Z(G)$ and thus $gx = xg$ holds for all g, hence $gxg^{-1} = x$.

Note that $\#G/H < \#G$, more precisely: If $\#G = p^r \cdot m$ with $p \nmid m$, then $\#G/H = p^{r-1} \cdot m$. Thus, G/H has a p-Sylow group by induction assumption.

$$S \subset G/H \text{ with } \#S = p^{r-1}.$$

Now consider the surjective group homomorphism

$$\pi : G \to G/H, \ g \mapsto [g]$$

with kernel $\ker(\pi) = H$ of order $\# \ker(\pi) = p$. Therefore, $\pi^{-1}(S) \subset G$ is a subgroup of order $\#(\pi^{-1}(S)) = p^{r-1} \cdot p$ since we have the pairwise disjoint decomposition into the residue classes

$$\pi^{-1}(S) = \bigsqcup_{[g] \in S} [g] = \bigsqcup_{[g] \in S} g \cdot H$$

and each of them has cardinality $\#(g \cdot H) = \#H = p$. Therefore, $\pi^{-1}(S) \subset G$ is a p-Sylow group in G. $\square$

The 2nd Sylow Theorem

Proof (of the 2nd Sylow theorem—Theorem 17.23)

Let again $\#G = p^r \cdot m$ with $p \nmid m$. Now let $H \subset G$ be a p-subgroup. We have to show that there is a p-Sylow group $S \subset G$ such that $H \subset S$ holds.

For this, consider the following set (which is not empty according to Sylow's 1st Theorem)

$$\mathfrak{S} := \{S \subset G \mid \#S = p^r\}$$

the set of all p-Sylow groups in G. Note that for $g \in G$ the conjugation

$$\text{conj}_g : G \to G, \ x \mapsto gxg^{-1}$$

is a group isomorphism. In particular, for $S \in \mathfrak{S}$, we have $\#(gSg^{-1}) = \#S = p^r$, so that again $gSg^{-1} \in \mathfrak{S}$ holds. More precisely: The group G acts by conjugation on the set $\mathfrak{S}$ of the p-Sylow groups:

$$G \times \mathfrak{S} \to \mathfrak{S}, \ (g, S) \mapsto gSg^{-1}. \tag{17.3}$$

Now let $T \in \mathfrak{S}$ be an arbitrarily chosen p-Sylow group. Then the stabilizer of T in this action is:

$$G_T = \{g \in G \mid gTg^{-1} = T\}.$$

Apparently, $T \subset G_T$. In particular, we have

$$\#T \mid \#G_T. \tag{17.4}$$

Let furthermore

$$\mathcal{O}(T) := G \cdot T = \{S \in \mathfrak{S} \mid S = gTg^{-1} \text{ for some } g \in G\}$$

be the orbit of T. According to Lemma 17.14 and (17.4), the following divisibility relation follows:

$$\#\mathcal{O}(T) = (G : G_T) = \#G/G_T = \frac{\#G}{\#G_T} \text{ divides } \frac{\#G}{\#T} = \#G/T = m,$$

and thus $\#\mathcal{O}(T)$ is coprime to p.

We now consider the (double) restriction of the conjugation action (17.3)

- to the subgroup $H \subset G$ (instead of G itself) and
- to the subset $\mathcal{O}(T) \subset \mathfrak{S}$ (instead of $\mathfrak{S}$ itself):

$$H \times \mathcal{O}(T) \to \mathcal{O}(T), \ (h, S) \mapsto hSh^{-1}, \tag{17.5}$$

(note that $S \in \mathcal{O}(T)$, hence $S = gTg^{-1}$ for some $g \in G$ and hence also $hSh^{-1} = (hg)T(hg)^{-1} \in \mathcal{O}(T)$ is true).

Given an $S \in \mathcal{O}(T)$ we denote by

$$\omega(S) := \{hSh^{-1} \mid h \in H\} \subset \mathcal{O}(T)$$

the (smaller) orbit of S under this restriction of the action. Due to Lemma 17.14 we have

$$\#\omega(S) = \#(H/\text{isotropy group of } S \text{ with respect to (17.5)})$$

is a divisor of $\#H = p^r$ and thus either 1 or a power p^k with $k \geq 1$.

Now $\mathcal{O}(T)$ decomposes into the orbits $\omega(S)$, more precisely:

$$\mathcal{O}(T) = \bigsqcup_{S \in R} \omega(S) \tag{17.6}$$

where $R \subset \mathbb{O}(T)$ is a representative system for the action (17.5). Now, if all $\omega(S)$ were of an order p^k each with some $k \geq 1$, then the cardinality of the right side of (17.6) would be a multiple of p. But we have seen above that $p \not| \#\mathbb{O}(T)$ holds.

It follows that there is an $S \in R \subset \mathbb{O}(T) \subset \mathfrak{S}$ with a singleton orbit:

$$\{hSh^{-1} \mid h \in H\} = \omega(S) = \{S\}.$$

Hence there is a p-Sylow group $S \subset G$, which remain invariant under conjugation with the elements in the given p-subgroup $H \subset G$:

$$hSh^{-1} = S \text{ for all } h \in H.$$

$\square$

Claim 17.26 In the above situation, the following applies: $H \subset S$.

Proof Consider the **normalizer** of S in G:

$$N := \{g \in G \mid gSg^{-1} = S\} \subset G.$$

According to the above, $H \subset N$ and furthermore $S \lhd N$ (by definition of N). Then

$$H \cdot S := \{h \cdot s \in N \mid h \in H, \ s \in S\} \subset N$$

is a subgroup, because

$$(h_1 \cdot s_1) \cdot (h_2 \cdot s_2) = h_1 h_2 \cdot \underbrace{h_2^{-1} s_1 h_2}_{\in S} \cdot s_2 \in H \cdot S$$

and

$$(h \cdot s)^{-1} = s^{-1} h^{-1} = h^{-1} \cdot \underbrace{h s^{-1} h^{-1}}_{\in S} \in H \cdot S.$$

Furthermore, one has the group isomorphism

$$H/H \cap S \overset{\cong}{\to} H \cdot S/S, \ h \cdot (H \cap S) \mapsto (h \cdot 1) \cdot S,$$

as one can easily show **(Check it out!)**.

However, the left group is either trivial or a p-group (because H is a p-group) and the right group fulfills

$$\#(H \cdot S/S) = \frac{\#(H \cdot S)}{\#S} \text{ divides } \frac{\#G}{\#S} = \#(G/S) = m,$$

and its cardinality is therefore coprime to p. Thus, this order can only be 1 and it follows that $H/H \cap S = \{1 \cdot (H \cap S)\}$ is singleton. Thus, $H \cap S = H$ and therefore $H \subset S$. This is what we wanted to show. $\square$

We have now shown assertion (1) in the 2nd Sylow theorem. If one looks at the proof more closely, we have actually shown a stronger version of it:

(1 *stronger*) If $H \subset G$ is any p-subgroup and $T \subset G$ is any p-Sylow group, then there exists a $g \in G$, such that

$$H \subset gTg^{-1}$$

holds.

Proof Because this is exactly how we conducted the proof above. Given an arbitrary p-Sylow group T, we found an $S = gTg^{-1} \in \mathcal{O}(T)$ such that ultimately $H \subset S$ holds. $\square$

However, this also immediately proves (2), if one chooses $H = P'$ and $T = P$ and takes into account that $\#(gPg^{-1}) = \#P = \#P' = p^r$. $\square$

The 3rd Sylow Theorem

Proof (of the 3rd Sylow theorem—Theorem 17.24) Let again $\mathfrak{S}$ be the set of all p-Sylow groups in G—for our fixed p. This time, we choose an arbitrary $T \in \mathfrak{S}$ and consider the conjugation action (17.3) restricted to T itself:

$$T \times \mathfrak{S} \to \mathfrak{S}, \ (t, S) \mapsto tSt^{-1}.$$

Let's write similarly to our notation before

$$\omega(S) := \{tSt^{-1} \in \mathfrak{S} \mid t \in T\} \subset \mathfrak{S}$$

for the orbit of $S \in \mathfrak{S}$ under this action. It now holds:

1. the orbit of T itself is $\omega(T) = \{T\}$, a singleton (clearly),
2. the orbit of $S \neq T$ is **not** a singleton: Otherwise, $tSt^{-1} = S$ would be true for all $t \in T$ and we could show just like in claim 17.26 (for $H := T$ and S) that this implies $T \subset S$.

Hence, we can apply the orbit equation (Theorem 17.15) and obtain

$$s_p := \#\mathfrak{S} = 1 + \sum_{S \in R_{\geq 2}} \#\omega(S)$$

(with $R_{\geq 2} \subset \mathfrak{S}$ denoting a representative system of all non-single-element orbits. The only single-element orbit is $\omega(T)$). Now, due to Lemma 17.14 we know that for each non-single-element orbit:

$$p \mid \#\omega(S) = \#T/\#(\text{isotropy group})$$

holds. Therefore, $s_p \equiv 1 \bmod p$, as was to be shown.

Additionally, consider the action (17.3):

$$G \times \mathfrak{S} \to \mathfrak{S}, \ (g, S) \mapsto gSg^{-1},$$

now on the whole group G. It is transitive according to the 2nd Sylow theorem, so it has exactly one orbit $\mathcal{O}(T) = \{gTg^{-1} \mid g \in G\}$ for any $T \in \mathfrak{S}$. Therefore, again according to Lemma 17.14 with the stabilizer group $G_T \subset G$:

$$s_p = \#\mathfrak{S} = \#\mathcal{O}(T) = \frac{\#G}{\#G_T} = \frac{p^n m}{\#G_T}.$$

Since we already know that $s_p \equiv 1 \bmod p$ holds, $p^n \mid \#G_T$ must also hold. This implies that $s_p \mid m$. $\square$

Red thread to the previous chapter
In this chapter, we have

- defined group actions $G \times M \rightarrow M$ of a group G on a set M ,
- seen that this corresponds to a group homomorphisms $G \rightarrow S(M)$ (and thus one can also easily introduce additional requirements when M itself carries an algebraic structure and the action should take this structure into account),
- that such an action leads to a decomposition of the set M into pairwise disjoint orbits, leading to the orbit equation theorem 17.15,
- that the conjugation action of a group G on itself is an interesting example, leading to the class equation theorem 17.18,
- that for $\#G = p^r \cdot m$ with $p \nmid m$ subgroups $S \subset G$ with $\#S = p^r$ are called the p-Sylow groups in G,
- about which one has **the three Sylow theorems**.

Exercises: 17.1 Let $L|K$ be a finite Galois extension and p a prime number that divides the degree of the field $[L : K]$.

(a) Show that there is an intermediate extension $K \subset Z \subset L$ such that

$$[L : Z] = p^m \text{ and } p \nmid [Z : K]$$

holds for a $m \in \mathbb{N}$. (Hint: Why is this exercise in the chapter about Sylow groups?)
(b) In the case of $K = \mathbb{Q}$, $L = \mathbb{Q}(\zeta_7)$ with a primitive 7th root of unity ζ_7 and $p = 3$ determine such an intermediate extension by specifying a primitive element $\alpha \in L$ for it.

17.2 How many groups up to isomorphism of order 2026 are there? (Hint: Also consider Appendix B.)

17.3 Let $H := \{z \in \mathbb{C} \mid \mathrm{Im}(z) > 0\}$ be the upper half-plane and $\mathrm{SL}_2(\mathbb{R})$ is the group of real 2×2 matrices with determinant 1. The map

$$\rho : \mathrm{SL}_2(\mathbb{R}) \times H \rightarrow H, \quad \left(\begin{pmatrix} a & b \\ c & d \end{pmatrix}, z \right) \mapsto \frac{az + b}{cz + d}$$

defines a group action of $\mathrm{SL}_2(\mathbb{R})$ on H.

(a) Specify the orbits of ρ.
(b) Specify the stabilizer group of $i \in H$.

17.4 Given are a prime number p and another prime number r which divides $p - 1$. Let $\mathbb{F}_p$ denote the field with p elements. Show:

(a) There is an element $\gamma \in \mathbb{F}_p^\times$ of order r.
(b) We fix such a γ. Consider the group

$$G := \left\{ \begin{pmatrix} 1 & 0 \\ \alpha & \beta \end{pmatrix} \mid \alpha \in \mathbb{F}_p \text{ und } \beta \in \langle \gamma \rangle \right\}.$$

Here, $\langle \gamma \rangle \subset \mathbb{F}_p^\times$ is the (multiplicative) subgroup generated by γ. Determine the number of p- and r-Sylow groups of G.

17.5 Let G be a finite group and $P \subset G$ a p-Sylow group for a prime number $p | \#G$. Furthermore, let s_p be the number of p-Sylow groups in G. Consider the **normalizer** of P in G:

$$N_G(P) := \{g \in G \mid gPg^{-1} \subset P\}.$$

(a) Show that $N_G(P) \subset G$ is a subgroup and that $P \subset N_G(P)$ is a normal subgroup in $N_G(P)$.
(b) Specify the statement: The group G acts on the set Σ of p-Sylow groups by conjugation and $N_G(P)$ is a stabilizer group of this action.
(c) Show that the index $(G : N_G(P))$ satisfies $(G : N_G(P)) = s_p$.

17.6 (The statements of this exercise were partially shown in the proof of the Sylow theorems or assumed there as "easy to show"): In the group G, let $N \lhd G$ be a normal subgroup and $H \subset G$ a subgroup. Show:

(a) The subset

$$H \cdot N = \{hn \in G \mid h \in H \text{ and } n \in N\} \subset G$$

is a subgroup.
(b) The subset $H \cap N \subset H$ is a normal subgroup in H.
(c) There is a natural group isomorphism

$$\varphi : H/H \cap N \to H \cdot N/N, \ [h] = h \cdot H \cap N \mapsto [h] = h \cdot N.$$

Solvability of Polynomial Equations

18

18.1 Solvable Groups

And now?

We stick to the idea that one might better understand a (finite) group G if one finds a suitable normal subgroup $H \lhd G$ and then considers the exact sequence

$$1 \to H \to G \to G/H \to 1$$

– see the *And now?* in Sect. 17.2. If we now assume that **abelian** groups are easier to understand, we can hope that we find $H \lhd G$ such that G/H is abelian. However, it might be too much to ask that H itself should also be an abelian subgroup in the possibly non-abelian group G.

But: One can successively hope again that H has a normal subgroup $H' \lhd H$ such that H/H' is abelian, and then so on ....—as the normal subgroups become smaller and smaller, this process will stop after a finite number of steps—if it works at all: This will not always be the case, not all groups have such normal subgroups (or normal subgroups inside the normal subgroup, or ...). This leads to the next definition.

Definition 18.1 A group G (not necessarily finite) is called **solvable**, if there is a **finite descending filtration**

$$G = G_0 \supset G_1 \supset G_2 \supset \ldots \supset G_N = \{1\} \tag{18.1}$$

such that for all $j = 0, \ldots, N-1$

1. each $G_{j+1} \lhd G_j$ is a normal subgroup and
2. G_j/G_{j+1} is an abelian group.

© The Author(s), under exclusive license to Springer-Verlag GmbH, DE, part of Springer Nature 2024
M. Hien, *Abstract Algebra*, Mathematics Study Resources 7,
https://doi.org/10.1007/978-3-662-67974-6_18

Remark 18.2

1. A trivial remark is that every abelian group G is solvable since one can then take

$$G = G_0 \supset G_1 = \{1\}$$

   as such a filtration.
2. Note that the subgroups G_j in (18.1) are **generally not normal subgroups in** G, they are only normal subgroups inside their respective predecessor G_{j-1}.

The question about abelian factor groups can also be well understood through the following concept:

Definition 18.3 Let G be a group. Then the subgroup generated by all commutators $[g, h] : = ghg^{-1}h^{-1} \in G$ with $g, h \in G$ is called the **commutator subgroup of** G and is written as

$$[G, G] : = \langle [g, h] \mid g, h \in G \rangle \subset G$$

.

Note that $[G, G] \lhd G$ is a normal subgroup and $G/[G, G]$ is abelian,

---

Check it out!
Check this out!

In fact, $G/[G, G]$ is the largest abelian quotient in the following sense:

Lemma 18.4 (Universal Property of the Commutator Subgroup) *The commutator subgroup $[G, G]$ is a normal subgroup and the factor group $G^{ab} : = G/[G, G]$ is abelian. Thus,*

$$\pi : G \to G^{ab}$$

is a group homomorphism into an abelian group.

   Conversely, if $\varphi : G \to A$ is a group homomorphism into an abelian group A, there exists a unique group homomorphism ψ, such that the following diagram is commutative:

Proof Every element in $[G, G]$ is a finite product of commutators $[g, h]$. For an arbitrary $x \in G$ we have

$$x[g, h]x^{-1} = xghg^{-1}h^{-1}x^{-1}$$

$$= (xgx^{-1})(xhx^{-1})(xg^{-1}x^{-1})(xh^{-1}x^{-1}) = [xgx^{-1}, xhx^{-1}]$$

again a commutator. Therefore, $[G, G] \lhd G$ is a normal subgroup. Furthermore, the classes $[g], [h] \in G^{\mathrm{ab}}$ satisfy

$$[g] \cdot [h] = [gh] = [gh \cdot \underbrace{h^{-1}g^{-1}hg}_{\in [G,G]}] = [hg] = [h] \cdot [g],$$

so that G^{ab} is abelian.

If $\varphi : G \to A$ is a group homomorphism and A is abelian, then $\varphi([g, h]) = [\varphi(g), \varphi(h)] = 1$ (we write A multiplicatively), so $[G, G] \subset \ker(\varphi)$. Thus,

$$\psi : [g] \mapsto \varphi(g)$$

is well-defined, fulfills the desired properties, and must necessarily have this form. □

Definition 18.5 Let G be a group. Then the descending filtration

$$G =: G^{(0)} \supset G^{(1)} := [G^{(0)}, G^{(0)}] \supset \ldots \supset G^{(j+1)} := [G^{(j)}, G^{(j)}] \supset \ldots$$

(a priori not necessarily finite) with normal subgroups $G^{(j+1)} \lhd G^{(j)}$ is called the **commutator series of** G.

One can now read off the solvability of a group from the commutator series:

Lemma 18.6 *A group G is solvable if and only if there is a natural number $N \in \mathbb{N}$ such that $G^{(N)} = \{1\}$ (we then say that the commutator series terminates after a finite number of steps).*

Proof If the commutator series terminates after a finite number of steps, it is a finite descending filtration as required in Definition 18.1.

Conversely, if a filtration as in Definition 18.1 is given (in the notation there) one can successively prove with Lemma 18.4 that $G^{(k)} \subset G_k$ holds for all k. In particular, $G^{(N)} = \{1\}$ since $G_N = \{1\}$ was the trivial group. To be more precise, let us apply Lemma 18.4 in the first step with $\varphi : G \to G/G_1 =: A$ (abelian by definition) and obtain the commutative diagram, where φ and π are the natural projections:

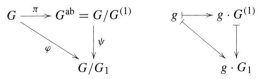

Since ψ is well-defined, the inclusion $G^{(1)} \subset G_1$ must hold.

In the second step, one considers analogously

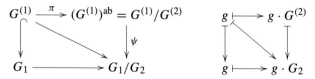

(where G_1/G_2 is abelian by definition). Again, it follows that $G^{(2)} \subset G_2$ must hold. Now we continue successively in this way—or reformulate the whole proof more elegantly as a proof by induction of the claim that $G^{(k)} \subset G_k$ for all k. $\square$

18.2   Solving Polynomial Equations by Radicals

Now let's consider an irreducible polynomial $f(X) \in \mathbb{Q}[X]$ and **assume that f is solvable by radicals**—Definition 14.5. Then there is a finite sequence

$$\mathbb{Q} =: K_0 \subset K_1 \subset K_2 \subset \ldots \subset K_N$$

of field extensions such that in each step an element $\mu_j \in K_j$ exists with

$$K_{j+1} = K_j[\sqrt[m_j]{\mu_j}]$$

for some $m_j \in \mathbb{N}$ and such that finally $Z \subset K_N$ for the splitting field $Z \subset \mathbb{C}$ of f.

Let $m := \prod_{j=0}^{N} m_j \in \mathbb{N}$ and $\zeta := \exp(\frac{2\pi i}{m}) \in \mathbb{C}$. Consider the fields $L_j := K_j[\zeta]$. Then $L_{j+1} = L_j[\sqrt[m_j]{\mu_j}]$ still holds, only that these extensions $L_{j+1} \mid L_j$ are all Galois.

According to Lemma 14.9, we can construct these fields in such a way that $L_N \mid \mathbb{Q}$ is Galois as well. So we have the field extensions as in (14.2):

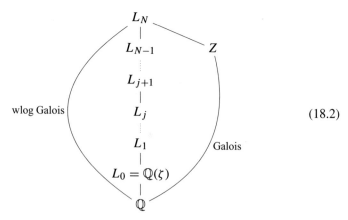

(18.2)

Claim 18.7 In this situation, $\mathrm{Gal}(f) = \mathrm{Gal}(Z \mid \mathbb{Q})$ is solvable.

Proof Firstly,

$$\text{Gal}(f) = \text{Gal}(Z|\mathbb{Q}) \cong \text{Gal}(L_N|\mathbb{Q})/\text{Gal}(L_N|Z).$$

We have already anticipated the actual argument: With the help of Lemma 14.11 we see that $\text{Gal}(L_N|\mathbb{Q})$ is solvable—in Lemma 14.11 we just hadn't used the term *solvable* yet. This leads to the assertion after applying the following lemma. □

Lemma 18.8 *If G is a solvable finite group and $H \lhd G$ is a normal subgroup, then G/H is also solvable.*

Proof Let

$$G = G_0 \supset G_1 \supset \ldots \supset G_N = \{1\}$$

be a finite filtration with $G_{j+1} \lhd G_j$ and each quotient G_j/G_{j+1} being abelian.
   Consider the surjective projection

$$\pi : G \to G/H =: G'.$$

The images $G'_j := \pi(G_j) \subset G'$ for each $j \geq 0$ are subgroups. Furthermore, each $G'_{j+1} \lhd G'_j$ is normal in its precursor since for a given $j \geq 0$, we have $g' = \pi(g) \in G'_j$ (note that in the special case $j = 0$ the surjectivity of π is used) and $x' = \pi(x) \in G'_{j+1}$ gives

$$g'x'g'^{-1} = \pi(gxg^{-1}) \in \pi(G_{j+1}) = G'_{j+1}.$$

Finally,

$$G_j/G_{j+1} \to G'_j/G'_{j+1}, \ g \cdot G_{j+1} \mapsto \pi(g) \cdot G'_{j+1}$$

is a surjective group homomorphism and since the group on the left hand side is abelian by assumption, so is the one on the right hand side. □
   In summary, we have obtained the following:

Theorem 18.9 *Let $f(X) \in \mathbb{Q}[X]$ be irreducible of degree $n \geq 1$ and assume that f is solvable by radicals. Then the Galois group $\text{Gal}(f)$ is a solvable group.*

Remark 18.10 One can also show the converse: If $\text{Gal}(f)$ is solvable, then f is solvable by radicals. However, we would still need to develop some preparations about extensions with abelian or cyclic Galois groups (Kummer theory) which we will omit here.

We can deduce some consequences about the question of finding solution formulas for polynomial equations of degree ≥ 5.

Lemma 18.11 *For $n \geq 5$ the symmetric group $S(n)$ is not solvable.*

Proof We first prove the following claim □

Claim 18.12 Let $n \geq 5$. If $G \subset S(n)$ is a subgroup that contains every 3-cycle and $H \lhd G$ is a normal subgroup with abelian factor group G/H, then H also contains all 3-cycles. $\square$

Proof Let $(a\,b\,c)$ be an arbitrary 3-cycle which by assumption lies in G. Then we find $e \neq d$ with $e, d \notin \{a, b, c\}$ since $n \geq 5$. Consider $\sigma = (a\,c\,e)$ and $\tau = (a\,b\,d)$. Then

$$\tau\sigma\tau^{-1} = (\tau(a)\,\tau(c)\,\tau(e)) = (b\,c\,e).$$

Furthermore,

$$\tau\sigma\tau^{-1}\sigma^{-1} = (b\,c\,e)(e\,c\,a) = (a\,b\,c).$$

Now consider $\pi : G \to G/H$. By assumption, G/H is abelian and G contains all 3-cycles, so in particular $\sigma, \tau \in G$. It follows that

$$\pi((a\,b\,c)) = \pi(\tau\sigma\tau^{-1}\sigma^{-1}) = \pi(\tau)\pi(\sigma)\pi(\tau)^{-1}\pi(\sigma)^{-1} = 1,$$

and thus $(a\,b\,c) \in \ker(\pi) = H$. $\square$

The lemma now is easy to see using the claim. If

$$S(n) =: G_0 \supset G_1 \supset G_2 \supset \ldots \supset G_N = \{1\}$$

were a finite chain with $G_{j+1} \lhd G_j$ and abelian G_j/G_{j+1} in the case that $S(n)$ was a solvable group, it follows from 18.12 successively that G_1 would have to contain all 3-cycles, but therefore also G_2 and so on—in contradiction to $G_N = \{1\}$. $\square$

Corollary 18.13 Let $f(X) \in \mathbb{Q}[X]$ be an irreducible polynomial of degree $n := \deg(f) \geq 5$ such that $\mathrm{Gal}(f) \cong S(n)$, then f is not solvable by radicals.

Example Our example considered in the intermezzo Chap. 11

$$f(X) = X^5 - 777X + 7$$

hence is not solvable by radicals.

Note For $n < 5$ the group $S(n)$ is solvable, by the way:

$$S(1) = \{1\}$$
$$S(2) \cong \mathbb{Z}/2\mathbb{Z} \text{ is abelian,}$$
$$S(3) \supset A(3) \supset \{1\} \text{ is a filtration as desired,}$$
$$S(4) \supset A(4) \supset H \supset \{1\} \text{ likewise,}$$

where $A(n)$ denotes the alternating group and in the case $n = 4$ the subgroup H is defined to be

$$H := \{1, (1\,2)(3\,4), (1\,3)(2\,4), (1\,4)(2\,3)\} \subset A(4)$$

Check it out!
Check the assertions about the normal subgroup properties and the factor groups in the cases $n = 3$ and $n = 4$.

18.3   The General Equation of Degree n

And now?
In the last section, we want to briefly address the question of a *solution formula* for a general equation of degree n for $n \geq 5$. In the previous section, we saw that there can be no representation of the solutions by radicals for explicit examples of such.

How would one exactly pose the question of a general *solution formula*?

In the quadratic case, we are talking about the equation $ax^2 + bx + c = 0$ with $a \neq 0$ (otherwise it is not quadratic), so we can divide by a and obtain the equation with the corresponding solution:

$$x^2 + bx + c = 0 \iff x_{1/2} = \frac{-b \pm \sqrt{b^2 - 4c}}{2}.$$

The general equation in degree n (again without loss of generality assumed to be monic) is to find a formula for the solutions of

$$x^n + a_1 x^{n-1} + a_2 x^{n-2} + \ldots + a_{n-1} x + a_n = 0$$

and, more precisely, we would hope for a solution formula as a radical extension with the parameters $a_1, \ldots, a_n$. The next definition clarifies how the expression *with the parameters* ... is to be understood.

Definition 18.14 Let $n \in \mathbb{N}$ and $\mathbf{k}$ be a field. Consider the quotient field

$$K := \mathbf{k}(u_1, \ldots, u_n) := \mathrm{Quot}(\mathbf{k}[u_1, \ldots, u_n])$$

of the polynomial ring over $\mathbf{k}$ with n variables $u_1, \ldots, u_n$—the so-called **function field in n variables over k**. Then the polynomial

$$f(X) := X^n - u_1 X^{n-1} + u_2 X^{n-2} + \ldots + (-1)^n u_n \in K[X] \qquad (18.3)$$

is called the **general polynomial of degree n over k**.

And now?
1. The distribution of signs will be explained below.
2. A decomposition field $Z|K$ of f is thus somehow a **solution field for the general polynomial**.

3. In the quadratic case (and let's say $k = \mathbb{Q}$) one has $K = \mathbb{Q}(b, c)$ and $f(X) = X^2 - bX + c$. Thus,

$$Z = K[\sqrt{b^2 - 4c}]|K$$

is a splitting field of $f(X) \in K[X]$, as follows from the quadratic formula. Conversely, if one knows that $Z = K[\sqrt{b^2 - 4c}]$ is a splitting field of f, it follows that every root $\alpha \in Z$ is of the form

$$\alpha = g(b, c) + h(b, c)\sqrt{b^2 - 4c}$$

for some $g(b, c), h(b, c) \in K$. This means that there is a solution formula. If you want to compute this formula explicitly, you now have to determine the $g(b, c)$, $h(b, c)$ contained therein and then you have obtained the quadratic solution formula.

If you consider a second version of the function field $L := k(t_1, \ldots, t_n)$ and the following polynomial over it

$$g(X) := \prod_{j=1}^{n}(X - t_j) \in L[X], \tag{18.4}$$

then, by expanding, one obtains a representation

$$g(X) = X^n - s_1 X^{n-1} + s_2 X^{n-2} + \ldots + (-1)^n s_n \tag{18.5}$$

with the coefficients $s_j \in L = k(t_1, \ldots, t_n)$, which even lie in the subring $k[t_1, \ldots, t_n]$. The justification for the distribution of signs in (18.3) comes from (18.5).

Definition 18.15 The polynomials just constructed

$$s_j(t_1, \ldots, t_n) \in k[t_1, \ldots, t_n]$$

are called the **elementary symmetric polynomials in n variables**.
These are **symmetric** in the sense that

$$s_j^{\sigma}(t_1, \ldots, t_n) := s_j(t_{\sigma(1)}, \ldots, t_{\sigma(n)}) = s_j(t_1, \ldots, t_n) \text{ for all } \sigma \in S(n)$$

In general, we will call any polynomial $h(t_1, \ldots, t_n) \in k[t_1, \ldots, t_n]$ **symmetric** if it has exactly this property.
Example: The polynomials s_j can easily be calculated:

$$s_1(t_1, \ldots, t_n) = \sum_{k=1}^{n} t_k, \quad s_2(t_1, \ldots, t_n) = \sum_{k<\ell} t_k t_\ell, \quad \ldots, \quad s_n(t_1, \ldots, t_n) = \prod_{k=1}^{n} t_k.$$

Now that we have these elements $s_j \in \mathbf{k}[t_1, \ldots, t_n]$ available we can consider the intermediate field $\mathbf{k}(s_1, \ldots, s_n)$ inside the field extension $\mathbf{k}(t_1, \ldots, t_n) \mid \mathbf{k}$ (i.e. the smallest intermediate field that contains $\mathbf{k}$ and all these polynomials s_j)—note that $\mathbf{k}(t_1, \ldots, t_n) \mid \mathbf{k}$ is an infinite transcendental extension. This leads to the following field extension

$$\begin{array}{c} \mathbf{k}(t_1, \ldots, t_n) \\ | \\ \mathbf{k}(s_1, \ldots, s_n) \end{array} \qquad (18.6)$$

and obviously $\mathbf{k}(t_1, \ldots, t_n)$ is a splitting field of the separable polynomial $g(X) \in \mathbf{k}(s_1, \ldots, s_n)[X]$ from (18.5), because $g(X)$ splits into the linear factors (18.4) and $\mathbf{k}(t_1, \ldots, t_n)$ is generated by the roots. Therefore, (18.6) is a finite Galois extension.

Theorem 18.16 The Galois group of the extension (18.6) is canonically isomorphic to

$$\mathrm{Gal}(\mathbf{k}(t_1, \ldots, t_n) \mid \mathbf{k}(s_1, \ldots, s_n)) = S(n),$$

where the elements of $S(n)$ act on $\mathbf{k}(t_1, \ldots, t_n)$ by permuting the variables (and thus can directly be read as an element in the Galois group).

The polynomial
$g(X) = X^n - s_1 X^{n-1} + s_2 X^{n-2} + \ldots + (-1)^n s_n \in \mathbf{k}(s_1, \ldots, s_n)[X]$ is irreducible.

Proof Let's write $L = \mathbf{k}(t_1, \ldots, t_n)$ and $M = \mathbf{k}(s_1, \ldots, s_n)$. Since L is the splitting field of $g(X) \in M[X]$, every $\sigma \in \mathrm{Gal}(L|M)$ must permute its roots $t_1, \ldots, t_n$. Conversely, every permutation $\sigma \in S(n)$ induces a field automorphism.

$$\sigma : \mathbf{k}(t_1, \ldots, t_n) \to \mathbf{k}(t_1, \ldots, t_n), \; t_j \mapsto t_{\sigma(j)} \,,$$

which is the identity on $\mathbf{k}(s_1, \ldots, s_n)$. This implies that each of the t_j must be a root of the minimal polynomial of t_1 over M, so $g(X)$ is this minimal polynomial and therefore irreducible. $\square$

Corollary 18.17 If $h(t_1, \ldots, t_n) \in \mathbf{k}(t_1, \ldots, t_n)$ is symmetric (i.e. invariant under permutations of the variables), there exists a uniquely determined $\ell \in \mathbf{k}(s_1, \ldots, s_n)$ with

$$h(t_1, \ldots t_n) = \ell(s_1(t_1, \ldots, t_n), \ldots, s_n(t_1, \ldots, t_n)).$$

Proof Since h is symmetric, it lies in the fixed field

$$\mathbf{k}(t_1, \ldots, t_n)^{S(n)} = \mathbf{k}(t_1, \ldots, t_n)^{\mathrm{Gal}(\mathbf{k}(t_1, \ldots, t_n) \mid \mathbf{k}(s_1, \ldots, s_n))} = \mathbf{k}(s_1, \ldots, s_n),$$

which was to be shown. $\square$

And now?

Have we now shown that the general polynomial (18.3) is not solvable by radicals for $n \geq 5$? Almost, but not quite. Because there is a subtle difference, whether one

- considers the polynomial $g(X)$ from (18.5) in $\mathbf{k}(s_1, \ldots, s_n) \subset \mathbf{k}(t_1, \ldots, t_n)$, or
- the polynomial $f(X)$ from (18.3) in $\mathbf{k}(u_1, \ldots, u_n)$.

In the first case, $\mathbf{k}(s_1, \ldots, s_n)$ is the intermediate field in $\mathbf{k} \subset \mathbf{k}(t_1, \ldots, t_n)$ which is generated by the elementary symmetric polynomials, in the second case the $u_1, \ldots, u_n$ are free variables.

However, the two situations are closely related to each other, as we will now see.

Let's consider the general polynomial

$$f(X) := X^n - u_1 X^{n-1} + u_2 X^{n-2} + \ldots + (-1)^n u_n \in K[X]$$

in n variables over the field $K = \mathbf{k}(u_1, \ldots, u_n)$.

Let $Z | \mathbf{k}(u_1, \ldots, u_n)$ be a splitting field of $f(X)$. Then, f has roots there, let's call them $\alpha_1, \ldots, \alpha_n \in Z$ (we do not yet know a priori that these are pairwise different) and then

$$Z = \mathbf{k}(u_1, \ldots, u_n)[\alpha_1, \ldots, \alpha_n] = \mathbf{k}(\alpha_1, \ldots, \alpha_n) \,,$$

the latter because $u_j = s_j(\alpha_1, \ldots, \alpha_n) \in \mathbf{k}[\alpha_1, \ldots, \alpha_n]$ is contained in this field.

And now?

We have, in a way, turned the tables compared to the consideration in the last *And now?*:

Before we had:

$$\overbrace{\mathbf{k}(t_1, \ldots, t_n)}^{\text{free variables}}$$

$$\underbrace{\mathbf{k}(s_1, \ldots, s_n)}_{\text{special elements}} \qquad g(X) \in \mathbf{k}(s_1, \ldots, s_n)[X]$$

Now we have:

special elements

$$\mathbf{k}(\overbrace{\alpha_1, \ldots, \alpha_n})$$

$$\mathbf{k}(\underbrace{u_1, \ldots, u_n}) \qquad f(X) \in \mathbf{k}(u_1, \ldots, u_n)[X]$$

free variables

We fully understand the **Before** situation by Theorem 18.16, but we want to understand the **Now** situation. However, these are isomorphic, as we will now show:

Let $t_1, \ldots, t_n$ be free variables again and consider the ring homomorphism ("inserting α_j"):

$$\Psi : \mathbf{k}[t_1, \ldots, t_n] \to \mathbf{k}[\alpha_1, \ldots, \alpha_n], \; h(t_1, \ldots, t_n) \mapsto h(\alpha_1, \ldots, \alpha_n).$$

Apparently, $\Psi(s_j(t_1, \ldots, t_n)) = u_j$, so Ψ induces a ring homomorphism by restriction.

$$\psi : \mathbf{k}[s_1, \ldots, s_n] \to \mathbf{k}[u_1, \ldots, u_n],$$

and we also have

$$
\begin{array}{ccc}
\mathbf{k}[t_1, \ldots, t_n] & \xrightarrow{\;\Psi\;} & \mathbf{k}[\alpha_1, \ldots, \alpha_n] \\
\uparrow & & \uparrow \\
\mathbf{k}[s_1, \ldots, s_n] & \xrightarrow{\;\psi\;} & \mathbf{k}[u_1, \ldots, u_n]
\end{array}
$$

And now?
With this, we have connected the **Before** situation with the **Now** situation.

Now, ψ is an isomorphism since the variables $u_1, \ldots, u_n$ are free variables and one can simply construct an inverse isomorphism by

$$\mathbf{k}[u_1, \ldots, u_n] \to \mathbf{k}[s_1, \ldots, s_n], \; u_j \mapsto s_j$$

("inserting s_j"). Thus, ψ also induces an isomorphism of the respective quotient fields which we again call ψ:

$$\psi : \mathbf{k}(s_1, \ldots, s_n) \xrightarrow{\cong} \mathbf{k}(u_1, \ldots, u_n).$$

But also Ψ is a ring isomorphism as we are going to show now.

And now?
Note that one cannot simply define a reverse isomorphism as we did for ψ by the attempt $\alpha_j \overset{??}{\mapsto} t_j$ because the $\alpha_1, \ldots, \alpha_n$ are not free variables. One would then have to show that this induces a well-defined ring homomorphism.

Apparently, Ψ is surjective, because the elements α_j are in the image. For injectivity we consider the kernel. We have

$$h(t_1, \ldots, t_n) \in \ker(\Psi) \Leftrightarrow h(\alpha_1, \ldots, \alpha_n) = 0.$$

Now if we "symmetrize" h by considering

$$H(t_1, \ldots, t_n) := \prod_{\sigma \in S(n)} h^\sigma(t_1, \ldots, t_n) = h \cdot \prod_{\sigma \neq 1} h^\sigma,$$

then H is symmetric and thus lies in the fixed field

$$\mathbf{k}(t_1, \ldots, t_n)^{S(n)} = \mathbf{k}(s_1, \ldots, s_n).$$

In particular, since ψ is an isomorphism,

$$0 = \Psi(H) = \psi(H) \Rightarrow 0 = H = h \cdot \prod_{1 \neq \sigma} h^\sigma \Rightarrow h = 0.$$

Therefore, Ψ also induces an isomorphism of the quotient fields, which is compatible with ψ. Thus, we obtain

$$
\begin{array}{ccc}
\mathbf{k}(t_1, \ldots, t_n) & \overset{\Psi}{\underset{\cong}{\longrightarrow}} & \mathbf{k}(\alpha_1, \ldots, \alpha_n) \\
\uparrow & & \uparrow \\
\mathbf{k}(s_1, \ldots, s_n) & \overset{\psi}{\underset{\cong}{\longrightarrow}} & \mathbf{k}(u_1, \ldots, u_n)
\end{array}
$$

and apparently $f(X) = g^\psi(X)$. Since $f(X)$ is irreducible and separable, so is $g(X)$ and the Galois groups are isomorphic via Ψ:

$$\mathrm{Gal}(\mathbf{k}(t_1, \ldots, t_n) \mid \mathbf{k}(t_1, \ldots, t_n)) \cong \mathrm{Gal}(\mathbf{k}(t_1, \ldots, t_n) \mid \mathbf{k}(t_1, \ldots, t_n)) \overset{\text{Theorem 18.16}}{\cong} S(n).$$

So, we have now shown:

Theorem 18.18 The general polynomial $f(X) \in \mathbf{k}(u_1, \ldots, u_n)[X]$ is irreducible and its Galois group is

$$\mathrm{Gal}(f) = \mathrm{Gal}(Z \mid \mathbf{k}(u_1, \ldots, u_n)) \cong S(n)$$

(Z is a splitting field of f) isomorphic to the symmetric group $S(n)$.
    With Lemma 18.11 we deduce:

Theorem 18.19 The general polynomial of degree n is not solvable by radicals for $n \geq 5$.

Remark 18.20
1. One could also phrase the theorem like this: For $n \geq 5$ there is no *solution formula.*
2. This theorem was already known before Galois (by Abel, Ruffini). The results from Sect. 18.2, especially corollary 18.13, are actually deeper than the last theorem. This was not known to Abel, Ruffini, but then discovered by Galois.

Thread to the previous chapter
In this chapter, we have seen

- how to define the concept of a **solvable group**,
- that the group $S(n)$ is not solvable for $n \geq 5$,
- that if an irreducible $f(X) \in \mathbb{Q}[X]$ is solvable by radicals, the Galois group is also solvable,
- that this can be used to show that there are irreducible $f(X) \in \mathbb{Q}[X]$ with $\deg(f) \geq 5$ that are not solvable by radicals,
- that the general equation of degree $n \geq 5$ is not solvable by radicals, so there can be no solution formula (as in the quadratic case) from degree 5 onwards.

Exercises: 18.1 Show that A_n is the commutator subgroup of S_n.

18.2 Let G be a finite group and $H \triangleleft G$ a normal subgroup. Let $m, n \in \mathbb{N}$ be such that $H^{(m)} = \{1\}$ and $(G/H)^{(n)} = \{1\}$ holds. Show that $G^{(m+n)} = \{1\}$ holds.

18.3 Let **k** be a field. Consider the group

$$G := \left\{ \begin{pmatrix} a & b \\ 0 & c \end{pmatrix} \in GL_2(\mathbf{k}) \mid a, b, c \in \mathbf{k} , \ ac \neq 0 \right\}.$$

Show that $G^{(1)} = [G, G]$ is abelian and deduce that G is solvable.

18.4 The polynomial $f(X) = X^4 + 2X + 2 \in \mathbb{Q}[X]$ (see Exercise 16) is solvable by radicals. Determine a corresponding tower of field extensions as in (18.2) and compute the associated normal series for $\mathrm{Gal}(f)$ with respect to the Fundamental Theorem of Galois Theory.

Proof of the Existence of an Algebraic Closure

We omitted the proof of theorem 12.2 in the main text because it would have represented a break in the procedure there. We are now catching up on the proof.

Proof
(of Theorem 12.2)

To do this, we carry out some steps and need additional definitions. Let K be an arbitrary field. $\square$

Theorem A.1 There is an algebraic field extension $L|K$ such that every polynomial $f(X) \in K[X]$ of degree $\deg(f) \geq 1$ has a root in L. $\square$

Proof
This is a generalization of Theorem 10.1 in which we only require this for **one** given polynomial $f(X)$. For finitely many $f(X) \in K[X]$ this would also not be a problem because we could then successively construct a splitting field of the product of the finitely many polynomials using theorem 10.1. Now one uses a (brilliant) trick. For any $f(X) \in K[X]$ (without loss of generality irreducible, otherwise we can consider the irreducible factors one after the other) we could have taken $L = K[X]/(f(X))$ to be the answer. The trick is now to use a polynomial ring **in infinitely many variables** X_i, indexed by an index set I (i.e., a variable X_i for each element $i \in I$, where I is any set) instead of the polynomial ring in one variable only. We first have to define this ring:

Definition inside the Proof A.2 Let $I \neq \emptyset$ be an (index) set and K a field. The **polynomial ring** $K[X_i | i \in I]$ **over** K **in the variables** X_i, $i \in I$ is defined as follows: First, let

$$\mathbb{N}_0^{(I)} := \{\underline{m} : I \to \mathbb{N}_0, i \mapsto m_i \mid m_i = 0 \text{ for all but finitely many } i \in I\}$$

M. Hien, *Abstract Algebra*, Mathematics Study Resources 7, https://doi.org/10.1007/978-3-662-67974-6

denote the set of mappings from I to the non-negative integers $\mathbb{N}_0$ which can only take a finite number of values not equal to zero. With this,

$$K[X_i | i \in I] := K^{(\mathbb{N}_0^{(I)})}$$

$$:= \{F : \mathbb{N}_0^{(I)} \to K \mid F(\underline{m}) = 0 \text{ for all but finitely many } \underline{m} \in \mathbb{N}_0^{(I)}\}.$$

We will write the elements $F \in K[X_i | i \in I]$ in the form

$$F : \mathbb{N}_0^{(I)} \to K \,,\, \underline{m} \mapsto F(\underline{m}) =: a_{\underline{m}}\,,$$

that is, we write any element as

$$(a_{\underline{m}})_{\underline{m} \in \mathbb{N}_0^{(I)}} \in K[X_i | i \in I].$$

Lemma in Proof A.3 If one defines an addition and multiplication in this situation by

$$(a_{\underline{m}})_{\underline{m}} + (b_{\underline{m}})_{\underline{m}} := (a_{\underline{m}} + b_{\underline{m}})_{\underline{m}}$$

$$(a_{\underline{m}})_{\underline{m}} \cdot (b_{\underline{m}})_{\underline{m}} := \left(\sum_{\underline{k}+\underline{\ell}=\underline{m}} a_{\underline{k}} b_{\underline{\ell}} \right)_{\underline{m}}$$

where $\underline{k} + \underline{\ell} : i \mapsto k_i + \ell_i$ then $K[X_i | i \in I]$ becomes a commutative ring with unity. Its

- zero element is $0 := (a_{\underline{m}})_{\underline{m}}$ with $a_{\underline{m}} = 0$ for all $\underline{m} \in \mathbb{N}_0^{(I)}$,
- unit element is $1 := (a_{\underline{m}})_{\underline{m}}$ with

$$a_{\underline{m}} := \begin{cases} 1 \text{ if } \underline{m} \in \mathbb{N}_0^{(I)} \text{ is the element such that } m_i = 0 \text{ for all } i \in I, \\ 0 \text{ otherwise.} \end{cases}$$

And now?
An element $\underline{m} \in \mathbb{N}_0^{(I)}$ is therefore a mapping that assigns an index $m_i \in \mathbb{N}_0$ to each $i \in I$ (and in such a way that only finitely many of the values can be non-zero). For such an $\underline{m}$ we also define the **monomial**

$$X^{\underline{m}} := \prod_{\substack{i \in I \\ m_i \neq 0}} X_i^{m_i}$$

and this can be read as the polynomial $X^{\underline{m}} \in K[X_i | i \in I]$, namely as

$$X^{\underline{m}} : \mathbb{N}_0^{(I)} \to K \,,\, \underline{n} \mapsto \begin{cases} 1 \text{ if } \underline{n} = \underline{m}, \\ 0 \text{ otherwise.} \end{cases}$$

With this definition, we have the following way of rewriting elements $F = (a_{\underline{m}})_{\underline{m}} \in K[X_i | i \in I]$ as described above using the monomials $X^{\underline{m}}$ as

$$F(X) = \sum_{\underline{m} \text{ such that } a_{\underline{m}} \neq 0} a_{\underline{m}} X^{\underline{m}}$$

and that is indeed the form one would expect from a polynomial.

One has an (injective) ring homomorphism

$$\iota : K \to K[X_i | i \in I] \,, \; a \mapsto \left(\underline{m} \mapsto \begin{cases} a \text{ if } m_i = 0 \text{ for all } i \in I, \\ 0 \text{ otherwise.} \end{cases} \right)$$

This polynomial ring (together with ι) has the following **universal property** which describes the question of *insertion of elements in a ring R over K*. More precisely:

Lemma in Proof A.4 Let R be any commutative ring with unity which admits a ring homomorphism $\varphi : K \to R$ (one then calls R a K-algebra via φ). If elements $z_i \in R$ are given for each $i \in I$, there exists a uniquely determined ring homomorphism $\psi : K[X_i | i \in I] \to R$ such that $\psi(X_i) = z_i$ and such that the following diagram commutes:

$$K \xrightarrow{\quad \iota \quad} K[X_i | i \in I]$$
$$\varphi \searrow \quad \downarrow \psi$$
$$R$$

Proof
This is actually clear: Due to the homomorphism property ψ should have the requirement $\psi(X_i) = z_i$ and the commutativity of the diagram,

$$\psi \left(\sum_{\underline{m}} a_{\underline{m}} X^{\underline{m}} \right) = \sum_{\underline{m}} a_{\underline{m}} \prod_{i \in I, m_i \neq 0} z_i^{m_i}$$

must hold. This shows the uniqueness. But also the existence, because one can define ψ exactly in this way. $\square$

Finally, here comes the announced brilliant trick: Which index set I should we choose for our problem? After all, we want to somehow have zeros for all polynomials $f(X) \in K[X]$ with $\deg(f) \geq 1$, the latter is a set, so we take

the index set I to be the set $I := \{f \in K[X] \mid \deg(f) \geq 1\}$.

We therefore consider the polynomial ring $K[X_f | f \in I]$ in infinitely many variables X_f, namely one variable X_f for each polynomial $f \in K[X]$ of degree ≥ 1.

For a fixed element $f \in I$ one then has the inclusion

$$K[X] \hookrightarrow K[X_h | h \in I]$$
$$g(X) \mapsto g(X_f) \,,$$

that is, one reads $g(X)$ as a polynomial in the one variable X_f (associated to the previously fixed $f \in I$)—that is to say, the other variable X_h with $h \neq f$ just happen not to occur for this special polynomial.

In particular, one can choose $g(X) := f(X)$ considering this inclusion and obtains the elements $f(X_f) \in K[X_h | h \in I]$. Now let

$$\mathfrak{a} := \langle f(X_f) \mid f \in I \rangle \lhd K[X_h | h \in I]$$

be the ideal generated by these (infinitely many) elements, that is, the ideal of all finite sums

$$\mathfrak{a} = \left\{ \sum_{\nu=1}^{N} g_\nu(X_h | h \in I) \cdot f_\nu(X_{f_\nu}) \mid g_\nu \in K[X_h | h \in I] \right\}$$

of arbitrary multiples of finitely many of these $f(X_f)$ in the polynomial ring.

Claim: We have $\mathfrak{a} \subsetneq K[X_h | h \in I]$,

Proof

Otherwise, we would have $1 \in \mathfrak{a}$ so the constant 1 would have the form

$$1 = \sum_{\nu=1}^{N} g_\nu(X_h | h \in I) \cdot f_\nu(X_{f_\nu})$$

with finitely many $f_\nu \in I$ and $g_\nu \in K[X_h | h \in I]$. The finitely many f_ν however, have a splitting field $M|K$ and if we choose a root $\alpha_\nu \in M$ for each f_ν, we can use the universal property and obtain

$$K \xrightarrow{\ \iota\ } K[X_h | h \in I]$$

Inclusion $\searrow \quad \downarrow \psi$

$$M$$

with $\psi(X_{f_\nu}) = \alpha_\nu$ and $\psi(X_h) = 0$ for all other $h \notin \{f_1, \ldots, f_N\}$. Then, however, we deduce that

$$M \ni 1 = \psi(1) = \psi \left(\sum_{\nu=1}^{N} g_\nu(X_h | h \in I) \cdot f_\nu(X_{f_\nu}) \right) =$$

$$\sum_{\nu=1}^{N} \psi(g_\nu(X_h | h \in I)) \cdot \psi(f_\nu(X_{f_\nu})) =$$

$$\sum_{\nu=1}^{N} \psi(g_\nu(X_h | h \in I)) \cdot \underbrace{f_\nu(\alpha_\nu)}_{=0} = 0 ,$$

a contradiction. $\square$

Consequently, there exists a maximal ideal $\mathfrak{m} \lhd K[X_h|h \in I]$ (one could have expected something like this: Zorn's Lemma will somehow have to come into play in such a proof—this is the spot where it does!) with $\mathfrak{a} \subset \mathfrak{m}$.

We now define

$$L := K[X_h|h \in I]/\mathfrak{m}.$$

This is a field and using ι we have an inclusion $K \hookrightarrow L$, so it is a field extension of K.

Claim: Every polynomial $f(X) \in K[X]$ of degree $\deg(f) \geq 1$ has a root in L.

Proof

This is now (almost as) clear (as in the case of Theorem 10.1). For an arbitrary f we have the element

$$\alpha_f := [X_f] = X_f \bmod \mathfrak{m} \in L$$

and $f(\alpha_f) = f(X_f) \bmod \mathfrak{m} = 0 \in L$ since $f(X_f) \in \mathfrak{a} \subset \mathfrak{m}$ holds. $\square$

Claim: $L|K$ is an algebraic extension.

Proof

Let $\alpha \in L$ be an arbitrary element, thus of the form

$$\alpha = [g(X_h|h \in I)] = g(X_h|h \in I) \bmod \mathfrak{m}.$$

We write the representative g in the form

$$g(X_h|h \in I) = \sum_{\underline{m}} a_{\underline{m}} X^{\underline{m}},$$

where only a finite number of the coefficients are non-zero $a_{\underline{m}} \neq 0$. So only **finitely many** of the variables X_h in the finitely many monomials $X^{\underline{m}}$ of g really enter its definition. Let $X_{f_1}, \ldots, X_{f_N}$ be these monomials - no other monomials are summands of g . Set

$$\beta_v := [X_{f_v}] = X_{f_v} \bmod \mathfrak{m} \in L \ .$$

Then obviously

$$\alpha \in K[\beta_1, \ldots, \beta_N] \subset L.$$

Furthermore, the elements $\beta_v \in L$ are algebraic over K because $f_v(\beta_v) = 0$. Thus, we have

$$[K[\beta_1, \ldots, \beta_N] : K] < \infty$$

and hence $K[\beta_1, \ldots, \beta_N]|K$ is an algebraic field extension, in particular α algebraic over K. $\square$

Thus, Theorem A.1 is proven. $\square$

We can now prove Theorem 12.2: We consider the chain of algebraic field extensions

$$K =: L_0 \subset L_1 \subset L_2 \subset L_3 \subset \cdots$$

where each L_{j+1} is an algebraic field extension of L_j as in Theorem A.1 such that every polynomial $f(X) \in L_j[X]$ has a root in L_{j+1}.

Now define

$$\Omega := \bigcup_{j=0}^{\infty} L_j.$$

This is a field since the field axioms yield conditions for two elements $x, y \in \Omega$ (e.g. that $x + y = y + x$ applies) or three elements x, y, z (for the distributivity) or one element x (for the existence of x^{-1}, whenever $x \neq 0$) and thus they hold since then $x, y, z \in L_N$ for a large enough N.

Furthermore, $\Omega|K$ is   algebraically   closed   because   if $f(X) = a_n X^n + \ldots + a_0 \in \Omega[X]$ is a polynomial of degree n, the finitely many $a_j \in \Omega$ are already contained in some L_N. But then $f(X) \in L_N[X]$ has a root in $L_{N+1} \subset \Omega$.

The extension $\Omega|K$ is algebraic, because every $\alpha \in \Omega$ already lies inside some L_N and $L_N|K$ is algebraic (after successive application of Theorem 2.24 (finitely often)).

With this, Theorem 12.2 is finally proven. $\square$

Tricks and Methods to Classify Groups of a Given Order

And now?
We want to collect a few results that are helpful in conjunction with the Sylow theorems when one wants to work on groups with predetermined orders.

B.1 Standard Arguments and Examples

We list some standard arguments, the first ones are just repetitions:

B.1.1 Number of p-Sylow groups:

If $\#G = p^n \cdot m$ for a prime number p with $p \nmid m$, then the number s_p of p-Sylow groups satisfies

$$s_p \equiv 1 \bmod p \text{ und } s_p \mid m$$

and this usually allows only a few possibilities.

Example If $\#G = pq$ with two prime numbers p, q and $q < p$. Then $s_p \in \{1, p+1, 2p+1, \ldots\}$ and also $s_p | q$. This is only possible for $s_p = 1$. Similarly, $s_q \in \{1, q+1, 2q+1, \ldots\}$ and $s_q \mid p$, so $s_q \in \{1, p\}$. If e.g. we know by some assumption that $q \nmid p - 1$, then necessarily $s_q = 1$ must hold.

Explicit Example $\#G = 77 = 7 \cdot 11$. Then $s_{11} = 1$ and since $7 \nmid 10$ also $s_7 = 1$ holds.

B.1.2 Sylow Groups as Normal Subgroups

A p-Sylow group $H \subset G$ is a normal subgroup if and only if $s_p = 1$.

B.1.3 Sylow Groups of order p

If $\#G = p \cdot m$ with $p \nmid m$, then every p-Sylow group H has the order p and is therefore cyclic, so

$H \cong \mathbb{Z}/p\mathbb{Z}$ (note that the group on the right hand side is written additively).

Note Note that H then is abelian in particular. However, this does not yet say anything about whether $H \subset G$ is a normal subgroup or not.

Explicit Example In the above example, $\#G = 77$ let $P_7 \subset G$ be the only 7-Sylow group and $P_{11} \subset G$ the only 11-Sylow group. Then

$$P_7 \cong \mathbb{Z}/7\mathbb{Z} \text{ and } P_{11} \cong \mathbb{Z}/11\mathbb{Z}.$$

Furthermore, both are normal subgroups in G (because they are the only Sylow groups).

B.1.4 Intersection of Sylow Groups

Note the following:

1. If $P \subset G$ is a p-Sylow group and $Q \subset G$ a q-Sylow group for two different prime divisors $p \neq q$ of $\#G$, then $P \cap Q = \{1\}$, because
   - every element $g \in P$ has an order $\operatorname{ord}(g) = p^s$ for some $s \in \mathbb{N}_0$,
   - every element $g \in Q$ has an order $\operatorname{ord}(g) = q^r$ for some $r \in \mathbb{N}_0$,
   and to fulfill both at the same time is only possible if $s = r = 0$.
2. If P and P' are two p-Sylow groups of G with $\#G = p^n \cdot m$ for the prime number p, then $P \cap P'$ is either trivial, i.e., $\{1\}$, or a proper p-subgroup of order p^s for $s \leq n$. Both cases can occur!!!
   There is, however, a special case: If $\#G = p \cdot m$ with $p \nmid m$ (i.e., the prime number p only appears in power 1), then for two different p-Sylow groups P and P', the strong statement $P \cap P' = \{1\}$ holds (since then every element $g \in P$ other than the neutral one 1 is already a generator of the cyclic group P of prime order.

B.1.5 Commuting Sylow Groups I

Let $p \neq q$ be two prime divisors of $\#G$ and assume we know that $s_p = 1$ and $s_q = 1$ holds in this group, that is, the respective Sylow groups $S_p \triangleleft G$ and $S_q \triangleleft G$ are normal subgroups. Then S_p and S_q **commute** in G in the sense that

$$a \in S_p, b \in S_q \implies ab = ba \in G.$$

Proof

What needs to be shown is that $aba^{-1}b^{-1} = 1$ holds. But

$$aba^{-1}b^{-1} = a \cdot \underbrace{\underbrace{ba^{-1}b^{-1}}_{\in S_p \triangleleft G}}_{\in S_p} = \underbrace{\underbrace{(aba^{-1})}_{\in S_q \triangleleft G} \cdot b^{-1}}_{\in S_q} \in S_p \cap S_q = \{1\}.$$

In this case, the map

$$\iota : S_p \times S_q \hookrightarrow G, \ (a, b) \mapsto ab \tag{B.1}$$

is an injective group homomorphism. Note that the homomorphism property follows because the two subgroups commute inside G.

Furthermore, the image $S_p \cdot S_q := \iota(S_p \times S_q)$ is again a normal subgroup in G, because for $g \in G$ and $a \in S_p$, $b \in S_q$ we obtain:

$$gabg^{-1} = gag^{-1}gbg^{-1} \in S_p \cdot S_q.$$

Explicit Example Any group G with $\#G = 77$ has the two commuting normal subgroups S_7 and S_{11}, hence

$$S_7 \times S_{11} \to G, \ (a, b) \mapsto ab$$

is an injective group homomorphism, which due to $\#(S_7 \times S_{11}) = 77 = \#G$ is also surjective thus a group isomorphism. Furthermore, due to argument B.1.3, it follows that $S_7 \cong \mathbb{Z}/7\mathbb{Z}$ and $S_{11} \cong \mathbb{Z}/11\mathbb{Z}$. We have thus shown:

Claim B.1 There is exactly one group of order 77 up to isomorphism, namely the cyclic group $\mathbb{Z}/7\mathbb{Z} \times \mathbb{Z}/11\mathbb{Z} \cong \mathbb{Z}/77\mathbb{Z}$.

Note: The latter isomorphism follows from the Chinese Remainder Theorem. A posteriori, every group G with 77 elements is cyclic. This was not known before applying all these arguments.

B.1.6 Commuting Sylow Groups II

Let $\#G = p^r \cdot q^s \cdot m$. As we have just seen, it is advantageous if two Sylow groups S_p and S_q in G commute, because then one has (B.1). In the argument B.1.5, the assumption was that $s_p = 1 = s_q$ should hold. If one only knows $s_p = 1$, one can consider the following considerations:

Because of $s_p = 1$, $S_p \triangleleft G$ is a normal subgroup and thus G as well as any subgroup of G acts on S_p by conjugation. In particular, this applies to the subgroup $S_q \subset G$ and the action induces the group homomorphism:

$$\varphi : S_q \to \text{Aut}(S_p), \ a \mapsto \text{conj}_a. \tag{B.2}$$

Now one can ask what kind of group homomorphisms of this kind can exist in this given situation a priori. To this end, one observes that

- for each $a \in S_q$ we have $\text{ord}(\varphi(a)) \mid \text{ord}(a) = q^k$ for some $k \leq s$,
- for each $a \in S_q$ we also have $\text{ord}(\varphi(a)) \mid \#\text{Aut}(S_p)$.

Check it out!
Justify the first statement!

In some cases, important conclusions can be drawn from this: If, for example, one knows that $q \nmid \#\text{Aut}(S_p)$, then each $a \in S_q$ must therefore be mapped to $\varphi(a) = 1$, thus $\varphi = \text{const}_1$. However, this means that $\text{conj}_a(b) = b$ for all $a \in S_q$ and all $b \in S_p$ and thus $ab = ba$. Then one has again

$$S_q \times S_p \hookrightarrow G, \ (a, b) \to ab$$

as an injective group homomorphism.

Note that one knows $\text{Aut}(S_p)$ well in the special case when $\#G = p \cdot q^s \cdot m$ (so p only appears in power 1). Then $S_p \cong \mathbb{Z}/p\mathbb{Z}$ is cyclic of prime number order, let us write $S_p = \langle \gamma \rangle$ by choosing a generator $\gamma \in S_p$ (any $\gamma \in S_p \setminus \{1\}$ will do) and one has an isomorphism

$$\text{Aut}(S_p) = \text{Aut}(\langle \gamma \rangle) \xleftarrow{\ \cong\ } (\mathbb{Z}/p\mathbb{Z})^\times$$
$$(\gamma \mapsto \gamma^a) \xleftarrow{\hspace{2cm}} [a] \tag{B.3}$$

where on the right hand side we find the unit group of the prime field $(\mathbb{Z}/p\mathbb{Z})^\times = \mathbb{F}_p^\times$. The latter is cyclic (theorem 15.9) of order $p - 1$.

So if by chance $q \nmid (p - 1)$ applies, then we know that $\varphi(a) = 1$ must hold for all $a \in S_q$, so φ has to be trivial. Then S_p and S_q commute in G.

Explicit Example Let $\#G = 2 \cdot 7 \cdot 13 = 182$. Then

$$s_7 \equiv 1 \bmod 7 \text{ und } s_7 \mid 2 \cdot 13 = 26 \Longrightarrow s_7 = 1$$
$$s_{13} \equiv 1 \bmod 13 \text{ und } s_{13} \mid 2 \cdot 7 = 14 \Longrightarrow s_{13} \in \{1, 14\}.$$

Hence there is exactly one 7-Sylow group S_7 and this is a normal subgroup. Let S_{13} be a (we do not know right now if it is unique) 13-Sylow group. We then have the group action by conjugation of S_{13} on S_7, and since S_7 is cyclic of order 7, we obtain:

$$\varphi : S_{13} \to \text{Aut}(S_7) \cong \mathbb{Z}/6\mathbb{Z}.$$

This group homomorphism must be trivial (because every element $a \in S_{13} \setminus \{1\}$ has order $13 \nmid 6$). Therefore, S_7 and S_{13} commute and we have an injective group homomorphism

$$\mathbb{Z}/91\mathbb{Z} \cong \mathbb{Z}/7\mathbb{Z} \times \mathbb{Z}/13\mathbb{Z} \cong S_7 \times S_{13} \hookrightarrow G$$
$$(a, b) \mapsto ab.$$

B.1.7 Non-commuting Sylow Groups

Let's consider the following situation: $\#G = p \cdot q^s$ for two prime numbers $p \neq q$. If then $p \nmid q^\ell - 1$ holds for all $\ell = 1, \ldots, s$, it follows that $s_p = 1$ and G has a single p-Sylow group $S_p \triangleleft G$.

If $q \nmid p - 1$ also holds, then as above, it follows that $S_p \times S_q \cong G$, $(a, b) \mapsto ab$, because the two Sylow groups in G commute.

However, if $q \mid p - 1$ applies, one cannot conclude this. Nevertheless, the consideration of (B.3) is then helpful. So we have the action (B.2) of (a q-Sylow group) S_q on the (only) p-Sylow group S_p i.e.

$$\varphi : S_q \to \mathrm{Aut}(S_p), \ a \mapsto \mathrm{conj}_a.$$

Now let's take a closer look at (B.3):

$$\mathrm{Aut}(S_p) = \mathrm{Aut}(\langle \gamma \rangle) \xleftarrow{\ \cong\ } (\mathbb{Z}/p\mathbb{Z})^\times \xleftarrow{\ \psi\ }_{\cong} \mathbb{Z}/(p-1)\mathbb{Z}$$

$$(\gamma \mapsto \gamma^k) \longleftarrow\!\!\dashv [k]$$

$$\qquad\qquad [\kappa^\ell] \longleftarrow\!\!\dashv \ell \qquad\qquad\qquad\text{(B.4)}$$

$$(\gamma \mapsto \gamma^{\kappa^\ell}) \longleftarrow\!\!\dashv \ell$$

when we let $[\kappa] \in (\mathbb{Z}/p\mathbb{Z})^\times$ be a chosen **generator of this cyclic group**—thus an element, which as in remark 16.11 one usually calls a **primitive element in $\mathbb{F}_p$**.

These isomorphisms are thus explicitly given (after the corresponding choices of generators). If one combines (B.2) and (B.3), one thus has

$$S_q \xrightarrow{\ \varphi\ } \mathrm{Aut}(S_p) = \mathrm{Aut}(\langle \gamma \rangle) \xleftarrow[\cong]{(\mathrm{B.3})} \mathbb{Z}/(p-1)\mathbb{Z}$$

$$(\gamma \mapsto \gamma^{\kappa^\ell}) \longleftarrow\!\!\dashv \ell$$

If one knows the group structure of S_q (as in the explicit example below), one can determine **all possible** group homomorphisms φ. Let's say, the set of possible homomorphisms is

$$\mathcal{M} := \{\varphi : S_q \to \mathrm{Aut}(S_p) \cong \mathbb{Z}/(p-1)\mathbb{Z} \mid \text{group homomorphism}\}.$$

Then one has:

- For a given group G with $\#G = pq^s$ and $s_p = 1$, the procedure above yields exactly one of the possible homomorphisms $\varphi \in \mathcal{M}$ (the procedure above refers to the conjugation action of the group S_q on the set S_p).
- Conversely, for a given $\varphi \in \mathcal{M}$ one can construct a group such that this φ appears as this conjugation action—see the next Defining Lemma and Lemma B.4.

Defining Lemma B.2 Let N, S be groups and let

$$\varphi : S \to \mathrm{Aut}(N)$$

be an action of S on N. We define a multiplication on the product $N \times S$ of the two sets by

$$(n_1, s_1) \cdot (n_2, s_2) := (n_1 \cdot \varphi(s_1)(n_2), s_1 \cdot s_2)$$

and obtain a group structure (with neutral element $(1, 1)$). The set $N \times S$ together with this group structure is called **the semidirect product** of N and S **with respect to** φ and is denoted as

$$S_\varphi \ltimes N \text{ oder } N \rtimes_\varphi S$$

Inside the semidirect product we can find the two groups S and N we started with as the subgroups:

$$S \hookrightarrow N \rtimes_\varphi S , \ s \mapsto (1, s)$$
$$N \hookrightarrow N \rtimes_\varphi S , \ n \mapsto (n, 1).$$

Then $N \triangleleft N \rtimes_\varphi S$ is a normal subgroup.

Note B.3

1. If the action φ is clear from the context, it is often omitted in the notation i.e. one only writes $N \rtimes S$. But: The group structure on $N \rtimes S$ depends on the choice of the action φ.
2. The notation $N \rtimes S$ is not symmetric, as in general S (in contrast to N) is not a normal subgroup in $N \rtimes S$. The direction of the symbol $\rtimes$ is easy to remember—it reminds us that N becomes a normal subgroup in the semidirect product:

$$N \rtimes S \Leftrightarrow N \triangleleft (N \rtimes S)$$

3. The inverse of an element $(n, s) \in N \rtimes_\varphi S$ is calculated as

$$(n, s)^{-1} = (\varphi(s^{-1})n^{-1}, s^{-1}).$$

With this, one can now construct groups of certain order and predescribed normal subgroups:

Lemma B.4 *Let p be a prime number and furthermore*

- *S_p a p-group of order p^r and*
- *U a group of order m with $p \nmid m$.*

If $\varphi : U \to \mathrm{Aut}(S_p)$ is an action of U on S_p, the semidirect product

$$G_\varphi := S_p \rtimes_\varphi U$$

is a group of order $n = p^r \cdot m$ *inside of which* $S_p \lhd G_\varphi$ *is the unique p-Sylow group of* G_φ. *The conjugation action of* U *(read as a subgroup of* G_φ) S_p *coincides with the given action* φ.

Proof

By construction, G_φ is a group and has $S_p \hookrightarrow G_\varphi$, $n \mapsto (n, 1)$ as a p-Sylow group, which is a normal subgroup, thus the only p-Sylow group. It remains to show that the conjugation of the subgroup $U \subset G_\varphi$ on S_p corresponds to the action φ, which is shown by

$$(1, s) \cdot (n, 1) \cdot (1, s)^{-1} = (\varphi(s)n, s) \cdot (\varphi(s^{-1})(1), s^{-1}) = (\varphi(s)n, 1)$$

□

Conversely, one has (formulated here in a more specific situation):

Lemma B.5 *Let G be a finite group with $\#G = p^r q^s$ for prime numbers $p \neq q$ and let's assume that G has only one p -Sylow group $S_p \lhd G$. If then S_q is a chosen q -Sylow group and*

$$\varphi : S_q \to \mathrm{Aut}(S_p) \,, \ s \mapsto (\mathrm{conj}_s : t \mapsto sts^{-1})$$

denotes the conjugation operation, then $G \cong S_p \rtimes_\varphi S_q$ holds.

Proof
The map

$$\psi : S_p \rtimes_\varphi S_q \to G \,, \ (n, s) \mapsto n \cdot s$$

is a group isomorphism—the essential property to prove is that this is a **group homomorphism**: In G we obtain for any $n, n' \in S_p$ and $s, s' \in S_q$, that

$$(ns) \cdot (n's') = n \cdot sn's^{-1} \cdot ss' = n \cdot \mathrm{conj}_s(n') \cdot ss' = (n \cdot \varphi(s)n') \cdot ss'.$$

From this we deduce that $\psi((n, s) \cdot (n', s')) = \psi(n, s) \cdot \psi(n', s')$.

Obviously, ψ is injective, since $1 = \psi(n, s) = ns$ can only hold for $n = 1 = s$ because of $S_p \cap S_q = \{1\}$. Since the two groups have the same number of elements, ψ is thus an isomorphism. □

Check it out!
Generalize Lemma B.5 to the situation where $\#G = mq^s$ with $q \nmid m$ and instead of considering a p-Sylow group, assume that G has a normal subgroup $N \lhd G$ with $\#N = m$: Then $G \cong N \rtimes_\varphi S_q$ also holds for suitable φ and the proof proceeds almost verbatim as above, if one replaces S_p with N.

And now?

One could try, in special cases of a given group order $\#G = n$, to determine all these groups up to isomorphism by the two above lemmas.

Let's say $\#G = p \cdot q^s$ and $p \nmid q^k - 1$ for $k = 1, \ldots s$. Then, due to the 3rd Sylow theorem, we know that $s_p = 1$ and thus G has a unique p-Sylow group $S_p \lhd G$. In addition, S_p is cyclic and we choose a generator $\gamma \in S_p$. Thus, Lemma B.5 can be applied to each of these groups and we must

1. find all possible q-groups of order q^s—then let S_q be any such,
2. find all possible operations

$$S_q \xrightarrow{\;\varphi\;} \operatorname{Aut}(S_p) = \operatorname{Aut}(\langle \gamma \rangle) \xleftarrow[\cong]{(B.3)} \mathbb{Z}/(p-1)\mathbb{Z}$$

$$(\gamma \mapsto \gamma^{\kappa^\ell}) \longleftarrow\!\!\!\mid \ell$$

—then $G := S_p \rtimes_\varphi S_q$ is such a group, and according to Lemma B.5 **every** group of order pq^s is then of this type.
3. We are not done yet: It may happen that several of the groups constructed above are isomorphic to each other. So we need to find all isomorphic groups from the list of groups we constructed and remove all but one of them from the list.

In general, all steps are quite difficult and often very complex. For example, the question of when two semidirect products $S_p \rtimes_\varphi S_q$ and $S_p \rtimes_\psi S_q$ are isomorphic to each other (important for step (3)), is very difficult for general groups—here one can understand this because $\#S_p$ and $\#S_q$ are coprime, but this is also another lemma that one must first prove in order to have it available.

It is almost impossible to provide general recipes on a few pages that would be applicable to many classes of examples. Therefore, I will not continue to go for this now but i will provide some few examples.

Let us note an explicit example that the question of the number of isomorphism classes of groups of a certain order can involve a lot of work. Considering the group order $2304 = 2^8 \cdot 3^2$ (where one might think that this could perhaps be carried out with not too much work with the above methods), there are exactly $15\,641\,993$ isomorphism types—see Eick, B., Horn, M. "The construction of finite solvable groups revisited." *Journal of Algebra* 408 (2014): 166–182.

A Simple Example Consider $\#G = 6 = 2 \cdot 3$, then obviously $s_3 = 1$. Let $S_3 \lhd G$ be the only 3-Sylow group. For the 2-Sylow groups we have $s_2 \equiv 1 \bmod 2$ and $s_2 | 3$, so we only know that $s_2 \in \{1, 3\}$. Is $s_2 = 1$, then has G has exactly one 2-Sylow group and as in B.1.5 it follows that $G \cong S_2 \times S_3 \cong \mathbb{Z}/6\mathbb{Z}$ is the cyclic group with 6 elements. If $s_2 = 3$ and S_2 is one of the three 2-Sylow groups, then G arises from $S_2 \cong \mathbb{Z}/2\mathbb{Z}$ and $S_3 \cong \mathbb{Z}/3\mathbb{Z}$ as a semidirect product by virtue of the only non-trivial action

$$\varphi : \mathbb{Z}/2\mathbb{Z} \to \mathrm{Aut}(\mathbb{Z}/3\mathbb{Z}) , \ 1 \mapsto (x \mapsto 2x).$$

Hence $G \cong \mathbb{Z}/3\mathbb{Z} \rtimes_\varphi \mathbb{Z}/2\mathbb{Z}$ with the group multiplication

$$(a,b) \cdot (c,d) = (a + \varphi(b)(c), b + d) = (a + 2^b \cdot c, b + d).$$

> **Check it out!**
> We have learned about the dihedral groups in Definition 17.8. In particular, we have the dihedral group D_6 with 6 elements. Convince yourself that $\mathbb{Z}/3\mathbb{Z} \rtimes_\varphi \mathbb{Z}/2\mathbb{Z} \cong D_6$, $(a,b) \mapsto \rho^a \cdot \sigma^b$ (in the notation from the definition of D_6) is a group isomorphism.
> Since also $\#S(3) = 3! = 6$ and the latter is not abelian, $S(3) \cong D_6$ must hold. Give an isomorphism!

B.2 Explicit Examples

B.2.1 Groups of Order 2021

The number $2021 = 43 \cdot 47$ is the product of two prime numbers. For the number of Sylow groups, the following holds:

$$s_{43} \equiv 1 \bmod 43 \text{ and } s_{43} \mid 47 \implies s_{43} = 1$$
$$s_{47} \equiv 1 \bmod 47 \text{ and } s_{47} \mid 43 \implies s_{47} = 1,$$

so there is exactly one Sylow group S_{43} and S_{47} each and the map

$$S_{43} \times S_{47} \to G , \ (a,b) \mapsto ab$$

is a group isomorphism, cp. withargument B.1.5.

Since both groups have a prime number order they are each cyclic and thus $S_{43} \cong \mathbb{Z}/43\mathbb{Z}$ and $S_{47} \cong \mathbb{Z}/47\mathbb{Z}$. Therefore, up to isomorphism, there is exactly one group of order 2021, namely the cyclic group

$$G \cong \mathbb{Z}/43\mathbb{Z} \times \mathbb{Z}/47\mathbb{Z} \cong \mathbb{Z}/2021\mathbb{Z}.$$

B.2.2 Groups of Order 2023

We have $2023 = 7 \cdot 17^2$. For the number of Sylow groups, the following holds:

$$s_7 \equiv 1 \bmod 7 \text{ and } s_7 \mid 17^2 \implies s_7 = 1$$
$$s_{17} \equiv 1 \bmod 17 \text{ and } s_{17} \mid 7 \implies s_{17} = 1,$$

so here too, there is exactly one Sylow group S_7 and S_{17} each and the map

$$S_7 \times S_{17} \to G , \ (a,b) \mapsto ab$$

is a group isomorphism. Again, $S_7 \cong \mathbb{Z}/7\mathbb{Z}$ is true, because $\#S_7 = 7$ is a prime number, but

$$\#S_{17} = 17^2$$

is not a prime number. So we still need to understand which groups of order 17^2 exist.

Lemma B.6 *If p is a prime number and G is a group of order p^2, then G is abelian and*

$$\text{either } G \cong \mathbb{Z}/p^2\mathbb{Z} \text{ or } G \cong \mathbb{Z}/p\mathbb{Z} \times \mathbb{Z}/p\mathbb{Z}$$

Proof

Every element $g \in G$ with $g \neq 1$ satisfies $\text{ord}(g) \in \{p, p^2\}$. If an element of order $p^2 = \#G$ exists, G is cyclic with this element as a generator.

So let's now assume that all $1 \neq g \in G$ have order $\text{ord}(g) = p$. Let $1 \neq g$ be such an element.

Claim B.7 The cyclic subgroup $\langle g \rangle \subset G$ is a normal subgroup. $\square$

Proof

The set (a priori still without group structure) $G/\langle g \rangle$ has p elements. Let $h \in G \setminus \langle g \rangle$, then $h[g] \neq [g] \in G/\langle g \rangle$. The elements are $g^i h^j$ are pairwise different for all $i, j \in \{0, \dots, p-1\}$ since

$$g^i h^j = g^r h^s \Longleftrightarrow g^{i-r} = h^{s-j}$$

for $i - r, s - j \in \{-(p-1), \dots, p-1\}$. Now if $s - j \neq 0$, then $s - j$ is coprime to p is and there are $a, b \in \mathbb{Z}$ with $a(s-j) + bp = 1$, hence

$$(h^{s-j})^a = h^{1-bp} = h$$

(here it is used that $\text{ord}(h) = p$) and thus also

$$h = (h^{s-j})^a = g^{(i-r)a} \in \langle g \rangle$$

in contradiction to the choice of h. Therefore, we have that

$$G = \{g^i h^j \mid i, j \in \{0, \dots, p-1\}\}.$$

If now $\langle g \rangle$ were not a normal subgroup, there would exist a $\gamma \in G$ with $h = \gamma g \gamma^{-1} \notin \langle g \rangle$. According to the above, every element of G has the form $g^i h^j$ for some $i, j \in \{0, \dots, p-1\}$. In particular, the element γ^{-1} is of this form

$$\gamma^{-1} = g^i h^j$$

and thus it follows that $h = \gamma g \gamma^{-1} = h^{-j} g^{-i} g g^i h^j \Rightarrow h = g$, a contradiction. $\square$

So, $\langle g \rangle \lhd G$ is a normal subgroup and we again choose an element $h \in G \setminus \langle g \rangle$. The subgroup $\langle h \rangle$ is itself cyclic of order p, so it is isomorphic to $\mathbb{Z}/p\mathbb{Z}$. It acts on $\langle g \rangle$ by conjugation:

$$\langle h \rangle \rightarrow \mathrm{Aut}(\langle g \rangle)\,,\ x \mapsto \mathrm{conj}_x.$$

Since $\langle h \rangle \cong \mathbb{Z}/p\mathbb{Z}$ and $\mathrm{Aut}(\langle g \rangle) \cong \mathrm{Aut}(\mathbb{Z}/p\mathbb{Z}) \cong \mathbb{Z}/(p-1)\mathbb{Z}$ hold (we have already considered this in (B.4)), this action must be trivial, so $\mathrm{conj}_x = \mathrm{id}_{\langle g \rangle}$, and thus as in the argument B.1.6 above, the map

$$\langle g \rangle \times \langle h \rangle \rightarrow G\,,\ (a, b) \mapsto ab$$

is a group isomorphism. Thus, $G \cong \mathbb{Z}/p\mathbb{Z} \times \mathbb{Z}/p\mathbb{Z}$ is proven. $\square$

We deduce that there are exactly **two** isomorphism types of groups of order 2023, namely $\mathbb{Z}/7 \times \mathbb{Z}/17^2\mathbb{Z} \cong \mathbb{Z}/2023\mathbb{Z}$ and $\mathbb{Z}/7\mathbb{Z} \times \mathbb{Z}/17\mathbb{Z} \times \mathbb{Z}/17\mathbb{Z}$.

B.2.3 Groups of Order 2022

And now?
This problem turns out to be relatively complex and also involves some rather *special* tricks, which in a certain way work *by chance* here. This is a common feature when trying to classify groups of a small order: It is often the case that for each order, different tricks and considerations must be taken into account which perhaps are of no use in other cases. I have written down the following arguments more for the sake of completeness: Whoever considers the cases 2021 and 2023, should also comment on 2022!

We have $2022 = 2 \cdot 3 \cdot 337$ and hence

$$s_2 \equiv 1 \bmod 2 \text{ and } s_2 \mid 3 \cdot 337 \Longrightarrow s_2 \in \{1, 3, 337, 3 \cdot 337\}$$
$$s_3 \equiv 1 \bmod 3 \text{ and } s_3 \mid 2 \cdot 337 \Longrightarrow s_3 \in \{1, 337\}$$
$$s_{337} \equiv 1 \bmod 337 \text{ and } s_{337} \mid 2 \cdot 3 \Longrightarrow s_{337} = 1.$$

At least we know that G has a unique 337-Sylow group $S_{337} \lhd G$. Furthermore, all Sylow groups are cyclic of the corresponding prime order.

So, there is a surjective group homomorphism

$$\pi : G \rightarrow G/S_{337}$$

into a group G/S_{337} of 6 elements. Of these, there are two isomorphism classes, the cyclic group $\mathbb{Z}/6\mathbb{Z}$ or the dihedral group D_6—see the **Check it out!** in Sect. B.1. In both cases, the group G/S_{337} has a subgroup (even a normal subgroup) of order 3, so a cyclic subgroup of order 3. Let's call this $V \subset G/S_{337}$, then

$$U := \pi^{-1}(V) \subset G$$

is a subgroup of G with $\#U = 337 \cdot 3$ elements and thus of index $(G : U) = 2$. It is even a normal subgroup, because it is the preimage of a normal subgroup. However, one could also use the following lemma, which might also be interesting otherwise.

Lemma B.8 *If G is a finite group and $U \subset G$ is a subgroup of index 2, then U is a normal subgroup in G.*

Proof

By assumption, there exists an $a \in G$ with $G/U = \{[1], [a]\} = \{U, aU\}$. In particular, $G \setminus U = aU$. Then for $u \in U$ we have $aua^{-1} \in U$ because otherwise $aua^{-1} \in aU$ thus $ua^{-1} \in U$ and therefore $a^{-1} \in U$, in contradiction to $a \notin U$.

If now $g \in G$ is arbitrary, then $g \in U$ or $g \in aU$. In the first case $(g \in U)$ $gUg^{-1} = U$ is trivial. In the second case, $g = av$ holds for some $v \in U$ and thus we obtain for aribtrary $u \in U$:

$$gug^{-1} = avuv^{-1}a^{-1} \in U$$

as seen above. Therefore, U is a normal subgroup. $\square$

We have seen that G admits a normal subgroup $U \triangleleft G$ of order $3 \cdot 337 = 1011$. Then we have the conjugation action

$$\varphi : \mathbb{Z}/2\mathbb{Z} \cong S_2 \to \mathrm{Aut}(U) \tag{B.5}$$

and G is thus isomorphic to $U \rtimes_\varphi \mathbb{Z}/2\mathbb{Z}$. If we consider the variant of the lemma B.5 from the subsequent ***Check it out!*** , we see that we can construct all such groups G as follows: If U is any group of order $3 \cdot 337$ and $\varphi : \mathbb{Z}/2\mathbb{Z} \to \mathrm{Aut}(U)$ an arbitrary action, then $U \rtimes_\varphi \mathbb{Z}/2\mathbb{Z}$ is a group with 2022 elements. So we need to investigate:

1. What kind of groups U of order $3 \cdot 337 = 1011$ are there?
2. Which group actions $\varphi : \mathbb{Z}/2\mathbb{Z} \to \mathrm{Aut}(U)$ exist for all these groups?

to (1): For the number of Sylow groups in a group U with $3 \cdot 337$ we have

$$s_3 \equiv 1 \bmod 3 \text{ and } s_3 \mid 337 \Longrightarrow s_3 \in \{1, 337\}$$
$$s_{337} \equiv 1 \bmod 337 \text{ and } s_{337} \mid 3 \Longrightarrow s_{337} = 1.$$

Hence U has a uniquely determined 337-Sylow group $S_{337} \triangleleft U$ (if we consider $U \subset G$ as a subgroup of a group G as above, then $S_{337} \subset U \subset G$ will also be the 337-Sylow group in G). Is $S_3 \subset U$ denotes a 3-Sylow group, we again have the conjugation action (conjugation now to be read inside the group U):

$$\psi : \mathbb{Z}/3\mathbb{Z} \cong S_3 \to \mathrm{Aut}(S_{337}) \cong (\mathbb{Z}/337\mathbb{Z})^\times \cong \mathbb{Z}/336\mathbb{Z}.$$

For this, there are three possibilities, because $\mathbb{Z}/336\mathbb{Z}$ has exactly two elements of order 3 (namely 112, 224) and the neutral element of order 1. Thus, $\mathrm{Aut}(S_{337}) \cong \mathrm{Aut}(\mathbb{Z}/337\mathbb{Z})$ also has exactly two elements of order three (each the square of the other), let's say κ, κ^2 – if you are curious which element this is, one can determine κ: The class $[10] \in (\mathbb{Z}/337\mathbb{Z})^\times$ is a generator and thus one can take $\kappa = [10^{112}]$.

Hence, in order to give such a ψ, there are three possibilities (where $\kappa \in \mathrm{Aut}(\mathbb{Z}/337\mathbb{Z})$ is to be read as the map *multiplication with κ*):

$$\psi_0 : \mathbb{Z}/3\mathbb{Z} \to \mathrm{Aut}(\mathbb{Z}/337\mathbb{Z}) , \ 1 \mapsto \mathrm{id}$$
$$\psi_1 : \mathbb{Z}/3\mathbb{Z} \to \mathrm{Aut}(\mathbb{Z}/337\mathbb{Z}) , \ 1 \mapsto \kappa$$
$$\psi_2 : \mathbb{Z}/3\mathbb{Z} \to \mathrm{Aut}(\mathbb{Z}/337\mathbb{Z}) , \ 1 \mapsto \kappa^2 .$$

Therefore every group U of order 1011 is isomorphic to one of the groups

$$U_0 = \mathbb{Z}/337\mathbb{Z} \rtimes_{\psi_0} \mathbb{Z}/3\mathbb{Z} = \mathbb{Z}/337\mathbb{Z} \times \mathbb{Z}/3\mathbb{Z} \cong \mathbb{Z}/(3 \cdot 337)\mathbb{Z}$$
$$U_1 = \mathbb{Z}/337\mathbb{Z} \rtimes_{\psi_1} \mathbb{Z}/3\mathbb{Z}$$
$$U_2 = \mathbb{Z}/337\mathbb{Z} \rtimes_{\psi_2} \mathbb{Z}/3\mathbb{Z} .$$

These are, however, **not pairwise non-isomorphic:** Obviously, $U_0 \not\cong U_1$ and $U_0 \not\cong U_2$, because U_0 is abelian and U_1, U_2 are not. But we have (I write $*_i$ for the multiplication in U_i):

$$\text{multiplication in } U_1 : (x,y) *_1 (x',y') = (x + \kappa^y x', y + y')$$
$$\text{multiplication in } U_2 : (x,y) *_2 (x',y') = (x + \kappa^{2y} x', y + y') \qquad \text{(B.6)}$$

and it is easy to see that $U_1 \overset{\cong}{\to} U_2$, $(x,y) \mapsto (x, 2y)$ is a group isomorphism.

to (2): Now the task is to determine all the possible homomorphisms φ as in (B.5) individually for U_0 and U_1 (U_2 is isomorphic to U_1, so it does not yield any new groups).

In the case of U_0:
We know that $U_0 \cong \mathbb{Z}/(3 \cdot 337)\mathbb{Z} = \mathbb{Z}/1011\mathbb{Z}$ and one has

$$\mathrm{Aut}(U_0) \cong \mathrm{Aut}(\mathbb{Z}/337\mathbb{Z} \times \mathbb{Z}/3\mathbb{Z}) \cong (\mathbb{Z}/337\mathbb{Z})^\times \times (\mathbb{Z}/3\mathbb{Z})^\times \cong \mathbb{Z}/336\mathbb{Z} \times \mathbb{Z}/2\mathbb{Z} .$$

The group on the right hand side has exactly **three** elements of order 2, namely $(0, 1)$, $(168, 0)$, $(168, 1)$, hence $\mathrm{Aut}(U_0)$ also has exactly three elements of order 2, together with the identity exactly four elements of order ≤ 2. This can be simply written down:

$$U_0 = \mathbb{Z}/337\mathbb{Z} \times \mathbb{Z}/3\mathbb{Z} \to U_0 = \mathbb{Z}/337\mathbb{Z} \times \mathbb{Z}/3\mathbb{Z} , \ (x,y) \mapsto (\pm x, \pm y) .$$

We obtain four possible constructions:

$$\varphi_0 : \mathbb{Z}/2\mathbb{Z} \to \mathrm{Aut}(U_0), 1 \mapsto \big((x,y) \mapsto (x,y)\big)$$
$$\varphi_1 : \mathbb{Z}/2\mathbb{Z} \to \mathrm{Aut}(U_0), 1 \mapsto \big((x,y) \mapsto (-x,y)\big)$$
$$\varphi_2 : \mathbb{Z}/2\mathbb{Z} \to \mathrm{Aut}(U_0), 1 \mapsto \big((x,y) \mapsto (x,-y)\big) \qquad \text{(B.7)}$$
$$\varphi_3 : \mathbb{Z}/2\mathbb{Z} \to \mathrm{Aut}(U_0), 1 \mapsto \big((x,y) \mapsto (-x,-y)\big)$$

Thus we obtain (if we take U_0 to be the corresponding subgroup) **four possible constructions:**

$$G_{0,i} := U_0 \rtimes_{\varphi_i} \mathbb{Z}/2\mathbb{Z} \text{ for } i = 0, 1, 2, 3 .$$

We want to confirm that these are not isomorphic to each other: Note that the subgroup

$$\{((0,0),0), ((0,0),1)\} \subset G_{0,i}$$

in each case is a 2-Sylow group in $G_{0,i}$. Let's determine which subgroups arise from it by conjugation with an element (because then we see all 2-Sylow groups each in $G_{0,i}$). For $((x,y),a) \in G_{0,i}$ we have (let's write again $*_i$ for the multiplication in $G_{0,i}$):

$$((x,y),a) *_i ((0,0),1) *_i ((x,y),a)^{-1} = \begin{cases} ((0,0),1) & \text{for } i = 0 \\ ((2x,0),1) & \text{for } i = 1 \\ ((0,2y),1) & \text{for } i = 2 \\ ((2x,2y),1) & \text{for } i = 3. \end{cases}$$

Hence we understand the 2-Sylow groups in side each of the groups $G_{0,i}$, what they each look like and how many there are:

$i = 0$: $\{((0,0),0), ((0,0),1)\}$ $\qquad\qquad\qquad\qquad \Rightarrow s_2 = 1$
$i = 1$: $\{((0,0),0), ((2x,0),1)\}$ for $x \in \mathbb{Z}/337\mathbb{Z}$ $\qquad\quad \Rightarrow s_2 = 337$
$i = 2$: $\{((0,0),0), ((0,2y),1)\}$ for $y \in \mathbb{Z}/3\mathbb{Z}$ $\qquad\quad\ \Rightarrow s_2 = 3$
$i = 3$: $\{((0,0),0), ((2x,2y),1)\}$ for $x \in \mathbb{Z}/337\mathbb{Z}$, $y \in \mathbb{Z}/3\mathbb{Z}$ $\Rightarrow s_2 = 1011.$

This shows that the groups $G_{0,i}$ are pairwise non-isomorphic to each other.
In the case of U_1:
For $U_1 = \mathbb{Z}/337\mathbb{Z} \rtimes_{\psi_1} \mathbb{Z}/3\mathbb{Z}$ one has

$$\text{Aut}(U_1) = \{\eta : \mathbb{Z}/337\mathbb{Z} \rtimes_{\psi_1} \mathbb{Z}/3\mathbb{Z} \overset{\cong}{\to} \mathbb{Z}/337\mathbb{Z} \rtimes_{\psi_1} \mathbb{Z}/3\mathbb{Z}\}$$

and this is not so easy to determine. But we don't have to, it is enough to find all elements η of order 2 in it, because exactly these will induce a non-trivial group action $\varphi : \mathbb{Z}/2\mathbb{Z} \to \text{Aut}(U_1)$, $1 \mapsto \eta$, which we are looking for.

In general, we see that for an arbitrary $\eta \in \text{Aut}(U_1)$ the following holds: Since U_1 is the only 337-Sylow group

$$\mathbb{Z}/337\mathbb{Z} \cong S_{337} \hookrightarrow U_1 = \mathbb{Z}/337\mathbb{Z} \rtimes_{\psi_1} \mathbb{Z}/3\mathbb{Z}, \ x \mapsto (x,0)$$

and $\eta(S_{337})$ is again a 337-Sylow group, we know that $\eta(S_{337}) = S_{337}$ must hold, that is $\eta(1,0) = (x,0)$ for some $x \in (\mathbb{Z}/337\mathbb{Z})^\times$ (not equal to zero, because η is bijective).

Let's additionally consider $\eta(0,1) =: (a,b) \in U_1$. Note that for $(c,d) \in U_1$ (with representatives $c,d \in \mathbb{N}_0$—I'm sloppily writing c instead of $[c] \in \mathbb{Z}/337\mathbb{Z}$) we obtain that

$$\eta(c,d) = \eta((1,0)^c *_1 (0,1)^d) = \eta(1,0)^c *_1 \eta(0,1)^d = (cx,0) *_1 (a,b)^d \text{ (B.8)}$$

(note that the multiplication $*_1$ is given as in (B.6)). Since $\text{ord}(\eta) = 2$ has to hold, we have to have $(1,0) \overset{\eta}{\mapsto} (x,0) \overset{\eta}{\mapsto} (x^2,0) = (1,0)$. Therefore, it follows that $x = -1 \in (\mathbb{Z}/337\mathbb{Z})^\times$.

Now, let's express (B.8) with this $x=-1$ and with the representatives $d = 0, 1, 2$:

$$\eta(c, 0) = (-c, 0)$$
$$\eta(c, 1) = (-c, 0) *_1 (a, b) = (-c + a, b)$$
$$\eta(c, 2) = (-c, 0) *_1 (a, b)^2 = (-c, 0) *_1 (a(1 + \kappa^b), 2b) = (-c + a(1 + \kappa^b), 2b)$$
$$(B.9)$$

Thus, the element $\eta^2(c, d)$ is computed as follows:

$$\eta^2(c, 0) = \eta(-c, 0) = (c, 0)$$

$$\eta^2(c, 1) = \eta(-c + a, b) = \begin{cases} (c - a, 0) & \text{for } b = 0 \\ (c, 1) & \text{for } b = 1 \\ (c + \kappa^2 a, 1) & \text{for } b = 2 \end{cases}$$

$$\eta^2(c, 2) = \eta(-c + a(1 + \kappa^b), 2b) = \begin{cases} (c - 2a, 0) & \text{for } b = 0 \\ (c, 2) & \text{for } b = 1 \\ (c - \kappa^2 a, 2) & \text{for } b = 2 \end{cases}$$

Thus, we see that for any $\eta \in \text{Aut}(U_1)$ with $\text{ord}(\eta) = 2$ we have that $b = 1$ or ($b = 2$ and $a = 0$) and also

$$\eta(0, 1) = (a, 1) \text{ for some } a \in \mathbb{Z}/337\mathbb{Z}, \text{ or } \eta(0, 1) = (0, 2).$$

The second case cannot occur, because then $\eta(c, 1) = (-c, 2)$ and $\eta(c, 2) = (-c, 1)$ would follow and thus

$$\eta((1, 1) *_1 (1, 1)) = \eta(1 + \kappa, 2) = (-1 - \kappa, 1)$$
$$\eta(1, 1) *_1 \eta(1, 1) = (-1, 2) *_1 (-1, 2) = (-1 - \kappa^2, 1),$$

and since $\kappa^2 \neq \kappa$, this contradicts the fact that η must be a group homomorphism.

On the other hand side, one can easily verify that (B.9) for any arbitrary $(a, 1) \in U_1$ defines a group isomorphism $\eta \in \text{Aut}(U_1)$ of order 2 (when checking the homomorphism property, one uses that $1 + \kappa + \kappa^2 = 0$). So we have seen:

Claim B.9 The group automorphisms $\eta \in \text{Aut}(U_1)$ of order 2 are exactly those maps that are defined by (B.9) for a given $(a, 1) \in U_1$. Let's denote the automorphism associated with $(a, 1)$ as η_a.

Let's write out the mapping rule (B.9) for η_a again:

$$\eta_a(c, 0) = (-c, 0)$$
$$\eta_a(c, 1) = (-c + a, 1)$$
$$\eta_a(c, 2) = (-c + a(1 + \kappa), 2). \qquad (B.10)$$

In particular,

$$\eta_0(c, d) = (-c, d). \qquad (B.11)$$

We obtain:

Claim B.10 The non-trivial operations of $\mathbb{Z}/2\mathbb{Z}$ on U_1 are exactly the operations

$$\phi_a : \mathbb{Z}/2\mathbb{Z} \to \mathrm{Aut}(U_1) \,, \quad 1 \mapsto \eta_a \tag{B.12}$$

for some $(a, 1) \in U_1$.

So in this case, we obtain the semidirect products

$$G_{1,a} := U_1 \rtimes_{\phi_a} \mathbb{Z}/2\mathbb{Z}$$

for all $a \in \mathbb{Z}/337\mathbb{Z}$ and the direct product (for the trivial group action):

$$G_{1,\mathrm{triv}} := U_1 \times \mathbb{Z}/2\mathbb{Z}.$$

Certainly $G_{1,\mathrm{triv}} \not\cong G_{1,a}$, because in the left group the 2-Sylow group is a normal subgroup and this is not the case in the right ones.

However, the groups $G_{1,a}$ are all isomorphic to each other, as we will show with the help of the next lemma.

Lemma B.11 *Let Z and U be groups and two group actions*

$$\varphi_1, \varphi_2 : Z \to \mathrm{Aut}(U)$$

be given. If there is an automorphism $\psi \in \mathrm{Aut}(U)$ such that $\varphi_2 = \mathrm{conj}_\psi \circ \varphi_1$ holds (i.e. $\varphi_2(z) = \psi \circ \varphi_1(z) \circ \psi^{-1}$ holds for all $z \in Z$), then

$$\Psi : U \rtimes_{\varphi_1} Z \to U \rtimes_{\varphi_2} Z \,, \quad (u, z) \mapsto (\psi(u), z)$$

is a group isomorphism.

Proof
What needs to be shown is that Ψ is a group homomorphism, because the bijectivity follows from ψ.

We write $*_j$ for the respective multiplication in $U \rtimes_{\varphi_j} Z$. Then

$$(u, z) *_1 (u', z') = (u \cdot \varphi_1(z)u', zz') \overset{\Psi}{\mapsto} (\psi(u) \cdot (\psi \circ \varphi_1(z))(u'), zz'). \tag{B.13}$$

On the other hand side

$$\Psi(u, z) *_2 \Psi(u', z') = (\psi(u), z) *_2 (\psi(u'), z') = (\psi(u) \cdot (\varphi_2(z) \circ \psi)(u'), zz')$$

and this is the same as the result of (B.13) according to the assumption. $\square$

Back to our groups $G_{1,a}$. Let $a \in \mathbb{Z}/337\mathbb{Z}$ be arbitrary. Then inside $\mathbb{Z}/337\mathbb{Z}$ we find an element α with $2\alpha = a$ (because $2 \in (\mathbb{Z}/337\mathbb{Z})^\times$). Then $\eta_\alpha^{-1} = \eta_\alpha$ (because the η_α are of order 2), and furthermore we have the following identity in $\mathrm{Aut}(U_1)$:

$$\eta_\alpha \cdot \eta_a \cdot \eta_\alpha = \eta_0 \,,$$

which is easy to see with the help of (B.10). With this we can choose $\psi := \eta_\alpha$ in order to compare $\varphi_1 := \phi_a$ and $\varphi_2 := \phi_0$ and obtain by Lemma B.11 that $G_{1,a} \cong G_{1,0}$ for all $a \in \mathbb{Z}/337\mathbb{Z}$ holds.

So we have shown:

Theorem B.12 There are exactly 6 isomorphism classes of groups of order 2022, namely the four isomorphism types:

$$(\mathbb{Z}/337\mathbb{Z} \times \mathbb{Z}/3\mathbb{Z}) \rtimes_{\varphi_j} \mathbb{Z}/2\mathbb{Z} \text{ for } j = 0, 1, 2, 3,$$

where φ_j are defined as in (B.7), and the remaining two isomorphism types of the two groups

$$(\mathbb{Z}/337\mathbb{Z} \rtimes_{\psi_1} \mathbb{Z}/3\mathbb{Z}) \times \mathbb{Z}/2\mathbb{Z} \quad \text{and} \quad (\mathbb{Z}/337\mathbb{Z} \rtimes_{\psi_1} \mathbb{Z}/3\mathbb{Z}) \rtimes_\phi \mathbb{Z}/2\mathbb{Z}$$

for the group action $\phi = \phi_0 : \mathbb{Z}/2\mathbb{Z} \to \mathrm{Aut}(U_1)$, $1 \mapsto \eta_0$ as in (B.12)—for η_0 cp. also (B.11).

Check it out!

Similar to the last *Check it out!* in Sect. B.1, one can also a priori specify some known groups of order 2022: Of course, we have the cyclic group $\mathbb{Z}/2022\mathbb{Z}$. Then there are the following groups, which are constructed with the dihedral groups D_{2n} of order $2n$: $\mathbb{Z}/3\mathbb{Z} \times D_{2 \cdot 337}$, $D_{2 \cdot 3} \times \mathbb{Z}/337\mathbb{Z}$ or $D_{2 \cdot 1011}$. It is a nice exercise to determine which of the isomorphism types from theorem B.12 these belong to.

And now?

As mentioned above, it's a long and painstaking journey to compute all isomorphism types of groups of order 2022, but: If one considers 2021 and 2023, one should also try 2022! The last section can also be seen as an example of how complicated it can get if one tries to classify groups of a certain order.

Printed in the United States
by Baker & Taylor Publisher Services